ENERGY RESOURCE CONFLICT

Also Available from Bloomsbury Academic

Energy Resources: Examining the Facts
Jerry A. McBeath

Sustainable Planet: Issues and Solutions for Our Environment's Future [2 volumes]
Edited by Barbara J. McNicol

Conflict in the Modern Middle East: An Encyclopedia of Civil War, Revolutions, and Regime Change
Edited by Jonathan K. Zartmam

ENERGY RESOURCE CONFLICT

ORIGINS AND GLOBAL IMPACT

Adrah N. Parafiniuk and Zachary A. Smith

BLOOMSBURY ACADEMIC
NEW YORK • LONDON • OXFORD • NEW DELHI • SYDNEY

BLOOMSBURY ACADEMIC
Bloomsbury Publishing Inc
1385 Broadway, New York, NY 10018, USA
50 Bedford Square, London, WC1B 3DP, UK
29 Earlsfort Terrace, Dublin 2, Ireland

First published in the United States of America 2025

Cover design: James Dare
Cover image © Maksim Safaniuk/Dreamstime.com

A catalog record for this book is available from the Library of Congress.

ISBN: HB: 978-1-4408-7180-1
ePDF: 978-1-4408-7181-8
eBook: 979-8-2161-8380-8

Typeset Newgen KnowledgeWorks Pvt. Ltd., Chennai, India

To find out more about our authors and books visit www.bloomsbury.com
and sign up for our newsletters.

To my family, my wife Diana, and my children Tatiana, Elijah, Sascha, and Sova.
Thank you for your patience, as the days rolled into years.
Thank you for your support and love.

To my friends, I apologize for the times missed and thank you for the support and encouragement needed to get through this. I look forward to sharing this work with you.

To my editors: Nicole Azze, Anne Crossman (Chapter 1), and Max Haub (Part II) thank you for bringing such coherence and clarity to this book.

To Zachary Smith who has been an inspiration as a mentor with a lifetime of writing and teaching. Due to his fight with cancer and untimely death, he was not able to contribute to this book in a way I know he would have liked. However, his words of encouragement were deeply appreciated. Cheers to making this world a little better in your lifetime.

May he rest in peace.

CONTENTS

Introduction

This book aims to describe how conflicts over energy resources have shaped human societies and civilizations. It hopes to take the reader on an adventure through ancient history and into modern times. Access and control of energy resources have always been attached to power, wealth, and status; they can also be a matter of life and death. Millions of people have died in the battle over energy resources; wars have been fought to gain control over energy resources. This is just as true today in the era of oil and solar energy as it was in the time of hunting mammoths. This book takes the reader behind the numbers into the stories that illuminate the how and why of energy resource conflict. Readers will learn the terms, people, events, and places that shaped energy resource conflict over time and today. In Part I, the first four chapters cover the origins and causes of resource conflict, current impacts, government and NGO responses, and future implications. Chapter 5 includes perspectives of academics and experts in the field of energy. Part II includes informative topical entries arranged alphabetically, as another way the reader can access the important concepts of this book.

Almost all energy derives from the sun. Fossil fuels—oil, coal, and natural gas—are derived from ancient plants that were created from sunshine and are now stored inside the earth. The first heat source, wood, is also a gift of the sun. Today, solar and wind power are beginning their ascension toward major energy sources. While solar power is more obviously connected, wind too is connected to the sun as the heated air of the equatorial zones mixes with the less heated air of the poles causing major global wind patterns. The earth itself was formed from the solar nebula about 4.5 billion years ago. The planets, including earth, could be called the children of the sun. Nuclear energy, alone, does not derive its energy from the sun.

The sun's energy, absorbed and transformed by plants, was again transformed into the herbivores, the megafauna, that drove human expansion across the planet. Humans chased the herds from Africa to Europe through Asia and on to North and South America. After humans hunted most megafauna to extinction, groups of humans began to struggle over the right to hunt specific animals in particular places. This was the beginning of energy resource conflict. As the abundance of easily obtained wild animal protein decreased, humans began to develop the ability to grow their own food, harnessing the power of the sun once again to feed growing civilization. Herding peoples had learned, by taming and domesticating wild animals, that some of the variability and unpredictability of the hunt could be avoided. Herders and farmers would spend millennia clashing over lands and fighting for power; the conflict over resources between these two groups continues in some areas to this day.

Civilizations grew and encompassed the farmers and the herders. (There were some notable exceptions such as the Mongols who were definitely "team herders.") These large growing civilizations were always in need of more resources, be it timber or food, for the growing cities. Empires in the Fertile Crescent, around the Mediterranean, in China, across Africa, and in both North and South America had their growth limited by their access to energy resources. Island people engaged in an acceleration of this process, in quick succession: native herbivores were hunted to extinction, non-native crops and livestock were introduced, the growing human population quickly ran out of resources, and some of the people needed to move on to the next island where the process repeated itself. These conflicts for resources were limited to land, wood, and food until the Industrial Revolution.

The Industrial Revolution, beginning in England, was originally powered by water and horses. Water mills allowed for machines to be run that created textiles faster and with less human labor; these textiles (cloth) were some of the first internationally manufactured imports. Horses could power machines when the rivers were frozen. Horses, oxen, camels, llamas, and other beasts of burden provided power and transportation across the planet. A new source of energy applied to technology at the end of the eighteenth century displaced them as the most important source of power.

Coal was known to the Chinese as early as 5,000 years ago, but not until James Watt and Matthew Boulton improved upon the steam engine did coal become the energy resource that propelled the Industrial Revolution. The Watt steam enginepowered ships, trains, and factories. Coal deposits occur across the globe and coal is plentiful in most areas, although coal is less plentiful in Africa (except South Africa) and the Middle East. Technology powered by coal transformed human existence.

Countries that produced coal and applied the use of steam engines transformed from agricultural, animal-powered societies to industrial powerhouses that could move faster and create more than ever before. Coal-powered steam engines moved industrializing cultures into conflict with traditional animal-powered cultures at a breakneck pace. People in Africa and the Americas that were once at a safe distance from European settlements found out quickly that steamships and trains could quickly move in settlers and armies. These settlers and armies came equipped with weapons that were powerful and plentiful due to steam-powered factories that were able to produce them quickly and affordably. The application of coal power caused new conflicts across the planet and old conflicts were intensified.

Steam-powered vehicles began appearing on roads in the 1800s, but by the beginning of the 1900s, the internal combustion engine (ICE) began to be the engine of choice. The ICE ran on gasoline derived from oil instead of coal. The age of the automobile commenced when Henry Ford began producing vehicles cheap enough for the average worker. Oil discoveries in Pennsylvania and Texas became oil booms. "Rock oil," as it was called, was replacing whale oil in lamps, but the need for this fuel in automobiles turned rock oil into black gold. Discoveries in Persia (Iran today) by the British and the Dutch East Indies (Indonesia today) by the Dutch created the first super majors (very large oil companies), BP (British Petroleum), and Shell. In the United States, Standard Oil (today split into ExxonMobil and Chevron) consolidated control over the oil markets in the eastern states and Texas. Gas derived from oil-powered cars and eventually airplanes. Much blood was spilled over the control of oil fields.

Japan bombed Pearl Harbor because it needed access to the oil fields of the Dutch East Indies and only the US Navy had the power to stop them. The oil giants carved new countries out of the Arabian Peninsula in order to control new oil fields. They used any means necessary to gain access to oilfields across the planet. These international oil companies (IOCs) many times left a wake of blood and destruction across the countries in which they worked. Nations fought back and nationalized oil in countries such as Venezuela, Iran, Saudi Arabia, and Russia. Today, over 70 percent of oil reserves are controlled by nationalized oil companies (NOCs).

An accidental effect of the industrialized, modern, coal- and oil-powered world was global warming, caused by the release of carbon dioxide (CO_2). New renewable energy resources such as solar and wind have become less expensive than fossil fuels, but the powerful companies that are profiting from the carbon economy are not willing to allow them to ascend and surpass fossil fuels without a fight. Fossil fuel interests are funding politicians and crafting policy that disadvantages renewable energy and favors the old fossil fuel power stations and transportation. The energy resource conflict continues with an important change: fossil fuel interests are fighting against the use and implementation of new technology to preserve profits. The fate and the heat of the planet may lie in the balance.

This book documents human conflict over energy resources from the time of ancient man hunting mammoths with primitive spears to the modern emergence of solar, wind, and hydrogen power. The fight for energy resources continues in new and surprising ways. Access to more affordable and powerful energy has transformed mankind and will continue to define human civilization into the future.

PART I

The History and Future of Energy Resource Conflict

CHAPTER 1

Origins and Causes

OVERVIEW

This chapter looks at human conflict over energy resources. It discusses broad historical conflicts over resources and how they shaped human migration, geopolitics, and wars. It introduces the major sources of energy from food sources, continues with the introduction of fossil fuels, and finally introduces nuclear and renewable energy resources. The advantages of each energy source are discussed, as well as the limits and hurdles for the widespread adoption of various types of renewable energy. The chapter then looks at some important examples of historical conflicts over oil resources within specific countries, as oil was discovered. (The next chapter goes into detail about international conflicts over resources.)

The historical view presented here demonstrates the cycle of human invention, overuse and stagnation, conflict over declining resources, and finally migration or invention. Humans first migrated for resources until those resources were exhausted across the planet or until an invention revealed a new resource that could replace the last one. Two examples of this are the worldwide overhunting of megafauna, which caused mass extinctions, and the replacement of burning wood for heat and energy with coal. This two-fold path resonates today, as humans are at an important juncture in history. It is yet unknown if we will continue to use available fossil fuels that cause global warming and climate change, or if we will transition quickly enough to renewable and zero-emissions energy sources in time to avoid the worst possible consequences. The history covered in this chapter is informative and demonstrates patterns and possibilities for the future.

Energy for humans can be measured in calories and joules (or BTU, British thermal unit). Humans have transformed from a simple species that required only food and warmth to survive, to an advanced civilization that requires energy for everything from cell phones to satellite launches. The measure of how much energy is required per person can vary incredibly, depending on lifestyle and country. Calorically, all humans consume similar amounts, but diets differ greatly among individuals and across cultures. A meat-heavy diet requires a much greater input of energy and water. The physical size of an individual can also change the equation. Heat in cold climates was once obtained by burning wood for warmth and using animal furs and skins to insulate and protect humans from the elements. Today's heat comes from nuclear, solar, wind, fossil fuels, and, for some, still wood. The progression of humans' heat and energy sources has transformed our planet and our atmosphere. Where we get our energy in the future will decide the fate of many species, billions of people, and the coastal cities where they live. Before we dive into the details, a quick review of human energy sources will be enlightening.

Humans were first gatherers and scavengers. The human ability to use simple tools, such as a well-placed heavy rock, allowed them to consume calories that other predators had left behind. Early human ancestors are believed to have been able to crack open the bones of large megafauna and drink the marrow (Ferraro et al. 2013).

Humans evolved in Africa alongside megafauna species such as elephants and rhinos. These animals feared humans, and that made hunting them difficult. Modern humans migrated out of Africa in the past 100,000 years (Robertshaw and Bellwood 2013), first along the coasts of Southern Europe, the Indian subcontinent, Southern Asia, and Eurasia, then later across Oceana, and eventually to the Americas. They found megafauna that had no fear of humans. (Ancient hominids, distantly related to modern humans, migrated out of Africa 2 million years ago.) Human hunters decimated megafauna across the planet, driving hundreds of species to extinction (Alroy 2001). From the iconic mastodons and mammoths to giant sloths and car-sized armadillos, the largest species were hunted to extinction. This created cascade effects that completely transformed ecosystems and led to other extinctions. Jungles once heavily browsed by megafauna became overgrown. This led to intense forest fires that converted forests to plains, which in turn created microclimate conditions that led to less rain and more arid plains where tropical forests once stood. The hunting of megafauna led to an increase of human populations and their spread across the planet in search of game.

At this point, human populations turned to farming for consistent and storable calories. Farming began in the Fertile Crescent in the Middle East about 10,000 BCE. It independently evolved in China, India, and Mesoamerica in the same time period. As humans settled into farming civilizations, the domestication of animals created an alternative to the unreliability of the hunt. While farming civilizations also had domestic animals, pure herding cultures continued to live nomadic lifestyles as herds had to be moved, due to the seasons and the availability of forage. These new cultures displaced native species and altered ecosystems to best serve human needs and the needs of their crops and domesticated animals.

Wood became a needed staple of civilizations, and it was imported from across all the lands with human populations. Wood built boats, furniture, weapons, buildings, and tools. It heated castles and huts alike. As human populations grew, local wood grew scarcer. Empires needed a constant supply to send to their armies and their cities. Wars were fought over access to woodlands, from the famous cedars of Lebanon to the Northern forests of Europe. People from kings to local tribes laid claim to dwindling woodlands as a source of game. Empires that could not keep a supply of wood were in dire straits (Oosthoek 1998). In the era of European exploration and colonization, trees suitable for ship masts were particularly important. Without the wood to create ships, European colonists could not access and defend their colonies. The era of wood for energy lasted until it was replaced by coal at the beginning of the Industrial Revolution. Even today forests are fought over, as they are seen as either a resource or a habitat and refuge for animal species.

Coal has been known for at least 5,000 years, but for most of that time, it was scarcely used. Wood was usually available and was more pleasant to burn, as it had no noxious odor and no chemically laden smoke that accompanies most burning coal. With the perfection of the steam engine, coal became much more valuable as steam engines were used to pump water out of mines and to move heavy materials. In the 1800s, coal mining expanded first in England and soon across all industrializing nations. Factories, ships, and trains ran on coal, and it became a necessary commodity for all countries hoping to be part of the Industrial Revolution. At that time, the long-term climate effects of burning coal and other fossil fuels were unknown. The widespread use of coal power plants to create electricity in the twenty-first century is the largest single contributor of CO_2 into the atmosphere. Climate change in turn threatens the long-term viability of the planet for humans (at current population numbers) and most other animal species.

Coal is hard to replace because it is widely dispersed, and there is enough coal to last for at least the next 300 years, at current rates of use. There are also sunk costs (existing investments that cannot be recuperated), such as coal-mining equipment and infrastructure, coal transportation systems, and actual coal power plants. And there are economic interests that benefit from coal-powered electricity. These interests are entrenched in political systems. Coal plants currently provide 20 percent of the power created in the United States, but closer to 40 percent of the power worldwide.

Coal revolutionized both factories and large-scale transportation systems, but liquid fuels revolutionized home heating, private transportation, and weapons of war. Oil discovered in the United States was first used to power oil lamps. In the 1900s, most homes and cities were illuminated either by oil made from whales and other animals or by plant-based oils. Petroleum changed that quickly, as it was much less expensive. Oil companies that used predatory pricing were able to create local and national monopolies across the industrializing world. The oil industry concentrated the wealth of the world on relatively very few people.

The invention of car and the combustion engine made oil even more important. Gasoline is one of the more energy-dense materials on earth, and it lets people move at unprecedented speeds. The car, with its ability to move without rail, was quickly adapted across the industrializing world. Industrialized and industrializing countries created troop carriers, tanks, ships, and eventually airplanes and aircraft carriers that were fueled by oil in the form of diesel or gasoline.

Today the oil and car industry make up a large part of the world economy. Industry leaders are loath to embrace legislation that would cause sweeping changes. Political leaders are not always willing to engage in change that has an unknown economic outcome. The risk inherent in large-scale industrial change is not in the financial interests of established enterprises. Transitioning to new energy sources would allow startup companies to gain market share that they would not be able to obtain if the status quo were enforced. All these risks and fears have created an environment that makes the transition to alternative-powered vehicles extremely difficult. Overall, electric cars make up 20 percent of the new automobile market, but countries such as Norway and states like California are changing that equation. In 2021, 86 percent of new cars sold in Norway were electric (IEA 2022). In 2022, 18 percent of new cars sold in California were electric. Like all new technology, growth will be exponential, not linear. Electric vehicle (EV) sales increased by 55 percent between 2021 and 2023. Despite political and industrial resistance, the transformation to electrified and eventually hydrogen vehicles is inexorable.

Natural gas has been in common use for more than 100 years. Its use was limited, due to availability and price spikes that could be 400 percent higher than average (US EIA 2024). Fracking technology has allowed companies across the United States and the planet to recover natural gas in old oil fields and in rock formations that had limited extraction potential in the past. Prices in 2018 were less than one-third of the price in 2008. The new natural gas boom has allowed many coal-powered plants to switch over to natural gas, which creates less pollution and less CO_2 emissions. Natural gas now provides 39 percent of the power in the United States. Natural gas power plants are proliferating across the planet at breakneck speed, especially in Florida, Texas, and the Appalachian states. This is dangerous, as it does not address continued CO_2 emissions.

While natural gas is not a long-term solution to climate change, it can cushion damaging effects of coal-powered plants. Some see it as a bridge between intermittent renewables like solar and wind, because gas-fueled power plants can quickly increase and decrease electrical output. That is not true for coal, which cannot quickly adjust to electrical needs and takes 10–20 hours to reach capacity. Nuclear plants take days to ramp up to full output. A gas-fired power plant can be at full output in 2–3 hours. Combined with a solar-wind-battery system, natural gas can provide a needed or precautionary backup.

Nuclear power was developed after the Second World War. The United States developed the first nuclear bomb. The power unleashed by the first nuclear explosion was both awe-inspiring and horrifying. Harnessing that energy for power can be seen as a miracle at best and a deal with the devil at worst. The ability to use less than a thousandth of the amount of fuel to make the same amount of energy appears miraculous. If there were no dangers or long-term spent fuel storage issues, there is no doubt that nuclear energy would power the planet.

Nuclear technology proliferated in the 1950s as American spies shared the US technology with the Soviet Union, and the United States shared technology with its allies. Today many countries get a good part of their electricity from nuclear power. France derives 75 percent of its electricity from nuclear power plants. The United States creates the most electricity (amounting to 18 percent of its power) from nuclear power plants of any country (EIA 2019). The reason that most countries do not depend mostly on nuclear energy has to do with the number of accidents at nuclear plants, including the 1979 Three Mile Island accident in the United States, the 1986 Chernobyl disaster in the Soviet Union (Ukraine), and the 2011 Fukushima disaster in Japan.

The Chernobyl nuclear plant disaster killed at least 46 people and left nuclear fallout all over Ukraine, Western Russia, and Eastern Europe. The disaster created an area the size of Luxembourg that will be uninhabitable by humans for 24,000 years. The city of Pripyat was only 16 years old, but people will never occupy it again. It remains a reminder of the dangers of nuclear power, as do the more than 200 villages that were permanently evacuated in modern-day Ukraine and Belarus.

Nuclear power does not create CO_2 emissions or pollution, but it does create spent nuclear fuel that remains radioactive and dangerous for thousands of years. In 2023, there were safer plant designs that may be able to overcome the problems that caused the nuclear reactor accidents, but public sentiment is unlikely to support new nuclear plants. Safety and spent-nuclear fuel storage are not the only difficulties facing nuclear energy. Nuclear plants are much more expensive than all other power sources. There is hope that smaller nuclear plants will be viable. There is also the hope that molten-salt reactors (MSRs) will prove successful. Unless less expensive and safer technology emerges quickly, use of nuclear power will continue to decrease across the planet.

Renewable energy has become the promised savior for human energy needs. Costs have always stopped that from happening, but that changed in the 2010s. New solar and wind power plants are less expensive than coal-, nuclear-, and gas-powered plants. Renewable energy power plants that use batteries to compensate for the intermittency of wind and solar are still less expensive. The difficulty of implementing renewable energy is that it is not available in every location, and that the financial interests of the fossil fuel industries are at stake. The people and the corporations that stand to lose are not going to sit back and allow this transition. They will fight it every step of the way.

Renewable energy was not originally focused on solar and wind power. Hydropower dominated the renewable energy space for over seventy years. Hydropower is created by damming a river and using the created pressure to spin turbines and create electricity. The costs were high, and the land was usually controlled or owned by governments. Governments were able to create large, impressive dams across the world. Those in the western United States are some of the most iconic: the Hoover Dam, the Grand Coulee Dam, and the Glen Canyon Dam are some of the most well-known dams. There are several downsides to these dams. They flood areas that people used to live in and they cover places of exceptional beauty, such as Glen Canyon. They also stop fish, especially salmon, from coming upriver to spawn, and that has devastated fish populations across the planet. Hydropower is limited to the number of major rivers that are suitable for damming. In the United States most hydroelectric dams were completed by the 1980s. There were few suitable spots left, and the damage caused by damming rivers was no longer acceptable. Hydropower still produces 6.2 percent of the power in the United States. It has been decreasing as the cost, both financially and environmentally, no longer makes sense when compared to solar (4.5 percent of worldwide power) and wind (7.5 percent of worldwide power).

The transition to renewables is hampered only by available resources and political will (Gilens and Page 2014). New technology tends to expand and grow exponentially, not linearly. Overall, this has proved true with wind and solar energy. Wind provides fifty times as much power as it did twenty years ago. Solar provides one hundred times as much power as it did twenty years ago. Geothermal power, while established in a few countries, is still in its infancy. This is also the case with tidal power. Both these technologies can expect to have similar growth patterns as solar and wind. Biofuels and bioenergy created by using plant matter to create fuel and run power plants also continue to grow. Until the people of the world stop building fossil-fueled power plants and gasoline-powered cars, CO_2 emissions will likely create a planet that is less livable for all species. Rising sea levels will also destroy many major cities across the planet until humanity fully embraces renewable energy.

THE HISTORY OF ENERGY TYPES

Whose Hunting Grounds?

Why would a species with little technology except fire and sharp stone tools head out in small groups into a complete wilderness, a wilderness full of animals larger than they, fiercer than they, and some that saw them as food? The only possible explanation is that human populations grew, and that led to overhunting and over-gathering. Disputes over who had the right to hunt a dwindling number of animals and to gather the remaining plant resources led to conflict and dispersal of the human population. This was the start of natural resource conflicts. The process was slow but constant. The spread of human settlement out of Africa can be traced archaeologically and through human DNA studies. Modern humans (*Homo sapiens sapiens*) first left Africa fewer than 100,000 years ago. Earlier human ancestors had already spread across Europe and Asia about 2 million years ago. Neanderthals and Denisovans later interbred with humans. Today most non-Africans have about 2 percent Neanderthal DNA.

The first wave of human migrants stayed close to the coast. Tidal pools provided easily accessible calories without much risk of loss of life. As settlements grew, humans expanded both along the coast and into the interior in search of new sources of food. Hunting with spears has been demonstrated to exist long before the evolution of modern humans. Spears were used at least 500,000 years ago. These early spears were wood, and wood does not preserve well, so it is hypothesized that they may have been used up to perhaps 5 million years ago. The use of these hunting tools allowed humans and their ancestors to eat animals that were much bigger and stronger than they. These abilities allowed human populations to spread across Africa, Europe, Asia, and Oceania 50,000 years ago. Humans did not settle in the Americas until about 15,000 years ago. Greenland and the Arctic Circle did not see human settlement until 4–6,000 years ago. The Polynesian people did not settle the last of the habitable Pacific Islands until about 900 years ago.

Human energy resources vary greatly. As omnivores, humans can live off an all-meat or all-vegan diet, and either way live healthy lives. Hunting megafauna became a vital part of human existence. This process of expansion into new territories, hunting new animal populations, growth of human populations, and conflict over limited resources continued for 69,000 years. It resulted in the extinction of most of the largest megafauna outside of Africa. The mammoths, mastodons, wild horses, and giant sloths that roamed Eurasia and the Americas were some of the many species lost to human hunting.

Neanderthal sites demonstrate that other hominid species were hunters. Cave art in Spain dated to 64,000 years old, along with stone spearheads and butchered animal bones, demonstrate that the Neanderthals were one of these groups (Greshko 2018). It appears from direct correlations between modern human settlement and large animal extinction that modern humans were more efficient at hunting. Hunting may also be a part of the spiritual practice of ancient man. This may help explain why humans continued to move in search of game before they adopted herding and farming as more sustainable ways of guaranteeing a constant food supply.

After the extinction of the largest of the megafauna, hunting smaller species continued. There is some disagreement over the definition of megafauna. Some scientists consider anything over 40 kg, but others consider 1,000 kg. More than ninety species over 40 kg went extinct with human arrival. Species such as the bison, sometimes referred to as the American buffalo, continued to survive in huge numbers until the nineteenth century. Many peoples were able to remain hunting-based cultures by finding a balance between how many animals to hunt and which animal to take based on sex and age. This form of selective hunting created a symbiotic relationship between the hunters and the hunted. The herds provided everything the tribes needed: food, clothing, and shelter. By limiting population numbers, hunters could stop the boom-and-bust cycle of herbivores that no longer had their historic predators. Saber-toothed tigers, cave lions, European leopards, and dire wolves went extinct due to a combination of human hunting of their traditional prey and hunting of the predators themselves.

By 10,000 BCE, there were 5 million humans across the planet, with little room left to exploit megafauna. Archeological evidence suggests that populations stayed stagnant for 2,000 years (Nielson 2016). With the extinction of most of the extremely large megafauna, humans could no longer count on hunting to supply additional calories, especially in the form of protein and fat. Hunting became more seasonal and less reliable. Humans, especially in highly populated areas, had to make the first big energy transition. After causing extinction and overhunting of medium and small game, humans had to adapt to new forms of extracting energy from the earth. Hunting and gathering always put limits on the human population, but herding and eventually farming gave them control over raising their own calories. This gave humans the energy and power for another large population increase.

From Hunt to Herd

Areas that were overhunted but still had available grasslands were transformed from hunting grounds to farming and grazing lands. Domesticating medium-sized fauna (originally sheep and goats) that were more docile evolved in one of two ways: humans raised baby animals that were orphaned by hunting or more docile animals that were drawn to human agriculture, then fed and eventually contained. Raising animals created new kinds of energy and water needs. Herding required large expanses of grassland and vigilance against predators.

Small-scale animal husbandry also seamlessly intertwined with farming, which we will address in the next section. Most cereal plants produce much more chaff than edible grains. The chaff can be eaten by herbivores. They, in turn, create fertilizer for the fields. In this way a small farm is a closed-loop system. There is no waste, as everything is part of the life cycle.

Nomadic herders moved their flocks with regularity so that there was always enough for their animals to graze on. As flocks and human families grew, herders needed more and more land. This led to conflict among herding tribes, as well as conflict between settled agrarian civilization and herding culture. The Eurasian Steppe that stretches from Eastern Europe to the Pacific Ocean in modern-day China provides an excellent and well-researched example. From the founding of the first civilization on the borders of the Steppe, there was conflict.

Chinese civilization was in constant conflict with the tribes of the Steppe from the very first cultivation of millet and rice 9,000 years ago until 1759, when the Manchu Dynasty finally subdued the tribes of Mongolia and dispersed them across the Steppe. The Great Wall of China was built specifically to fend off constant attacks, raids, and occasional conquests by these nomadic herding tribes of the Steppe. The Wall, of course, suffered the fate of all walls; it was ineffective, unable to contain the Mongols, who eventually ruled all of China.

Horses and dogs allowed ancient pastoral people to herd their animals. Dogs were the first domesticated animals, as they worked well with hunter societies. They defended against predators and assisted in hunts. Dogs kept the herds together and relieved humans of the constant work of physically herding animals together. According to genetic evidence, dogs were first domesticated from wolves in the Middle East (Pollinger et al. 2010). Burials with dogs date back at least 30,000 BCE but there is disagreement on how to classify these animals: wolves or dogs (Perri 2016). Regardless, these burials imply a close relationship between canines and humans, which of course continues to this day.

The herders of the steppe eventually became mounted warriors. Horses allowed humans to move at speed with their animals. Horses were domesticated across Eurasia starting 10,000 years ago. Genetic research demonstrates that mares were domesticated independently at various times and locations (Achilli et al. 2012). Their use in transport, labor, and warfare was so advantageous that any group that domesticated horses had a military edge, as well as a better chance of survival.

Herders engaged in local raids, often due to the high payoff of capturing the herd of a rival. As pastoral tribes grew more sophisticated, they became adept at full-scale warfare from the back of a horse. When conflict arose between farming communities and pastoral tribes (herders), the herders had an advantage, as they could come in quickly and leave just as quickly. They were also hard to locate. A farming community had to store its energy (grain) in a permanent place, but a pastoral tribe could move its energy (meat) anywhere that four hooves could go. This gave an advantage to the herders. As farming communities became part of larger civilizations and empires, walled cities and/or elevated cities were built to defend against the raiding tribes. This happened across the planet. The Hopi (Native American tribe) and their Anasazi ancestors, in what became the American Southwest, built cities high on mesas. The Chinese built walls and walled cities to keep tribes at bay. And rulers in the Mediterranean and Fertile Crescent cities erected walls beginning in Jericho about 9,000 BCE.

Resource constrictions became the main driver of conflict in human society. Drought and plant or animal disease could drive one civilization against another. All societies saw themselves as "the people," be they chosen, advanced, or civilized. They viewed others as barbarians and less evolved. This distinction was especially true in the settled farming communities. They looked down on the nomadic herders. In turn, the herders and farmers looked down on the hunters and gatherers. Cities allowed for specialization, but in herding and hunting cultures, everyone was engaged in the daily work for sustenance. Chiefs and priests worked alongside mothers and fathers. Advances in farming technology allowed one person to grow enough for more than one person. This permitted a society to have bricklayers and artists who could dedicate time to their crafts. This specialization allowed settled people to create large art projects and to build temples and buildings that convinced them of their superiority.

It had always been assumed that hunter-gatherer societies were not able to create many permanent structures and buildings. Discoveries in Turkey, near the southern border with Syria, dating back 12,000 BPY demonstrate that this assumption was wrong. At the archaeological site Göbekli Tepe, 20-foot-tall stone "Ts" rise out of the ground in a complex referred to as the world's first temple. While that may be unlikely, it may be the world's first large-scale stone temple. It was built by hunter-gatherer people thousands of years before either herding or farming transformed human culture. Its existence shows that hunter-gatherer communities lived in settled communities for at least part of the year. The human manpower required to move stones between 10 and 50 tons required hundreds of people and months of work. These discoveries, starting in the 1990s and continuing to the present day, question the notion that hunter and later herding cultures were less sophisticated than herders and farmers.

The nomadic lifestyle of herders generally did not encourage creation of permanent stone structures. The herders built structures of wood, earth, and animal skins, which would not stand the test of time. This reality gives archaeologists very little evidence to understand the cultures of ancient nomadic peoples. The herder cultures that are most well-known are those that fought with and sometimes conquered the settled people of the world. Written records attest to the continual attacks from nomadic people looking for food, treasure, or women. The armies of ancient empires were always kept busy fending off invasions from the frontier or the boundaries of the empire's controlled lands.

Alexander the Great (356–23 BCE) had to defeat the Scythians in order to continue his conquests. The Scythians were a nomadic herding culture, famous for their skill in battle and their women warriors. These women famously had their right breasts removed as young women (according to the ancient historian Herodotus), so that their sword/spear hands were not encumbered. The Scythians ruled a vast empire for over a millennium, from the eleventh century BCE until the second century CE. Their territory stretched from Eastern Europe into modern-day Pakistan and India. The agricultural communities that surrounded them were subservient and provided annual grain and tribute.

Although Alexander and his father Philip of Macedon were able to win a battle against these fierce foes, they were not able to subjugate the people or conquer this empire. Later, Alexander's viceroy in Thrace was defeated and killed trying to conquer the Scythians. It was not until the end of the third century BCE that the empire began to shrink and fragment from the combined pressures of Celts and Thracians from the west and Sarmatians from the east. A smaller Scythian culture survived until the second century CE when the Scythians, now more settled and agrarian, were finally defeated by invading Goths, another nomadic-herder culture from the north. These same Goths, along with other nomadic tribes, eventually sacked Rome, which led to the fall of the Western Roman Empire (Aurelius 2015).

For the Romans, the people of the Steppe were a continual nuisance, and the Romans did not always win the battles. They fought the Scythians, Sarmatians (culturally Scythian people), the Goths (German people who adopted the Scythian herding culture), and the Huns (a people who originated on the Eastern Steppe bordering China) among many others (Baumer 2012). It is important to note that these herders or pastoral nomads were seasonally nomadic, but they did have set territories. Most herding cultures wintered in the same area every year and moved their herds in the spring and the summer. It was only pressures from other groups of pastoral nomads that caused them to move to terra incognita.

For example, the ancestors of the Huns, the Xiongnu, were soundly defeated by the Chinese Han Dynasty in the second century BCE. This defeat and others that followed pushed the Huns westward, leading to their arrival on the Pontic Steppe that bordered the Roman Empire (Baumer 2012). These new powerful nomadic tribes forced other tribes to either fight the Romans or ask for protection and land within Roman borders. These tribes could cause trouble later from either inside or outside the empire.

After the fall of the Western Roman Empire, the Eastern Roman or Byzantine Empire kept the warring herding cultures of the Western Steppe in check until 1453 CE, followed by the Muslim Empires. They continued to make forays into the settled farming areas of the Mediterranean, Western Europe, India, and China. The empires in India and China limited their growth outside the Steppe. The area of the Eurasian Steppe, still ruled by herding cultures, was referred to as Tartary or Great Tartary by Western settled cultures. In the late period of the pastoral herding peoples of the Steppe, from about 500 CE to 1700 CE, one group was successful at winning battles and conquering entire empires. These people were the Mongols.

The Mongols were a group of unified tribes under a leader named Temujin, better known as Genghis Khan (the Great King). The Mongols rose out of the Steppe in the thirteenth century CE. They created the largest land empire in human history. They did this by being excellent mounted archers, as most herding warriors have been, and by adopting any new technology they encountered in their conquests or defeats.

Their traditional enemy was the Chinese Empire. They had been defeated by Chinese artillery that launched explosives. Attacking walled cities made victory difficult. The Chinese were the first to implement gunpowder in warfare. The Mongols gradually adapted and used explosive artillery to deadly effect. They also employed engineers who built siege engines on site, eliminating the need to haul equipment to the battlefield. The Mongols used the advantage of a herding society. Each warrior had three to five horses. By switching horses, a warrior could ride 100 miles in a day. Six out of ten mounted warriors were light cavalry, meaning they carried bows and wore lighter armor. The other four were heavy cavalry with more armor and spears as their main weapons. Their horses, mostly mares, provided milk and in desperate times, meat. The small sturdy horses of the Mongols could live off the land and did not require grain for food. They could also withstand cold weather. This allowed the Mongols to use frozen rivers as roads and to attack in the winter, which was unheard of at that time. These advantages allowed Mongols to conquer most of Eurasia within a century.

The Mongols defeated the Chinese Empire, the Indian Empires, the Muslim Empires of the Middle East, and the Europeans. The knights of Europe were too heavy and slow to defeat the lightly clad Mongols. The Mongols wore light chain mail or lamellar (plated) armor over silk shirts that stretched over armor-piercing arrows. This allowed for easy extraction of arrows, which limited deaths from infections. Only Europe was saved by the death of the Great Khan. At that time all the Mongol leaders had to gather and decide on their next leader. The Mongols had been moving through Western Europe after defeating huge allied armies of Europe in Poland and Hungary. Christendom believed the Great Khan's death was an answer to their prayers. It gave Western Europe time to prepare and develop new tactics. However, the Mongols became victims of their own success, and they settled down to rule the empires of India, China, Persia, and Russia. Eventually the Mongols became a part of the people they conquered. They no longer respected the rule of the Great Khan. They fought among themselves, and for this reason they never finished their conquest of Eurasia. Both Japan and Thailand remained independent, due to the Mongols' lack of naval success and a number of unfortunate storms that wiped out their landing armies before they could assemble a fighting force (Morgan 2007).

The Mongols created dynasties that ruled for hundreds of years across Eurasia, but eventually, as happens to all dynasties, new dynasties took over. These new rulers fought against the various herder tribes. The last powerful herding people, the Dzungar Khanate, were defeated and eliminated by China's Qing Dynasty in the Dzungar Genocide of 1755–8. They lost over 80 percent of their population and were the last herders or horse people to have power that rivaled the settled empires of their neighbors. These agricultural empires exist today as modern nation-states.

Fields and Food

The agricultural revolution forever changed human interaction with the natural world. It eventually removed humans from the food chain and allowed them to create worlds that, outside of weather, they could completely control. There are two theories of why humans gave up a hunter-gatherer lifestyle for one of sedentary farming and animal husbandry. Before we discuss these competing theories, it is important to understand some differences between the daily, weekly, and seasonal workloads of these two lifestyles.

According to anthropologists who study modern hunter-gatherer societies, they tend to work about twenty-five hours a week hunting and gathering. They have a lot of time for social gatherings and social time in general. Their diet is varied and seasonal. The vast majority (75–95 percent) comes from gathering. The rest comes from hunting. Most cultures reserve hunting activities for the men of the tribe. Gathering food is generally a group activity that involves all sexes and ages. Food preparation and cooking responsibilities vary, depending on tribe and tradition.

Ancient farming was a much more time-intensive process. Typical workdays stretched from sunrise to sunset during the growing season. However, a successful farmer might have an incredible amount of free time during the off-season. In colder climates, this off-season could include gathering and splitting firewood, as well as preparing tools and property. In warmer climates, farmers could have two separate growing seasons. Weather permitting, the hard work of farming guaranteed stable food supplies. It also allowed for specialization, as one farmer could grow enough food for more than one person.

Farming civilizations had a less healthy diet than hunter-gatherer societies. There was less meat in a farmer's diet and less variety. Grains stored very well, but they did not pack the nutritional punch of fruits, vegetables, roots, and fungi that a foraging lifestyle could provide.

Why did people choose to give up a healthier, easier way of life for farming? This question has befuddled ancient historians, anthropologists, and psychologists alike. One theory is that a growing population created scarcity of food to hunt and gather in some regions, and farming proved more stable and allowed populations to grow. Another theory involves beer. Ancient gatherers gathered seeds of grasses that were precursors of today's grains. If these grains were ground, wet down, and left in containers, they fermented. There is no doubt that mammals enjoy getting drunk. Elephants and primates (other than humans) have been observed stashing fruit and coming back to eat it after it has fermented. They will eat until they are falling over. (Authors' note: do not get close to a drunk elephant.) The theory here is that having a constant supply of grains created a steady supply of alcohol for humans. The growing of grapes soon added wine to the supply of alcohol. This school of thought contends that growing crops for alcohol slowly morphed into raising all food supplies. Originally crops were supplemented by gathering and hunting, but growing populations eventually limited the amount of hunting and gathering that could be done by a concentrated population.

The agricultural revolution changed humanity in so many ways that modern cultures could never have developed without it. Ancient and modern hunter-gatherer societies had roles and specialties: chief, hunter, medicine person, priest. However, agricultural settlements that grew into towns and cities lay fertile ground for increased specialization. These full-time specialists made advances in many fields, including mathematics, astronomy, architecture, agriculture, and weaponry.

This is not to say that ancient hunter-gatherers were not making their own advances and discoveries. Stonehenge is just one example of ancient peoples' creating amazing structures, temples, and art. The ancient site of Göbekli Tepe, mentioned previously, shows that pre-agriculture humans could build extraordinary stone structures as well as symbolic art. Cave paintings attest to the art and creativity of early humans and Neanderthals. Cities were able to create and maintain advances in a more permanent way.

Agricultural settlements tended to trade food, supplies, and ideas. As ideas intermixed, new technologies developed and improved. These memes (ideas, behaviors) spread from person to person and were passed down through teaching. They extended across all humanity and allowed the sum of humanity to be much greater than its parts. Great buildings arose across India, China, and the Fertile Crescent. Eventually, the entire Mediterranean was surrounded by cities and cultures that brought forth the alphabet, advanced shipbuilding and sailing techniques, and bureaucracies to control and manage the growing cities.

As settlements became towns and towns became cities, nearby resources shrank. Towns and cities were storehouses of wealth and food supplies. The cities became walled, and invaders learned the art of the siege. Empires grew by conquering neighboring towns. Despite occasional losses, cities tended to win more than they lost. The agricultural lifestyle that began in a few small settlements 10,000 years ago became dominant across the planet by 2000 BCE. There were still many hunter-gatherer peoples. Across the steppes and prairies, herding cultures ruled supreme.

Trade enabled people to have goods from hundreds and in some cases thousands of miles away. Memes, products, and technology were traded. Empires in China, the Mediterranean, and Oceana built their trade with sailing technology that allowed sailors to navigate even out of sight of land. The Phoenicians, Carthaginians, and Greeks colonized the Mediterranean. Their trading outposts swelled into cities and helped spread goods, ideas, and technology to all adjacent lands.

Despite successes, these growing towns were susceptible to regional weather variances, especially droughts, and in some cases early freezes. Cities were also targets of invaders who had an advantage in warfare technology. Any combination of droughts and invasions could spell disaster to entire empires and entire regions. The Mediterranean fell into a dark age following drought, as demonstrated by pollen taken from Dead Sea core samples. Invaders known simply as the Sea Peoples swept in. The Sea Peoples had iron technology in a Bronze Age world, and they decimated all the great empires of the Mediterranean region. Only a weakened Egyptian Empire limped on.

Despite these setbacks, the advantages of agriculture outstripped the disadvantages by a factor of 10. Every dark age of history, be it the Mediterranean, China, or Europe, was followed by a period of continued enlightenment. Agriculture allowed humans to put energy into anything their brains could dream up. The ability to dedicate teams of people to formulate projects, ideas, and questions allowed humanity to make great strides. It was only the conflicts over resources that sent humanity backward. Unfortunately, these conflicts continue today.

Forests and Wood

To build an empire, you must build it with something. That something was usually wood. To have wood, you need trees. To have trees, you need access to or control of forests. Empires have fought over this resource for millennia. The *Epic of Gilgamesh,* written in Mesopotamia in the third millennium BCE, describes a forest journey. The hero had to fight and kill the guard of the Gods who fought to stop humans from destroying the forest. The hero was Humbaba. He was destroyed, as were the famed forests of Lebanon. These forests are mentioned many times in the Hebrew/Christian Bible: Ezekiel, Psalms, 1 Kings, and many others. The Phoenicians built their fleet with these magnificent cedars that helped them become the area's first seafaring people, spreading their culture and the precursor of today's alphabet. The forests of the Middle East were the building blocks of ancient culture.

Today only small fragments of these forests are left. These strong, slow-growing cedars are the most well-known and revered, but they were only one of ten coniferous species that thrived in the once lush forests of Lebanon and the surrounding areas: Cilician fir (*Abies cilicica*), Cedar of Lebanon (*Cedrus libani*), Mediterranean cypress (*Cupressus sempervirens*), Syrian juniper (*Juniperus drupacea*), Grecian juniper (*J. excelsa*), Willdenow or stinking juniper (*J. foetidissima*), Prickly juniper (*J. oxycedrus*), Calabrian pine (*Pinus brutia*), Aleppo pine (*P. halepensis*), and stone pine (*P. pinea*) (Talhouk, Zurayk, and Khuri

2001). Stone pine was of particular importance as it produced an edible pine nut. It is still cultivated today for that very purpose. The forest has also produced game for human hunters since they first picked up spears. For these reasons, the forests were an integral part of human civilization. But overuse and destruction of local forests have caused the demise of empires. The loss of a resource of this importance forced a civilization to invade and fight neighboring peoples and empires. Without the resources of the forest, ancient civilizations could not survive.

The Hellenistic period (323 BCE–30 CE) saw a shift in power, depending on the availability of timber. Crete's 600 years of dominance (1900–1300 BCE) in the Bronze Age was abruptly cut short by erupting volcanoes and resulting tsunamis. Some historians theorize that it was the loss of the forest caused by the volcanoes and tsunamis that shifted power from Crete to mainland Greece. Crete's bronze-making abilities became dependent on timber imports (Hurtt et al. 2006). Alexander the Great (356–323 BCE) hailed from Macedon, a Northern Greek kingdom rich in timber. That timber wealth helped his ancestors and especially his father Philip to create a powerful kingdom that would eventually give him the strength and ability to conquer and rule lands across the Mediterranean, the Middle East, and India.

The Romans (27 BCE–400 CE; the Eastern Roman Empire continued to 1453 CE) quickly denuded the Italian Peninsula of her forests. They looked with great longing at the German forests that appeared almost endless. The Romans never conquered the land past the Rhine River, and so never controlled all the German forests. They conquered modern-day France, Southern England, Spain, and Portugal, but the forests that served them best were those in Lebanon and Syria. These lands were stripped of their forests until Emperor Hadrian (76–138 CE) claimed what remained of the forest as his own and put an end to their destruction. Eventually, the Western Roman Empire came to an end, after the Germanic tribes (Goths, Ostrogoths, Vandals, Visigoths, and Franks, among others) were uprooted from their lands by invading Huns, who pushed the tribes out of their traditional lands into the Roman Empire. Ironically, the Huns, a nomadic people, had only a modest need for timber.

As Europe settled into a millennium of warlords and small empires expanding and contracting but always warring, the Vikings put wood to good use and created strong light ships that could quickly row up rivers. They used these ships to raid and eventually settle across Europe, from Scandinavia in the north and Sicily in the south, from England in the west to Russia and Ukraine in the east. The woodlands of Scandinavia provided the resources to conquer people across the continent.

China used her forests to build the greatest navy in history, up to that time. Chinese emperors had tribute ships that were four stories tall and twice as long as the longest Viking ships of that era. These Chinese ships sailed as far west as the Middle East and Africa with large armies and holds filled with cannons, centuries before gunpowder technology came to Europe. They demanded and received tribute from the entire Eastern world. To this day, historians argue about how the Chinese could create wooden ships of that size and remain seaworthy. There is no argument, however, that the forests of the Chinese Empire turned China into a worthy sea power. It was only the internal politics of China, which concluded that its advanced culture did not gain enough from the voyages, that stopped

China from demanding tribute from more countries across the world. Chinese politics turned inward, and China let its tribute ships, as well as its rhino-hide-covered warships, rot in the harbors. Later the rising European kingdoms used their wood and Chinese gunpowder technology to conquer the known world.

The Europeans of the Age of Exploration (also known as the Age of Discovery, 1400–1600 CE) brought the conflict over resources to a much higher level. The Portuguese, Dutch, and Spanish used the forests of Europe to build ships that could circumnavigate the world. In their search for dominion over the lucrative spice trade, they discovered two continents that were unknown to Eurasia and Africa. They plundered the resources of these new continents and killed millions of inhabitants in an endless drive for gold, silver, and spices. They created a slave trade that enriched themselves and their countries. When they had worked the inhabitants of some Caribbean islands to death, they imported African slaves to work sugar, tobacco, and cotton plantations. They fought each other in endless wars over the wealth of plundered nations.

Millions of people from the Americas, Africa, Asia, and Europe died in this age, 1500–1900 CE. The conflicts over resources were often fought because of population pressures or a need to feed a people. This era of conquests killed millions in the pursuit of wealth and riches. These resources further empowered the home nations and allowed them to control ever-growing segments of the world. After the Dutch, Portuguese, and Spanish came the English and French. They were able to finally conquer Africa and almost all the kingdoms of Asia. In the last century of this era, three new countries, the United States, Germany, and Italy joined the imperialist movement and conquered the remaining free peoples of the world. Even the few countries that were still free, such as Japan and Thailand, were forced to trade with the imperialist nations or face destruction from the cannons of imperialist ships. China was never fully conquered, but European countries forced her to allow each imperialist country to have an area of influence where they could trade and settle their military and traders.

Rebellions were common, but it took many countries centuries before they won their own freedom. Some people who were forced to be part of countries drawn on a map by European bureaucrats still do not have a nation to call their own. Hundreds of tribes in Africa were forced into countries that split up their people or forced them to live next to traditional enemies. In the Middle East, the Kurdish people, who numbered in the millions, were separated into Turkey, Syria, Iran, and Iraq, where they are minorities with limited rights and political power. The political turmoil today can be traced to this era of conquest in search of riches. These riches of gold and silver gave way to the riches gained from international trade. At the end of this era, new resources emerged on the horizon. Technology was changing transportation and manufacturing. The world would demand new resources to fuel these changes.

Coal and Industry

We have seen how humanity fought over the most basic resources: animals to eat and land to grow grains and raise livestock. We have seen the herders of the plains battle the farmers of the cities. We have seen the Vikings and later the European nation-states fight over the resources of the world that would bring them wealth and power. The battles over international trade bring us into a more modern era in which technology takes leaps and bounds in only a century. This era brings change at a rate the human world has never seen. Steam power, originally derived from burning wood, changed how things were made and how goods and people traveled. Inventions in England in the mid-1800s led to steam-powered manufacturing that created products in mere hours, which once had taken days. It created steam-powered trains that could literally turn a month-long journey into days. Steam-powered ships no longer depended on the winds to sail. They no longer had to follow ocean currents across the world. Ship captains could simply pick any direction they chose. Steam-powered engines quickly made short work of any wood resources, but there was another resource even more energy-dense and plentiful. That was coal.

The coal mines of England produced coal for steam engines and heated homes of people in cities. The famous fog of London was actually smog formed from thousands of sooty coal fires heating and running the city. Coal-powered steam engines pumped water out of coal mines, spun wool, and moved goods across the planet.

Coal became the new gold. Those who owned the coal mines became rich coal barons. Countries without coal reserves were left behind. Even today, most of those countries that did not have natural coal reserves are less developed and less wealthy than those that do. Some of those countries later found oil reserves that brought wealth, but some still languish behind in their development.

Before the era of coal, labor came from humans and animals. The agricultural revolution freed up some people for specialization in the cities, and they created art, literature, and civilization, as we know it, in the towns of ancient cultures. Coal created what appeared to be almost free labor. But centuries of burning coal turned out to have devastating consequences, proving once again that most things that seem free have hidden costs.

This free labor created wealth at levels the world had never seen before. At first, the wealth was concentrated at the very top with the few who owned the land, the factories, and the coal mines. They lived in luxury far beyond the dreams of ancient kings and queens. This spawned a new struggle for resources, which continues to this day. Soon workers banded together to demand a fair piece of the pie. Wealthy owners raised their own armies to beat and kill those who went on strike to demand fair pay, days off, and safe working conditions.

Eventually new political systems evolved to fairly distribute the new wealth. Socialism and communism arose in the twentieth century. At one time these systems were embraced by more than half of the world's population. Through a century of protest and strikes, which left many workers dead, labor unions in capitalistic democracies earned some of the rights for which they had fought. A forty-hour work week, sick pay, vacation, safe working conditions, and abolishment of child labor became the law of the land in most countries. But these struggles continue. Many countries still do not have decent working conditions, job safety, and rights that are enjoyed by people in more developed countries.

Socialism is a system that redistributes some of the wealth to all citizens. It still functions across the planet to keep citizens safe, healthy, and educated. Almost all countries practice some form of socialism. Socialism in Northern Europe, such as Norway, Denmark, and Sweden, provides healthcare and education through college to all citizens, regardless of income level. American socialism provides K-12 education and programs for the poor and elderly. It creates public spaces for leisure and play. Socialist programs work within free-market capitalist economic systems. Many of the countries that have this combination are referred to as social democracies. The United States is currently the only economically developed country to not consider healthcare a right and provide it to her people.

Communism dominated the former Soviet Union, China, countries in Eastern Europe, countries in Africa, countries in Asia, and countries in South America. Today, only Vietnam, North Korea, China, and Cuba are still communist countries. China and Vietnam have allowed capitalism to thrive while retaining a communist ruling party. The Communist Party does not permit elections. The communist theorist Karl Marx thought that wealthy business interests always found ways to warp democracies to their own needs and desires. His idea of a communist government that eventually would lead to small communes with local control has never come to fruition. The communist leaders who overthrew the monarchies or capitalistic democracies never relinquished powers. In almost every case the leader of the revolution stayed in power until his death: China had Mao Zedong, the Soviet Union had Joseph Stalin, Cuba had Fidel Castro, and all these leaders remained until they died or were physically incapacitated by old age. The top-down economies and repressive dictators who accompanied communist revolutions stifled growth. While most citizens' lives usually improved with a communist revolution, future generations had little room for growth and limited options for work and wealth. Like the industrialists and landowners before them, the members who ruled the communist parties tended to keep much of the wealth for themselves. The Soviet Union eventually broke up, due to the increasing cost of supporting other communist countries while being involved in an arms race with the United States. Only North Korea and Cuba still have a pure communist system.

The industrial capitalism that sprung out of the use of coal also continues to have problems with the distribution of wealth and with interference in elections. Politicians in many countries need the wealthy to fund their election campaigns or need to be wealthy themselves. Many politicians will then be beholden to the interests of their wealthy backers. The 90 percent of people who are not wealthy are not able to obtain what they want from their politicians. The wealthy also own the media outlets where many people receive their news. This creates situations in which democracy is at risk.

Not only did coal change manufacturing and transportation, it also ushered in the age of electricity. In 2023 coal accounted for 37 percent of the power sent into people's homes worldwide. The percentage of energy coming from coal has significantly decreased. Coal at one time created almost all the electricity in the world. Worldwide demand for power continues to grow, so even as the United States and countries in Europe decommission coal plants, countries in Asia and Africa are building them. China and India in particular are responsible for more than half of all new coal plants. Renewable energy for electricity accounted for 30 percent of electricity in 2023, but coal use continued to slowly grow in 2023 as well (IEA 2024). Demand for coal should begin to decrease in 2025 according to estimates by the International Energy Association (IEA 2018).

Coal is abundant. Estimates are that the world could use coal at current rates for hundreds of years before running out. Unfortunately, coal creates pollution and soot with heavy metal concentrations that decrease health and longevity in communities around the planet. The mining of coal causes black lung disease that kills miners when they are still young. Mining accidents have killed tens of thousands over the years. CO_2 emissions from coal-fired power plants have led to global warming and other detrimental effects of climate change. The use of coal has already heated the earth by 1°C. It will, most likely, with the use of oil, lead to a planet that is 3–5 degrees warmer than today. Glaciers and ice caps will melt, and many coastal cities will be flooded. This will cause migration on a level never seen before.

Climate change will also alter weather patterns and affect plants and animals across the planet including massive forest fires and species die-offs in the now warmer ocean. Major storms are increasing and are likely to continue to increase. Exactly how catastrophic these effects will be is up for debate. If the entire world could quickly transition to using renewable energy for electricity and transportation, many of the worst possible effects could be avoided. However, many countries have heavily invested in coal (as well as gas and oil), and financial interests that have the ear of politicians are also unlikely to stop fighting for the profits that can be made off coal. It is unlikely that humanity will be able to transition soon enough to avoid the worst effects of climate change. New technologies are promising, and low or zero-carbon alternatives may be able to stem the tide of climate change. The question is, will they be implemented in time?

Oil and Transportation

We have discussed human conflict over energy resources from Stone Age man's fighting over wooly mammoth hunting grounds to the Industrial Revolution's insatiable thirst for coal. In between, we have seen global conflict over the herding land of the Eurasian Steppe, and civilizations clashing over the ability to control the most fertile fields. We have seen how the access to forests and the resources that forests provide, such as lumber and game, affected some of the largest empires on earth. The resources for agriculture, herding, and hunting were used to fuel humans and their animals. Humans and their draft animals—mules, horses, donkeys, elephants, alpacas, and oxen—were the engine that moved the world until wood-powered and then coal-powered steam engines began to supplant them. Coal created mass transportation that had not existed before. Trains replaced ox-driven covered wagons that moved people across the United States and other countries. Coal-powered ships replaced ships that were dependent on wind power. Even with all these advances, most people still relied on their feet or their horse to get from place to place. That would change dramatically over the 1900s as oil changed daily life.

As we have learned, coal revolutionized manufacturing, mass transportation, and electrified the world. In its turn, oil revolutionized personal transportation and warfare in ways that have lasted for over 100 years. Gasoline has an amazing amount of energy inside of it. Gasoline and diesel are both products of oil. Every day, refineries across the planet turn millions of barrels of oil into billions of gallons of gas and diesel. The sheer quantities are mind-boggling. The United States alone uses 391 million gallons per day (US EIA 2023c). The invention of the combustion engine was a long process that stretched from 1791 to 1892. Inventors from France, Germany, Switzerland, Italy, and the United

States all patented types of engines that led to the creation of functional vehicles. Some of the names of these inventors are synonymous with their inventions or the companies that they founded. Gottlieb Daimler, Karl Benz, and Rudolf Diesel are among them. Henry Ford changed the way cars were manufactured and brought the car from a toy of the super-wealthy to a daily tool of the common man.

One defining moment for the car industry came from the most unlikely of places: small towns in the Midwest. In 1875, the young state of Wisconsin in the Northern United States offered a $10,000 reward to anyone who could drive a horseless carriage (in those days a steam-powered vehicle) from Green Bay, Wisconsin, to Madison, Wisconsin. In a race that had only two vehicles finish, humans saw their future. It was another forty years before the sight of a car would be as common as a horse-drawn carriage that had dominated human life for thousands of years. Across the planet, animals had always taken people farther than they could go on their own. But soon fossil fuels would take them farther than they ever dreamed possible (Donald 1921).

Oil before the Car Before we continue to discuss oil and cars, it is important to note that oil was already an extremely vital resource before the invention of the combustion engine. Oil lamps were used to provide light in public and private spaces. Since ancient times, oil lamps were the most common way for people to see in the darkness. Traditionally, oil was derived from natural sources specific to each region: olive oil in the Mediterranean, cottonseed oil in the southern United States, and whale lard across the Arctic. The whaling boom of the early 1800s made whale oil, extracted from the blubber of whales, affordable and thus popular. It was clean-burning and long-lasting. The United States alone imported up to 15 million gallons a year in the 1850s. World production of whale oil topped 34 million gallons in 1914 (Tønnessen and Johnsen 1982). Whaling led to the near extinction of many whale species. Regulations and bans eventually saved what was left of the whale population, but it was the discovery of petroleum in Pennsylvania that may have saved whales from extinction by driving down the price of lamp oil. Petroleum could be converted to lamp oil much more cheaply than either coal or whale blubber.

Oil, more specifically petroleum (from the Latin: *Petra* means rock and *oleum* means oil), is created from dead zooplankton and algae that have been covered by sediment that eventually forms sedimentary rocks. Over millions of years, these rocks exert pressure. As these organisms are pushed deeper into the earth, they are exposed to intense heat. So dead sea life, plus heat, plus pressure, plus millions of years create petroleum. Due to these requirements for creating oil, most oil is deep within the earth, making it hard to get to. There are many exceptions to this, and there are naturally occurring oil springs and tar pits. These raw materials have been used since ancient times for building, lighting, and medicinal purposes.

Oil has been used in some areas since the beginning of recorded history. China used oil for fuel as early as the fourth century BCE. By the fourth century CE, China had many bamboo-drilled oil wells, and oil was used for a plethora of purposes. Ancient Baghdad paved streets using petroleum-based tar. The *Epic of Gilgamesh* describes a precursor of Noah making his boat watertight by using bitumen (heavy oil). The epic was written 4,000 years ago in 2100 BCE about Mesopotamian king Gilgamesh. Oil was used anywhere it was available. During what is called the Islamic Scientific Revolution (the eighth to the fourteenth centuries CE), Arab and Persian chemists could refine oil and distill it into more flammable and explosive substances that were used in warfare. Despite oil's use for millennia and for many purposes, it was limited to regional availability and was never a major worldwide resource until the nineteenth century. There was also no technology available for discovering oil deposits without drilling. Many discoveries were accidental. People discovered oil while drilling for water. Because oil was a niche market, it was not profitable to spend money drilling specifically for oil. However, some accidental finds turned out to be profitable.

As mentioned, this situation changed in the mid-1800s when oil was needed for lighting. An English chemist James Young took oil from an oil seep to create oil for lamps, as well as heavier oil for lubricating machinery in the burgeoning Industrial Revolution. England was leading the world during the Industrial Revolution and had the most need for oil at the time. Eventually, oil was also used for home heating. Many older homes still use furnaces that run on oil. The oil market now began to have real value. Entrepreneurs took ancient knowledge from regions known to have oil resources and invested money in drilling deep underground.

In the 1850s, Canadian inventor Abraham Gesner patented a process of distilling a type of flammable oil from coal. He called it kerosene, but many people called it "coal oil." As coal oil grew in popularity, people began to be interested in petroleum, specifically natural tar and oil seeps in Pennsylvania. The first man to use Pennsylvanian oil was Samuel Kier. He observed that "American Oil" from Tennessee appeared to cure his wife's consumption (tuberculosis). Kier started collecting his own oil and advertised it as "Kier's genuine petroleum! Or rock oil! A natural remedy, procured from a well 400 feet deep, and possessing wonderful Curative Powers in diseases" (Wells and Wells 2022). Other businessmen saw an even bigger opportunity. They created Pennsylvania Rock Oil Company and Seneca Oil and employed noted geologists to assist in their projects. Seneca Oil was led by Edwin Drake. After many failures and the invention of new drilling technology, success was finally achieved on August 27, 1859. This is a date that changed human history forever. Selling first to Kier at $20 a barrel, it was quickly discovered that lamp oil refiners would pay $40 a barrel.

Unbeknownst to those in America, these same processes and discoveries were happening in the Russian Empire. Oil was drilled for and discovered in Baku, Azerbaijan. Oksana Grigorieva improved on Gesner's process and helped the Russian Empire convert oil into kerosene. Rock oil was also extracted in Bóbrka, Galicia (Poland). In Ploieşti, Romania, the world's first refining plant opened in 1857. The next year thousands of oil lanterns lit up the streets of Bucharest (Auzanneau 2018). Petroleum began to revolutionize American life and the life of those living around the world, beginning with its uses for lighting and lubrication.

The small-time companies that drilled for oil in Pennsylvania were soon bought out or squeezed out by larger players. The Pennsylvania oil rush, the first of many such rushes, brought in powerful people who could see the potential for "rock oil" in lighting and industry. One of these players was John D. Rockefeller and his Standard Oil Company. The boom in oil supply also caused many busts. Oil prices could swing from $10 a barrel to

10 cents a barrel in a year. Despite an early attempt by Pennsylvania drillers to limit supply through the Oil Creek Association, too many producers cheated, and it was not effective until Rockefeller created a local monopoly by controlling the refineries. Rockefeller's local monopoly quickly spread across the Northeastern United States. By 1877, he controlled more than 90 percent of the refinery capacity in the country. He raised prices in cities where he had a monopoly, then used those profits to subsidize the low prices of his oil in cities where he did not have a monopoly. His low prices put other refiners out of business, and the cycle continued. As the United States and the world began to be illuminated at night, Rockefeller amassed a fortune that became the largest in world history (Auzanneau 2018).

Oil and the Car Early cars came in all shapes and sizes and were powered by a variety of sources. Electric, steam-powered, gas-powered, and diesel-powered vehicles all competed for superiority. Petroleum-powered cars did not become supreme until the beginning of the twentieth century with the rise of Ford Motor Company and the inexpensive Model T, built on what was then cutting-edge technology of the assembly line.

Electric vehicles did not produce any exhaust. They were popular in cities that were electrified (had electric power from either local generators or electric utility companies). Battery technology was in its infancy, and many batteries were wet (meaning they had fluid in them) and not fully sealed (meaning they could spill). People thought electric vehicles would eventually win the battle as the dominant means of transportation. But by the time Thomas Edison created an efficient battery that did not spill, the Model T had already achieved market dominance. Battery-powered vehicles became the main mode of public transport and most subway systems. The one in New York City still uses electric battery systems.

Steam-powered vehicles were also initially popular. They could run on coal or wood, but most ran on kerosene with white gas in the pilot. They had the disadvantage of often needing to be refilled with water. Their popularity in Boston went into sudden decline when the city closed all the water troughs for horses, due to an outbreak of a waterborne pathogen. These troughs were also used by the drivers of steam-powered vehicles to refill the water needed for a steam engine to function. Eventually, Henry Ford, using the Model T made at his assembly line, was able to produce combustion-engine cars at a price that put the Stanley Steamer and other steamers out of business for good.

The combustion engine was not a superior machine, but Ford created a vehicle that was affordable for all people. The Model T completely changed the urban and rural landscape of the United States and eventually the world. By the early 1900s, oil produced in the United States and the Russian Empire was being exported across the planet. What came to be known as gasoline was just a by-product of oil-refining processes. It had limited uses and was affordable. On the other hand, oil had a wide variety of uses. First, it illuminated the world. Before "rock oil" was discovered, only the wealthy could keep illuminated as long as they wanted. Oil lamps were expensive and mostly filled with whale oil that was dangerous to obtain and expensive to buy. Petroleum was used for almost everything imaginable: lighting, heating, lubricating, cleaning, protecting, and making cement, pharmaceuticals, gum, explosives, and fertilizers. Demand for oil grew as its uses grew. This created plenty of gasoline available for the exploding car market. In fewer than fifteen years (1900–15), the major American cities went from horse-powered to combustion-engine powered vehicles.

The demand for oil continued to rise as car ownership rose across the United States and the planet. Oil was also needed for weapons of war, and by the First World War, it was indispensable to the militaries of the world. As we will discuss in later chapters, entire wars were fought over access to oil. Today civil wars are regularly fought over access to oilfields. The political landscape of today is dominated by powers controlling access to oil and natural gas. The conflict over oil resources has been long and arduous. Oil's dominance in the energy landscape is unlikely to continue past 2040. Quick adoption of electric vehicles and zero-carbon power generation could hasten that decline as soon as 2030.

Nuclear: War and Energy

The discovery of nuclear energy, both for weapons and for energy use, is one of the most important and transformative discoveries in human history. The potential for energy use and for destruction is almost limitless. Nuclear weapons ended the Second World War. Today nuclear power provides over 10 percent of the world's energy needs, and it creates no CO_2 emissions (World Nuclear Association 2019). At one time, it seemed that nuclear power would create limitless power, but accidents at nuclear power plants caused fear across the world. Most countries do not use nuclear power due to expense and fear of accidents. Some countries are in the process of decommissioning all their nuclear reactors, while a few countries are developing new reactors. A new reactor just opened at the time of writing this book in the American state of Georgia.

We have seen the transformation in human society from hunting and gathering to farming, from human/animal power to fossil fuel power. We have seen the accidental side effects of the worldwide use of fossil fuels and global warming. At first glance nuclear energy appears to be a cure-all. However, it is much more complicated, expensive, and messy than it appears. Accidents, radiation, and problems in storing nuclear waste have created many difficulties for countries using this incredible energy source. Rogue nations have used the cover of nuclear energy to create nuclear weapons. Nuclear power remains important and controversial, as the world looks to fill its energy needs and solve the problem of climate change.

Nuclear Weapons A letter dated August 2, 1939, is undoubtedly the most important letter in the history of nuclear weapons and eventually nuclear power. American physicist Albert Einstein signed this letter and had it delivered to President Franklin Delano Roosevelt (FDR). It was written by Leó Szilárd in consultation with fellow Hungarian physicists Eugene Wigner and Edward Teller. All four of these men were Jews who had fled Europe from Nazi persecution of the Jewish people. The Einstein–Szilárd letter, as it is referred to, convinced FDR to begin the Manhattan Project. This project created the first nuclear explosion in human history. The project was staffed by the brightest scientists in the United States, many of them refugees from Nazi-controlled parts of Europe.

It could be argued that Nazi anti-Semitism allowed the United States, instead of Germany, to acquire nuclear weapons first. The racist and anti-intellectualism of the Nazis drove out the best and brightest from Europe's academic world. Einstein, who enjoyed fame at that time, was instrumental in encouraging the United States, Great Britain, Turkey, and other countries to let Jewish scientists into their countries. It is important to note that this was very difficult, as many countries were not allowing immigration, due to the Great Depression. Additionally, most universities including Harvard, Yale, and Princeton had Jewish quotas that limited admission of Jewish students. Few universities had Jewish professors, and those that did had very few. At this time, anti-Semitism was widespread across the Western world.

The Einstein–Szilárd letter inspired FDR and others in the American government to spend over $40 billion (in 2023 dollars). This massive project, initially called the Manhattan Engineering District, was eventually renamed the Manhattan Project. It brought together the brightest physicists from across the world. The Manhattan Project eventually employed more than 130,000 people, and in a relatively short time took a fringe theoretical idea into reality. This was a Herculean task, given that nuclear weapons in 1939 were as theoretical as Einstein–Rosen bridges, otherwise known as wormholes in space-time. That may have been one reason that the Nazis did not spend anywhere near as much time and money to create nuclear weapons. It was also helpful that the scientists who could have led the effort in Europe were now living in the United States, Great Britain, and other allied countries, due to Nazi persecution.

On July 16, 1945, the first nuclear bomb was detonated in the deserts of New Mexico. After over five years of nonstop work, the scientists in Los Alamos, New Mexico, had succeeded. The Manhattan Project was a success. This success would lead to two nuclear bombs being dropped on the Japanese cities of Nagasaki and Hiroshima and causing between 130,000 and 225,000 deaths. This resulted in the surrender of the Japanese Empire and the end of the Second World War. The morality of dropping a nuclear bomb on cities that were mostly populated by non-combatants is debated to this day. Casualty estimates of the planned invasion of the islands that make up Japan were between 1.7 and 4 million American soldiers and 5–10 million Japanese civilians.

Fortunately, the two nuclear bombs dropped on Japan have been the only nuclear weapons used in human warfare. Spies in the United States and Great Britain brought nuclear weapon technology to the Soviet Union and China. Eventually India, Pakistan, Great Britain, France, South Africa, North Korea, and most likely Israel developed their own nuclear weapons. Only South Africa has denuclearized. Even though many wars have been fought by countries with nuclear capabilities, these countries have not used them. The theory is that because both the United States and Russia (formerly the USSR) have nuclear capabilities, they deter each other from using the weapons, due to the fear that these traditional foes would retaliate. This also holds true for India and Pakistan. Israel has remained ambiguous about its nuclear arsenal, but international estimates put its arsenal in the 50–130 individual weapons range.

As of 2019, there were a total of 13,850 known nuclear weapons. Of those, 92 percent are controlled by the United States and Russia. This is down 75 percent from the height of the Cold War in the 1980s. Unfortunately, there are still enough weapons in existence to destroy the entire planet. Modern nuclear weapons can be up to 3,000 times more powerful than the bombs dropped on Hiroshima and Nagasaki. Fortunately, the most fissionable material is used for nuclear energy instead of nuclear weapons (Ploughshares 2019).

Nuclear Power Nuclear power is most easily understood as the creation of a slow and controlled nuclear reaction, usually using uranium as a fuel source. The heat released in a nuclear reaction is used to turn water into steam. Steam turns a generator, which creates electricity. A generator moves a magnet, usually an electromagnet, around a coil of metal wires, and this generates an electric current. Michael Faraday discovered this process for electricity generation in 1831. Any heat source can be used: wood, coal, oil, gas, solar, or flowing water. Nuclear reactions create no greenhouse gases (GHGs) and release no CO_2. One kilogram of uranium has between 2 and 3 million times the amount of energy as coal or oil. In these ways, nuclear energy does not cause global warming and climate change (US EIA 2023d).

But nuclear power releases radiation, and that can be deadly. It also creates nuclear hazardous waste, which in some cases can take over 10,000 years to dissipate before it is considered "safe." If this waste comes near humans, water sources, or in contact with the earth, it contaminates everything in the area. It can cause genetic mutations in all life forms and lead to birth defects in all animals, including humans. Nuclear energy uses 400 gallons of water per megawatt hour of electricity produced (Styles 2017). This adds up to 3 percent of all fresh water used in the United States. Accidents at nuclear power plants have killed people and can cause damage to the environment. Three accidents in particular have made nuclear energy a difficult political proposition: Three Mile Island (United States, 1979), Chernobyl (USSR, 1986), and Fukushima (Japan, 2011).

The anti-nuclear power movement was born after the Three Mile Island incident. The reactor meltdown caused by human error allowed radioactive gas to be released into the atmosphere near Harrisburg, Pennsylvania. While no deaths were recorded, there is ongoing debate about the effects of radiation on the people, plants, and animals that were downwind. The anti-nuclear energy movement got stronger with the Chernobyl nuclear accident in 1986. This was a much larger accident. It was directly fatal for twenty-eight people and indirectly fatal for at least fifteen more. The disaster occurred in Northern Ukraine near the now abandoned town of Pripyat. In 1986, Pripyat was a part of the USSR. The radiation released from the Chernobyl accident ironically was caused during a late-night safety test. The radiation spread across Europe and contaminated twelve countries, in varying degrees. This disaster temporarily ended nuclear reactor building in the United States. Of the 100 reactors online in the United States, only 2 have come online since 1996. The more recent one opened in the state of Georgia in 2023. However, climate change has caused some states to reconsider nuclear energy. Utah, at the time of writing this book, has plans for twelve SMRs (small modular reactors), which will generate only 50 MW each, compared to an average modern reactor of over 1,000 MW.

Despite these accidents and the growth of the anti-nuclear energy movement, the United States still has 100 reactors in operation, and they provide 20 percent of the electricity in the country, with a total output of 797 billion kilowatt-hours (US EIA 2023d). The United States produces almost twice as much nuclear energy as France, which has the next highest nuclear capacity at 419 billion kilowatt-hours. It is unclear if reactor capacity expansion and additions of new nuclear plants will keep pace with reactor retirements in the United States. However, many countries in Asia, especially China, India, and Korea, have plans to build new nuclear plants. The World Nuclear Association lists 50 plants that are currently under construction and another 100 ordered or planned, with a total capacity of 120,000 MWe (megawatt electrical). Three hundred more plants are proposed. In the United States, five reactors have been canceled during construction (with billions of dollars already invested), due to high costs, compared to natural gas and renewable energy plants. It is unclear how many of the 300 proposed plants worldwide will become operational. There is a need for carbon-free electricity, but costs and the possible dangers of nuclear plants make the future of nuclear energy precarious.

The most recent nuclear accident at Fukushima Daiichi Nuclear Power Plant, located in Japan near the towns of Ōkuma and Futaba, occurred on March 11, 2011. This accident was caused by a 9.0-magnitude earthquake and a tsunami set in motion by the quake. The accident released radiation in the surrounding area, and the Japanese government created a 30-km evacuation zone. The plant's radioactive coolant was released into the Pacific Ocean. Heightened radiation levels were reported as far away as California. The Japanese government confirmed only one death from radiation exposure, but more than 2,200 early deaths have been attributed to evacuation and resettlement of the population near Fukushima. The disaster appears to have hastened the decommissioning of several German nuclear plants that had over twenty years of predicted life remaining. Until March of 2011, Germany had obtained 25 percent of its electricity from its nuclear plants. Eight plants were immediately shut down, and the remaining plants were shut down by 2023, despite the fact that many of these plants could continue operating for decades longer. Fukushima demonstrated that even with advanced technology and fastidious focus on safety, natural disasters could leave countries vulnerable to the fallout from nuclear disasters (Appunn 2021).

Nuclear power is currently providing 10 percent of electrical power worldwide. The rise of natural gas and renewables has made nuclear power relatively more expensive. However, nuclear power has the advantage of being carbon-free and reliable. Neither natural gas nor renewables such as solar and wind can claim both these advantages. The growing economies of Asia may require an "all of the above" approach to energy if they hope to maintain growth without increasing global warming and the effects of climate change. If countries, either individually or through international treaty, decide to put a price on carbon, nuclear may once again be affordable, compared to natural gas and coal options. Renewable energy, using solar and wind, is constrained only by the price and capacity of battery systems that can augment electricity needs in times lacking sunshine or wind. If renewables can overcome battery hurdles, it may mean that nuclear power, with its high costs and associated risks of accidents, will no longer be competitive with renewable energy options.

Due to the long lifespan of nuclear power plants, nuclear power will continue to be part of the energy mix for years to come. However, as the German example illustrates, political pressure can prematurely end the use of nuclear plants, regardless of planned span. The conflict over the use of nuclear energy and its safety will continue, and democracies will always have the ability to increase or decrease the use of nuclear power, depending on the political zeitgeist of the time.

Renewables versus Fossil Fuels

Renewable energy is any energy that can be made without the input of a finite fuel source. These include geothermal, tidal, hydro, solar, and wind power. Geothermal power is currently used on a large scale only in Iceland and a few plants in the United States. Hydropower is the most widespread renewable power source. However, as the majority of cost-effective dams needed for hydropower have already been built, hydropower is not growing quickly. Solar and wind power are growing exponentially, and in many places they are the most affordable form of energy generation. Tidal power is promising but is in its infancy. It is unlikely to be a major producer of electricity until the 2030s. The biggest obstacles faced by people promoting renewables are energy storage and entrenched fossil fuel interests, which have found ways to control politicians and government regulators. It is clear that renewable energy is necessary if the world is to limit global warming and the worst effects of climate change. These can be avoided while still providing humanity with power.

In the following pages, we will look at the history of each of these five major forms of renewable energy: hydro, geothermal, tidal, solar, and wind power. Chapter 3 will look into the projections and possibilities of these technologies and how they can be used to slow or possibly reverse the effects of global warming. This section will also look at some of the obstacles to using renewable energy, including regional availability, energy storage, and fossil fuel corporations' political entrenchment.

Hydropower Hydropower has been used for millennia. Rivers have provided transportation and power. Indians, Greeks, Chinese, and Egyptians used hydropower over 2,000 years ago. Watermills ground grain and ground ore in a pre-electric world. Water powered ancient mining operations. In the Middle Ages, Islamic kingdoms used water to power dozens of devices. At the same time, Europeans continued milling and grinding grain. Some of those water mills are still in operation today. Things changed with the inventions of Faraday (1831) in England and Ányos Jedlik (1829) in Hungary. These inventions allowed water movement to be converted to electricity, eventually both AC (alternating current) and DC (direct current). These inventions enabled people to convert the flow of water into the flow of electricity.

Hydropower has disadvantages, as well as advantages. The process of building a dam on a river causes flooding of land behind the dam. In some cases, it can displace entire populations. Small towns, farmlands, and traditional peoples have lost their homes to permanent flooding from the building of hydropower dams. Dams are also dangerous to fish that spawn upriver. Dams have significantly reduced populations of salmon, shad, and herring, among many others, that need access to small creeks upriver of the dams. Dams also change the temperature of the water that flows out of them. Native species that do not tolerate colder water cannot thrive and survive. Sometimes invasive species can fill the space left by declining native populations. A hydroelectric dam also changes the actual physical characteristics of a river ecosystem. The constant and consistent flow of water out of the dam replaces the seasonal ebb and flow of the river. Sediment accumulates behind the dam. The sand and mud banks that were refreshed by seasonal flooding downriver disappear. These changes can be catastrophic for both fauna and flora. Many plants and trees have adapted to the natural ways of the river for reproduction and fertilization. This is another case where invasive, non-native species can replace native plants. These detrimental effects must be weighed against the advantages of emission-free electricity.

Hydropower is emission-free, renewable, and consistent. Wind and solar power depend on wind and sunshine. Neither is as consistent as hydropower. Wind and solar must have battery systems or backup gas-fired plants to guarantee continued and uninterrupted electric power. The technology required for hydropower has existed for over 100 years. Countries across the world have harnessed the power of rivers to electrify their cities and towns and create water storage for cities and crops. The newly formed lakes are used for recreation and fishing. Hydropower accounts for 16 percent of all the electricity produced in the world. That total is down from 20 percent at its height in 1974 (World Bank 2014). This is because hydroelectric dams are expensive, long-term projects that are not always profitable. It is why most dams are built, owned, and operated by governments. However, most of the possible large hydroelectric dams have already been built.

The IEA expects continued but slow growth in hydroelectric power. There is one important untapped resource in the hydroelectric space. Most dams do not produce electricity. In the United States, only 3 percent of dams produce electricity. Almost all dams must release water. The EIA estimates that these non-powered dams have the potential of over 12,000 MW. Hydroelectric energy will continue to be an important part of the renewable energy mix, but as a percentage of renewable energy, currently over 50 percent, it will decrease as other forms of untapped renewable energy come online.

Geothermal Power Geothermal power is energy and heating produced using the heat of the earth. This power comes in many forms. Homes can be individually heated by systems that pump room temperature water down into the earth and return hot water. This same system can be used to create hot water for showers, washing, and cooking. Places that have natural geothermal heat close to the earth's surface can take advantage of this heat at low costs. With its volcanic activity close to the surface, Iceland can heat almost all buildings using these methods. Alternatively, the stable temperature of the earth below 10 feet is between 50°F and 60°F and can be used to heat or cool homes.

For our purpose, the most important use of geothermal energy is to produce electricity. Iceland gets 30 percent of its electricity from geothermal plants, Kenya gets 51 percent, and the United States gets .4 percent. However, as the United States uses many times the power of Kenya and Iceland combined, it is the world leader in geothermal energy production, with 3,591 MW of installed capacity. This is not particularly impressive, as one average power plant produces 600 MW of energy. The next closest countries are Indonesia at 1,948 MW and the Philippines with 1,868 (Richter 2018). The vast majority of US geothermal capacity is in Nevada and California, as most other states have not invested much capital in building their geothermal capacity.

The biggest hindrance to building geothermal electrical output is the cost of finding geothermal power. Unless there is steam directly coming out of the ground, it is unknown exactly where geothermal energy is available. Drilling wells at least 500 feet deep can cost millions of dollars, and there is no guarantee of success. The uncertainty inherent in building geothermal electric plants has stopped most commercial investment in this area. Most other forms of electrical production have fixed costs and expected output for the investment. Conversely, once a successful well has been drilled, geothermal power is one of the most consistent renewable energy resources. Technology being developed by Jason Craig at the University of Nevada, Reno, looks promising (Craig 2021). It has already successfully identified two blind geothermal systems (systems that showed no outward signs of geothermal energy). If this new system can prove successful, it may help the United States obtain a possible 10 percent of power from geothermal resources. The United States is not alone in looking to increase its use of geothermal energy to create electricity. Germany has eight geothermal projects under construction and thirty-seven already in operation (Bundesverband Geothermie 2019). Interest in geothermal energy is growing, as countries look to wean themselves from fossil fuels. Implementation of carbon markets may change the economic calculus of building geothermal plants.

Wave and Tidal Power Tidal power is a continual, dependable resource that is in its technological infancy. Working with seawater, storms, and hurricanes can make the challenge of harnessing the vast powers ocean waves appear daunting. The ocean waves and tides are estimated to have enough power to provide for all the electrical needs of the world. Yet, the technology is 20–25 years away from providing a serious amount of electricity.

There are two large-scale tidal power plants. In France, the Rance Tidal Power Plant built in La Rance, in 1966, has an output of 240 MW. It was the largest tidal power plant in the world until the Sihwa Tidal Power Plant, located at Sihwa Lake, South Korea, was completed in 2011 with a total output of 254 MW. Tidal power plants have remained unpopular due to their expense. They were more expensive to build than coal plants, when coal dominated electricity production. Coal still produces the most electricity of any source, but it will soon be overtaken by natural gas. Tidal power plants are currently much more expensive than wind and solar. Many countries have spent time and money investigating tidal power, but few countries are seriously investing in it. The exceptions are South Korea and Scotland. South Korea proposed building a 1,320 MW tidal barrage plant off the islands near Incheon Bay. Scotland completed a yearlong project that demonstrated the effectiveness of tidal energy and produced over 3 GWh in one year (Osbourne 2019). These examples demonstrate that given political will, tidal power has been effective in producing carbon-free electricity. Tidal power joins geothermal, hydropower, and wave power as the most consistent forms of renewable energy. Its power is constant and predictable. The onset of national and international carbon pricing should place tidal power into the position of a rising renewable.

Wave power is even more difficult to tap than tidal power, as it is deeper in the ocean. There are more disadvantages to overcome, such as which mechanism to use to convert waves to power. They range from writhing snake-like contraptions to buoys attached to the ocean floor. Salt water is corrosive and ocean life attaches itself to metal in the water. Transporting electricity can also be a problem. Everything in the ocean must be able to stand up to hurricane-force wind and waves. At this point, wave power is in the investigative/experimental phase. It is unlikely that the massive power of ocean waves will become a part of the energy mix. Only Australia, Portugal, the United States, and Great Britain are testing wave technologies. Barring a huge technological breakthrough, the massive potential of wave power will remain only potential.

Wind Power Wind power, like hydropower, has existed for millennia. It has helped sail ships since at least 3,500 BCE. Egyptian pots and vessels from that period display images of sailing ships. Ancient Austronesians harnessed wind power to travel across the Pacific Ocean at least 3,000 BCE. Their ancestors came on rafts and settled in Australia and surrounding areas 25,000–50,000 years ago. Humans have used the power of wind to their advantage throughout history. In Persia, people ground grain and moved water as early as 900 BCE. This was the beginning of windmills. The technology spread to Northern Europe, where farmers drained swamps and lakes to create more farmland. Windmills were used across the planet for physical labor for thousands of years. It was not until the 1850s that technology advanced so that windmills could create electricity.

In the western United States, wind power was used to pump water out of the arid earth. Windmills were so named because they were primarily used to mill grain and did so across windy parts of the world. The inventions of metal blades increased the popularity and longevity of windmills for both grinding and pulling water out of wells. Long-standing windmills still dot the western American landscape, especially in areas used for livestock and where there is no electricity. From 1850 to 1970, more than 6 million small wind machines (colloquially still called windmills although they did not mill anything) were sold in the United States (Dodge 2014).

In 1888, Charles F. Brush built the first large windmill capable of generating electricity. The picket-fence style of mill was 17 m in diameter. It was bulky and did not provide much electricity. Advances in design and technology eventually turned wind turbines into feasible generators for farmers and ranchers across the planet. Some of these turbines could generate between 1 and 3 kW of electricity. The first heyday of wind turbines eventually came to a halt, due to advances in electrification. In the United States, electrification spread to the countryside in the form of New Deal projects, such as the Tennessee Valley Authority. These projects, both public and then private, continued until most of the country was electrified by the end of the 1940s. Personal electrification continued and still exists as a niche market for people who live in extremely remote areas, such as the Alaska Bush country and the Navajo and Hopi Reservations in Arizona.

Despite a long lull between the 1950s and 1990s, wind-turbine technology continued to advance. Post-Second World War experiments in Germany and Denmark improved designs and efficiency. Advances in safety and materials made wind turbines more resistant to wind gusts and storms. Dropping prices of fossil fuels post-war kept wind turbines from being commercially viable, compared to coal-generated power. That changed with the Arab Oil Embargo that came out of the Arab-Israeli War of 1973. Oil prices quadrupled overnight after the Arab members of OPEC (Organization of Petroleum Exporting Countries) involved in the war implemented an embargo on oil imports to the United States. These events spurred the US government to spend more money creating "home-grown" energy. Wind was a very local resource. The government began wind-turbine research in 1974.

After much trial and error, these projects began showing success. Unfortunately, by 1981, politicians were starting to view this line of research as a financial boondoggle, and it was defunded in 1989.

Research, technology, and policies combined to start developing wind farms in the 1980s and 1990s in California and Northern Europe. By the early 2000s, the need for carbon-free energy was clear. Wind energy has since grown steadily, with China, the European Union, and the United States, in that order, leading the way. In 2000, there were 17,000 MW of wind power installed worldwide, and by 2023 that figure jumped to 906,000 MW. The need to de-carbon energy in order to limit the rate of global warming has driven the demand for wind power. Wind power is the second most abundant form of renewable energy. Only hydropower is more abundant (see Figure 1.1).

Wind power has been dropping in price every year. The explosive growth of wind power will continue at an annual predicted increase of about 10 percent per year over the next five years, an average of 65,000 MW per year (Wood Mackenzie n.d.). These estimates depend on a multitude of factors and could be vastly underestimated if countries become serious about decreasing their CO_2 levels by 2030. The US government estimates wind power in the United States to be at 224 MW by 2030, compared to 94 MW in 2018 (US Department of Energy n.d.). Wind will likely provide 20 percent of all power by 2030 and 35 percent by 2050.

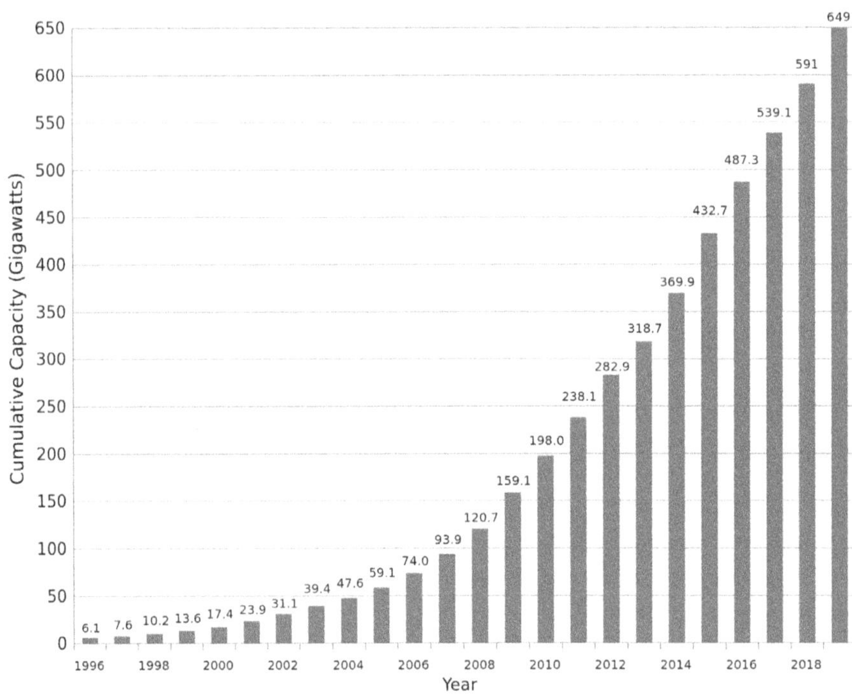

FIGURE 1.1 Global wind power cumulative capacity. Data: Global Wind Energy Council; Wikimedia Commons, https://commons.wikimedia.org/wiki/File:Global_Wind_Power_Cumulative_Capacity.svg.

Barrier to wind power is generally the physical appearance of wind turbines. Many people do not want to see them in their neighborhoods. It is a classic NIMBY (not in my back yard) problem. People like the idea of wind power, but not the giant wind turbines that block their view. This is especially true of offshore wind turbines, which are usually much larger than onshore ones. GE has a turbine that is 853 feet tall. For comparison, the Statue of Liberty is 305 feet tall. Onshore turbines are also large, an average 466 feet tall (Roberts 2018). Wind turbines can also decrease home values if the viewshed is impacted. There are additional concerns if turbines are built in environmentally sensitive areas, where construction could impact local wildlife. People who live close to wind turbines can be disturbed by the vibrations that emanate from the wind turbines. Wind installations have been more successful in less-populated rural areas. Wealthy neighbors have had success using the courts to block wind power projects. On the other hand, farmers have been able to increase their income by allowing energy companies to use their land for installing turbines. Wind turbines do not have a large footprint, so they are compatible with continued farming operations. Despite some local opposition, wind power is increasing every year, and it is the cheapest form of energy in areas that have the wind needed to support wind power. Large offshore wind-power projects have been approved, and they are an appealing option for coastal cities. The AWEA (American Wind Energy Association) estimates that the United States has 2,000 GW of offshore power potential. That is almost twice the annual energy that the United States consumes.

Wind power is poised to become an important part of the energy transition. Most large coastal cities will likely be powered by a combination of wind, solar, and battery-storage energy systems. But it is not known how long this transition will take. The economics of wind, combined with the growing calls to address climate change, indicate that many of these transitions will take place by the year 2030.

Solar Power Solar power is the fastest growing renewable energy source. It is estimated to overtake wind as the most widely used renewable energy source after hydropower. Solar power continues to cost less each year. Advances in technology and economies of scale have allowed solar power to move from a niche market to the growth leader. Solar prices decreased by 70 percent from 2008 to 2018. In the same period, its average annual growth rate was 50 percent. Total solar capacity has grown exponentially. It increased from 15,800 MW in 2008 to 1,200,000 in 2023, an increase of 7,600 percent. Installed capacity is still steadily increasing. If solar maintains this rate of growth, it will soon become the dominant source of electricity. Predictions vary widely as to when solar will become the dominant energy source, but given continually falling costs, advances in battery-storage technology, and the transition to a carbon-free energy system, there is no reason to doubt that solar will dominate by 2030.

People have always used solar energy. As we have discussed, the sun is the source of almost all energy on the planet. Wood, coal, gas, and oil are all energy derived from the sun and transformed into various forms. Solar power was originally used for direct heat. People have used the sun for drying and heating since time immemorial. Sunlight has been used to dry meat, fruit, and other foodstuff. Homes have been built to absorb solar heat during the day, so that it is released at night. We see this in adobe homes that are traditional in the desert southwest.

In 1977, producing 1 kW of solar electricity cost $77. In 2018, it cost $.03 per kilowatt hour, which is 2,567 times less expensive. The price to build solar power has dropped to as low as 2.3 cents per kilowatt-hour in sunnier places. These are the cheapest prices for energy. No energy at any time by any means has been this affordable. Woods Mackenzie believes that solar power will be cheaper than gas-powered electricity anywhere in the world (Stromsta 2019). Decreasing prices of solar power, combined with decreasing prices for battery systems, continue to increase adoption of solar systems. Solar systems pair well with wind systems, as wind is more abundant at night and less in the day. Wind also tends to be more abundant in the winter, as total hours of sun decrease in the Northern Hemisphere.

Barriers to solar power are political in nature. As described, prices continue to fall and will continue to fall in the foreseeable future. Electric companies resist changes to energy systems because they are already making money with fossil fuel power plants. New technology disrupts systems. Many energy providers are intertwined with the fossil fuel extraction industry. Most importantly, they own the plants that are producing electricity. Until building new solar is cheaper than running a gas or coal plant, these providers have little incentive to change.

Obstacles to Renewable Power Regional availability of renewable power sources is difficult in some areas. Any area with little sunshine and wind, no large rivers, no easily accessed geothermal energy, and no nearby ocean has trouble obtaining renewable energy. Fortunately, there are very few places like this. Almost all areas have one of these five types of renewable energy. However, to make these renewable energy systems work together, there is a need for smart grid systems. These are systems that can use intelligent technology that adjusts electricity flows from areas producing large amounts to areas that need it. Smart grid systems must also be coupled with battery storage. There is a need for governments to invest large amounts of money to update these antiquated electrical systems. It is estimated that a nationwide smart grid system in the United States would cost $300–500 billion, but the savings would be close to $2 trillion. These savings are due to more efficient transportation of electricity, avoidance of future repairs, and no money lost from outages that occur from storms and disrepair of old systems. A smart grid would also integrate with electric cars and could use millions of electric cars as a giant battery storage system.

Energy storage is what links solar and wind power to continual electrical generation. The constantly decreasing costs demonstrate that solar and wind plus storage will become the most logical and affordable option for almost every community. The LCOE (levelized cost of electricity) is the expected cost per kilowatt-hour to produce electricity over the predicted lifetime of a power plant. This is usually described as cents per kilowatt-hour. The LCOE for solar, wind, and battery systems is now dropping so quickly that utility companies based in the Great Lake States, such as Minnesota, Michigan, and Indiana (areas famous for their long, cold cloudy winters), are shutting down coal plants earlier than planned and making pledges to move to zero-carbon electricity. If this can be done in states with significantly less solar and wind resources than other states, it will be even easier in the windy states of the Great Plains and the sunny states of the southwest. This is not just a national phenomenon. Much of Germany has a similar climate to Minnesota, and Germany has been a leader in adopting both wind and solar power. While early adoption was not cost-effective there, it is now. With the LCOE of wind, solar, and battery systems continuing to decline, the biggest obstacles are political.

Fossil fuel political entrenchment is the biggest obstacle facing clean energy transition. This works in two ways. The first is a classic political problem. People from areas that produce fossil fuels will fight a change away from that fossil fuel. Coal-mining towns in West Virginia and Germany have put pressure on politicians to subsidize coal-mining operations. Germany stopped its coal subsidies only in 2019. The United States still heavily subsidizes coal, oil, and natural gas production. According to the International Monetary Fund, the United States is subsidizing fossil fuels by an estimated $757 billion. Worldwide, subsidies are over $7 trillion (Black et al. 2023).

The facts presented by the International Monetary Fund demonstrate that while fossil fuel companies loudly complain about the unfair subsidies that wind and solar receive, they are a tiny fraction of the amount of subsidy received by coal, oil, and gas. The $757 billion in explicit and implicit subsidies (such as forgone consumption revenue, not charging for environmental costs, and not taxing on extraction of resources) do not count the costs of pollution and the cost of ameliorating the effects of increased CO_2 production that may well run into the trillion in the long term. Renewable energy received $15.6 billion in subsidies in 2022, double of what was received before the Inflation Reduction Act. This is still a tiny fraction of the subsidies received by gas, coal, and oil, which amount to 4,852 percent more. This demonstrates how truly competitive wind and solar are with gas and coal. Forty-eight times the subsidies are not enough to stop the closing of fossil fuel plants or to stop the exponential growth of wind and solar. Government policy that evenly subsidizes all energy would quickly lead to the transition to carbon-free electricity. If the government would make every company responsible for sequestering the carbon that they put into the atmosphere directly and indirectly, the transition would happen in a few short years. There is less pressure to subsidize emerging industries, as these industries are less established, have less money, and in some cases, create fewer jobs.

The second way that fossil fuel companies are entrenched in politics is more direct. Companies such as Mobil, Chevron, Koch Industries, and Marathon spend hundreds of millions of dollars either directly supporting candidates or supporting lobbying and special-interest groups that advocate for fossil fuel companies. In the United States, this is an especially big problem. The US Supreme Court has ruled that companies can spend unlimited amounts of money supporting or attacking a candidate, as a form of free speech, as long as it is not coordinated with the candidate's campaign. The companies that are receiving billions in subsidies from the government can then spend a small fraction of that money to ensure they continue to receive more money. This cycle is very hard to break. It takes external market changes in order to break this grip on government. A new energy resource or a new technological innovation can sometimes bankrupt these companies, even with the subsidies they receive. This happened in 2018 when Peabody Coal went bankrupt, due to combined pressure of cheaper natural gas and cheaper renewable energy. Despite these massive subsidies, renewable energy surpassed coal in annual energy production in 2019. Natural gas still produces more energy than either coal or renewables, as of 2023. Unless US law is changed, it will take additional paradigm shifts in energy use and production to break the current stranglehold that the fossil fuel industry has on the US government and its energy policy.

LOCATION, LOCATION, LOCATION: ORIGINS OF THE OIL INDUSTRY

United States

On August 27, 1859, on a tributary of the Allegheny River aptly named Oil Creek, a man named Edwin Drake, on a rig locals nicknamed Drake's Folly, drilled the first successful oil well in the United States. From then on, the rush was on in the wilds of western Pennsylvania. Small oil boomtowns such as Oil City, Oilopolis, and Pithole sprang up across the area, along with bars, brothels, and gambling halls (Auzanneau 2018). It was dangerous work that sometimes ended in death from explosions, due to the flammable nature of the product, and murder over competing claims to the right to drill a particular area. The boom-and-bust cycle of a commodity depending on demand resulted in oil prices jumping as high as $40 a barrel and dropping to as low as 10 cents. This cycle destroyed many oil producers. Refiners who were big enough to hold onto their supply until the cycle reversed course had the best shot at riding out the lows of the market. Rockefeller, as discussed earlier, solved this problem by developing a virtual monopoly on the American petroleum industry. He used the power to drive out competitors with lower prices and to raise prices when he had no competition.

Eventually oil was discovered in other parts of the United States, from the Appalachians to California, from Alaska to Oklahoma and Texas. US oil production continued to grow, and for a time it dominated worldwide production. In 1909, US oil production was more than the rest of the world combined. Oil production continued to grow, and US oil was used to support the Allies in the First and the Second World Wars. Oil imports remained low until the 1970s, when new US oil discoveries could not keep up with demand (US EIA 2023a). The United States would remain heavily dependent on oil exports and continually increase imports, until the fracking boom of the 2010s. New techniques that included horizontal drilling and fracking sparked a new oil boom across the country. Fracking is a process of injecting water, sand, and a cocktail of chemicals into oil deposits that have gone dry using the drilling technology of the past. For the first time since 1949, the United States became a net exporter of oil in 2020 and continues to export more oil than it imports (US EIA 2023b).

This statistical view of oil exploration avoids the harsh environmental realities of oil exploration. Land has been stolen from individuals and tribes. Pollution due to oil wells, oil spills, and fracking has left permanent damage to the planet. The promise of easy money has led people and communities to do dark things to other people and other communities.

One of the darkest moments in US oil history was the plotting against American Indians in Oklahoma. They were murdered in order for white settlers in the area to steal their newfound oil wealth. These murders involved collusion with government officials and a blind eye by local law enforcement authorities. In some cases, sham marriages that produced children ended with the murder of the American Indian and the children, sometimes at the hand of the spouse. Many of these murders remain unsolved. This was all made possible by a US law passed in 1921 requiring all Osage tribal members who had head rights to oil to have a white financial guardian, regardless of age. This led to over sixty confirmed murders of men, women, and children. Millions of dollars were stolen by the "guardians" (Gann 2017). For them, robbery was not enough. Murder was their tool. Gann estimates the murders may have been in the hundreds, but due to poisoning and crooked officials, most of the murders were recorded as natural deaths. Historically, any site that is awash in oil is eventually awash in blood. This has been the history of oil across the planet. The Oklahoma case was particularly horrid, as racism against American Indians was widespread and embraced by the vast majority of Americans in the 1920s. Since "settling" the US-delivered genocide to the native people who lived here, it is logical that this genocide is justified with beliefs that American Indians were somehow less than human. This is why the case of the Osage ended up in possibly hundreds of murders instead of the normal assortment of accidental deaths while drilling, due to explosion and accidents, and occasional murders for oil rights, which was more common in places like Pennsylvania, Texas, and Alaska.

Russian Empire (USSR/Russia)

Oil was discovered in Baku in modern-day Azerbaijan, at that time part of the Russian Empire. While small oil operations existed for some time in the area, the Bibi-Heybat well, found in 1886, produced more oil every day than all other wells worldwide combined (Auzanneau 2018). At that time, the United States had 25,000 wells. The Nobel family (its most famous member, Alfred Nobel, invented TNT) moved into Baku in 1877 and began modernizing operations. The family created pipelines that ended the practice of humans pushing carts with barrels of oil. The Nobels bought and built refineries. They built the first tankers. They also had exceptional qualities of treating workers fairly, sharing profits, and educating the workers' children. Branobel, as their company was called, soon began to compete with Standard Oil for a share in the world oil market.

There soon was famous competition. Two of the Rothschilds from France found a way to skirt Russia's anti-Semitic land-ownership laws and created a partnership with companies that competed against Branobel. This new company was called Bnito. With Branobel, it built a 68-km pipeline that got through the Caucuses by using 400 tons of Alfred Nobel's TNT. Baku was described as a living hell. Workers toiled in filthy conditions where illness and disease were rampant and criminal enterprises flourished. Deaths from explosions were still common. Gasoline was still just a by-product of kerosene that was so valuable. Branobel's first tanker blew up, killed many, and caused the world's first (of many) oil spills.

Peru

Peru had an early start in oil, but never became a major player. In 1863, the first oil well was dug in Peru in the Zorritos field. The field is still producing. Then in 1875, the Pirin field was drilled in the Titicaca Basin. Despite the early start, Peru produces only 40,000 barrels per day (bpd) of oil, significantly down from 140,000 bpd in the mid-1990s. For comparison, Russia, the United States, and Saudi Arabia each produce over 11 million bpd. The United States leads the world with over 20 million bpd. Peru has been limited by geography. Much of its oil is estimated to be deep in the Amazonian jungle. It is fortunate for the native Peruvians that the technology and resources to drill in these pristine forests have not been sufficient. Peru's oil boom in the 1960s and 1970s was limited. Many areas were off limits. This has appeared to change, as today 70 percent of the Peruvian Amazon is leased for oil exploration or extraction. On paper this appears very dangerous for both the un-contacted and known tribes that still call the Amazon home.

Texaco, which was bought by Chevron, was accused of dumping in the lakes and rivers of the Ecuadorian Amazon in the 1990s, and it has been in court for decades over it. Spills have also been common in Peru. Eight thousand barrels were leaked into an Amazonian river in 2018. There are fears that confrontation between oil companies and indigenous people will continue. The indigenous people who fight against oil, gas, as well as palm oil plantations, often end up murdered. In 2017, more than 200 land defenders were murdered. Most of these murders were in Amazonian countries (Zachos 2018). Fortunately for those who live in the Amazon, low oil prices driven by growing US supply have made oil drilling in the Amazon less profitable. Drilling in the Amazon will depend on when the world reaches peak oil demand.

Peak oil demand is the point at which less oil is used each following year. It is likely to come as electric vehicle sales continue to grow. Peak oil demand is predicted to occur between 2025 and 2045. It will depend on a number of factors, including adopting rates of electric vehicles, government support for electric vehicle subsidies and infrastructure, and demand for plastics, which are created with oil.

Dutch East Indies (Sumatra)

Shell, one of the biggest oil producers on the planet, made important discoveries and an investment in what was then called the Dutch East Indies (modern-day Sumatra). The Telega Tiga was the first exploratory well drilled in Sumatra in 1883. Two years later the Telega Tunggal well was drilled and became the first successful well in Sumatra. This oil field was the starting point for what would later be the Royal Dutch Shell Company. Today it is called Shell. Aeilko Jans Zijklert dug those wells and acquired a partner. Zijklert died at age twenty-five in the year 1890, but his partner continued on and made discoveries throughout modern-day Indonesia.

Shell Transport and Trading Company was named after the ocean shells that Marcus Samuel sold for interior design in London. His sons began importing oil, using some of the early oil tankers. Shell combined with the Royal Dutch Company in the Dutch Indies.

It had been chartered in 1890, and by 1892 was processing oil in the lands controlled by the Dutch in what they called the East Indies.

In 1907, the two companies merged in order to compete with the behemoth of the industry Standard Oil. Sixty percent was controlled by the Dutch company and 40 percent by the British company (Venturi n.d.). Royal Dutch Shell maintained a government-supported monopoly until American companies were allowed before the Second World War. By 1939, oil production was 170,000 barrels per day. Japan wanted this oil, and in January 1942, after bombing the United States at Pearl Harbor in December, Japan invaded Indonesia. By March 1942, Japan had full control of all the oil fields in Indonesia. Japan held control until they were driven out in 1945. This will be covered in more detail in the next chapter. The Dutch East Indies declared independence in 1945 and went to war with the Dutch and British. The casualties on both sides were high: more than 100,000 Indonesians died, both armed and civilian. British and Dutch deaths were over 7,000 with estimates of 20,000 missing in action. The oil of Indonesia was the impetus to continue to fight for the islands that make up Indonesia (Kahin 2013). Indonesia still produces 833,000 barrels per day.

Persia (Iran)

In 1901, a wealthy English investor named William Knox D'Arcy received prospecting rights from the king of Persia, Mozaffar ad-Din Shah Qajar. The prospecting rights covered an area twice the size of Texas (Auzanneau 2018). The discovery of oil in 1908 led to the creation of the Anglo-Persian Oil Company (APOC) in 1909. A new company named British Petroleum (BP) began marketing the oil in 1916. The discoveries in Persia led to interest in oil possibilities across the Middle East. In May 1914, a young British politician named Winston Churchill convinced the British government to buy a 51 percent stake in APOC (Auzanneau 2018). The future prime minister saw that control of oil would dictate the winners and losers of future wars. Oil production in Iran has continued to grow over the last century with a few bumps along the way.

Starting in the 1920s, the Persian government moved to end a contract with APOC, as it felt it should have been receiving more than the 16 percent of profits it was getting. The vast majority of the wealth leaving the country went to Britain. Despite years of national and international court battles, the D'Arcy agreement held solid until the Shah legally canceled it in 1936. After that, Iran, as Persia was officially called since 1935, benefited from more favorable contracts.

In 1941, during the Second World War, the Soviet Union (USSR) and the British and Commonwealth of Nations invaded neutral Iran to secure oil supplies for both armies. They arrested Reza Shah and put his son Mohammed Reza Pahlavi in his place. They also rolled back some of the autocratic regulations of Reza Shah and allowed for a more democratic government. In 1945 after the war, the USSR remained in Iran and "sponsored" two People's Democratic Republics. APOC, still controlled by the British, refused to have its books audited. The Iranians assumed they were being cheated of their share of the profits. Most likely they were.

In 1951, Iran's new prime minister Mosaddegh ran on a policy of oil nationalization. Nationalization is a process by which a country takes full ownership of its oil. The British countered by creating a naval blockade and orchestrating an international boycott of Iranian oil. They froze Iranian cash and assets. They banned British nationals from working in Iran. They also refused to import British products needed in the oil industry. However, despite internal conflicts, Iran continued nationalizing its oil.

After interference in Iranian elections did not achieve the desired results, Britain asked the United States to back a coup d'état. President Harry Truman originally refused, but the British later convinced the new American president Dwight Eisenhower that communists were infiltrating Iran's government. The US Central Intelligence Agency (CIA) told the shah that he could work with them, or else he would be deposed. The United States and Britain planned to place the Shah in power and to get back control of the oil. These governments only orchestrated the first coup attempt, and Mosaddegh defeated it. He told his supporters who were out on the streets fighting for him to return to their daily lives. The CIA sent in infiltrators who posed as communist revolutionaries and they set about destroying capitalist equipment and looting shops. Next the CIA sent in fake Shah supporters to organize crowds who were angry at the violent destruction. These pushed out the "communists" and the Shah brought in the army. Mosaddegh fled after a tank fired at his residence. He later turned himself in to the army. The Shah spent that week in a hotel in Italy. He flew from Iran to the capital Tehran with CIA director Allen Dulles. After the coup, five US oil companies, along with Royal Dutch Shell and a French company, were allowed to join AIOC in Iran to drill for oil. Britain and the United States backed the Shah for the next twenty-six years.

In 1979, a religious revolution overthrew the pro-Western Shah and created an Islamic State. This revolution came from resentment over decades of British and US control, both covert and overt. The new Iranian government held US embassy staff and personnel hostage for 444 days. They were all eventually returned, but the United States broke formal relations with Iran.

Iran and the United States continue to have a contentious relationship over Iran's support of Shia militias and terrorism, as well as Iran's movement toward creating a nuclear weapon. Iranians hate the United States and Britain for overthrowing their democratically elected leader in order to create profit for international oil companies. Iran has nationalized its oil, and it remains a main source of income. At their height, Iranian exports were 2.5 million bpd. Estimates were as low as 1.5 million bpd in 2023, due to international sanctions. The United States has imposed sanctions on Iran for decades. Under President Barack Obama, the United States agreed to relieve sanctions in return for Iran's giving up its nuclear ambitions. Iran, the United States, the UK, Russia, China, France, Germany, and the European Union all signed this agreement. However, former president Donald Trump unilaterally pulled out of the agreement, and his objective was unclear. Iran appears to be restarting its nuclear program as the United States again tries to enforce sanctions.

Mexico

British entrepreneur Weetman Pearson discovered oil in Mexico in 1908. This caused a land rush by foreign oil companies. In 1917, a new Mexican Constitution declared that all oil was property of the Mexican government. It took many decades for this to become a reality. By 1925, the Mexican government required all oil companies to register with the government and limited its concessions to a fifty-year period. Mexico was second only to the United States in oil exports throughout the 1920s. In 1938, Mexico nationalized all oil fields, equipment, and machinery involved in extracting and transporting oil. Mexico Eagle Petroleum Company was a subsidiary of Shell, but it was set up as a Mexican company under Mexican law, so Shell had no legal standing for damages. President Roosevelt had recognized Mexico's right to nationalize its oil. Mexico paid companies for the equipment and infrastructure. This was the first instance of a country nationalizing its oil without facing international pressure and/or war.

Mexico's timing was perfect, as most countries were dealing with the Great Depression and unable to spend money defending oil companies. Europe was busy dealing with the rise of Hitler's Nazi Germany, and Roosevelt was friendly to many socialist ideas. Mexico created the company Petróleos Mexicanos, or Pemex. Pemex continues to control all oil production and distribution in Mexico. US business leaders were successful in organizing a boycott of Mexican oil and silver, as well as limiting exports of parts needed in oil production to Mexico. This did not last. After the Japanese attack on Pearl Harbor in 1941, Mexican oil was needed in the war effort against Japan, Germany, and Italy.

Mexico reached peak oil production in 2004 at 3.5 million bpd. By 2019, Mexico produced 1.75 million bpd (IEA 2024). If trends continue, Mexico is likely to become a net importer of oil.

FURTHER READING AND REFERENCES

Achilli, A., A. Olivieri, P. Soares, H. Lancioni, B. H. Kashani, U. A. Perego, S. G. Nergadze, V. Carossa, M. Santagostino, S. Capomaccio, and M. Felicetti. 2012. "Mitochondrial Genomes from Modern Horses Reveal the Major Haplogroups That Underwent Domestication." *Proceedings of the National Academy of Sciences* 109, no. 7: 2449–54.

Alroy, J. 2001. "A Multispecies Overkill Simulation of the End-Pleistocene Megafaunal Mass Extinction." *Science* 292, no. 5523 (June 8): 1893–6. https://doi.org/10.1126/science.1059342.

Appunn, Kerstine. 2021. "The History behind Germany's Nuclear Phase-Out." *Clean Energy Wire*, December 15. Accessed October 23, 2022. https://www.cleanenergywire.org/factsheets/history-behind-germanys-nuclear-phase-out.

Aurelius, M. 2015. *Meditations*. London: Penguin.

Auzanneau, Matthieu. 2018. *Oil Power and War: A Dark History*. Chelsea, VT: Chelsea Green.

Baumer, Christopher. 2012. *The History of Central Asia: The Age of the Steppe Warriors*, volume 1. London: I.B. Tauris.

Black, Simon, Antung A. Liu, Ian Parry, and Nate Vernon. 2023. "IMF Fossil Fuel Subsidies Data: 2023 Update." *International Monetary Fund*, August 2023. Accessed July 5, 2024. https://www.imf.org/en/Publications/WP/Issues/2023/08/22/IMF-Fossil-Fuel-Subsidies-Data-2023-Update-537281.

Bundesverband Geothermie. 2019."Geothermie in Zahlen." The Geothermal Congress, May 29. Accessed December 12, 2021. https://www.geothermie.de/geothermie/geothermie-in-zahlen.html.

Craig, Jason W., James E. Faulds, Nicholas H. Hinz, Tait E. Earney, William D. Schermerhorn, Drew L. Siler, Jonathan M. Glen, Jared Peacock, Mark F. Coolbaugh, and Stephen B. Deoreo. 2021. "Discovery and Analysis of a Blind Geothermal System in Southeastern Gabbs Valley, Western Nevada, USA." *Geothermics* 97: 102177.

Dodge, D. M. 2014. *Illustrated History of Wind Power Development*. Littleton, CO. Accessed July 5, 2021. https://web.archive.org/web/20160405043530/http://telosnet.com/wind/index.html.

Donald, J. S. 1921. "Race of First Steam Buggies to Madison for Prize Is Recounted." *Capital Times*, May 26. Accessed December 12, 2020. https://www.wisconsinhistory.org/Records/Newspaper/BA3778.

Ferraro, J. V., T. W. Plummer, B. L. Pobiner, J. S. Oliver, L. C. Bishop, D. R. Braun, P. W. Ditchfield, J. W. Seaman III, K. M. Binetti, J. W. Seaman Jr, and F. Hertel. 2013. "Earliest Archaeological Evidence of Persistent Hominin Carnivory." *PloS one* 8, no. 4, p.e62174.

Gilens, M., and B. I. Page. 2014. "Testing Theories of American Politics: Elites, Interest Groups, and Average Citizens." *Perspectives on Politics* 12, no. 3: 564–81.

Grann, David. 2017. *Killers of the Flower Moon: The Osage Murders and the Birth of the FBI*. New York: Doubleday.

Greshko, Michael. 2018. "World's Oldest Cave Art Found—and Neanderthals Made It." *National Geographic*, February 22. Accessed March 10, 2020. https://www.nationalgeographic.com/science/article/neanderthals-cave-art-humans-evolution-science.

Hurtt, G. C., S. Frolking, M. G. Fearon, B. Moore, E. Shevliakova, S. Malyshev, S. W. Pacala, and R. A. Houghton. 2006. "The Underpinnings of Land-use History: Three Centuries of Global Gridded Land-use Transitions, Wood-harvest Activity, and Resulting Secondary Lands." *Global Change Biology* 12, no. 7 (May 22): 1208–29. https://doi.org/10.1111/j.1365-2486.2006.01150.x.

International Energy Agency (IEA). 2018. "Global Coal Demand Set to Remain Stable through 2023, Despite Headwinds." *World Energy Outlook*. Accessed November 23, 2019. https://www.iea.org/newsroom/news/2018/december/global-coal-demand-set-to-remain-stable-through-2023-despite-headwinds.html.

International Energy Agency (IEA). 2022. "Global EV Outlook 2022." Trends in electric light-duty vehicles. Accessed June 28, 2023. https://www.iea.org/reports/global-ev-outlook-2022/trends-in-electric-light-duty-vehicles.

International Energy Agency (IEA). 2022. "Electricity 2024." Analysis and forecast to 2026. Accessed June 29. https://www.iea.org/reports/electricity-2024.

International Energy Agency (IEA). 2024. "Where Does Mexico Get Its Oil?" Accessed July 6, 2024. https://www.iea.org/countries/mexico/oil.

Kahin, George Mc T. 2013. *Nationalism and Revolution in Indonesia (No. 35)*. Ithaca, NY: Cornell University Press.

Morgan, David. 2007. *The Mongols*, 2nd edition. Malden, MA: Wiley-Blackwell.

Nielsen, Ron William. 2016. "Growth of the World Population in the Past 12,000 Years and Its Link to Economic Growth." *Journal of Economics Bibliography* 3, no. 1 (March): 1–12.

Oosthoek, K. Jan. 1998. *The Role of Wood in World History*. Environmental History Resources, October 15. Accessed June 22, 2019. https://www.eh-resources.org/the-role-of-wood-in-world-history/.

Osborne, Roger. 2019. *An Overview and Awareness Briefing for Offshore Renewable Energy, Wind, Wave, Flow, Hydrokinetic and Thermal Convertors*. OnePetro Offshore Technology Conference, April. https://doi.org/10.4043/29672-MS.

Perri, A. 2016. "A Wolf in Dog's Clothing: Initial Dog Domestication and Pleistocene Wolf Variation." *Journal of Archaeological Science* 68: 1–4.

Ploghshares. 2019. *World Nuclear Stockpile Report*. Ploughshares Fund. Accessed December 7, 2019. https://www.ploughshares.org/world-nuclear-stockpile-report.

Pollinger, J. P., K. E. Lohmueller, E. Han, H. G. Parker, P. Quignon, J. D. Degenhardt, and K. Bryc. 2010. "Genome-wide SNP and Haplotype Analyses Reveal a Rich History Underlying Dog Domestication." *Nature* 464, no. 7290: 898.

Richter, A. 2018. "Global Geothermal Capacity Reaches 14,369 MW—Top 10 Geothermal Countries." *ThinkGeoEnergy*, September 28. Accessed March 2, 2019. http://www.thinkgeoenergy.com/global-geothermal-capacity-reaches-14369-mw-top-10-geothermal-countries-oct-2018/.

Roberts, D. 2018. "These Huge New Wind Turbines Are a Marvel. They're Also the Future." *Vox*. Accessed December 8, 2019. https://www.vox.com/energy-and-environment/2018/3/8/17084158/wind-turbine-power-energy-blades.

Robertshaw, P., and P. Bellwood. 2013. *The Global Prehistory of Human Migration*. Chichester, UK: John Wiley and Sons.

Stromsa, Karl-Erik. 2019. "WoodMac: Solar Plants Cheaper Than Natural Gas Just about Everywhere by 2023." *Greentech Media*, A Wood Mackenzie Business, March 14, 2019. https://www.greentechmedia.com/articles/read/solar-plants-cheaper-than-natural-gas-just-about-everywhere-by-2023-woodmac.

Styles, John Henry, III. 2017. "Nuclear Power Plant Water Usage and Consumption." Stanford University, 2017. Accessed October 19, 2019. http://large.stanford.edu/courses/2017/ph241/styles2/.

Talhouk, S., R. Zurayk, and S. Khuri. 2001. "Conservation of the Coniferous Forests of Lebanon: Past, Present and Future Prospects." *Oryx* 35, no. 3: 206–15. https://doi.org/10.1046/j.1365-3008.2001.00180.x.

Tønnessen, J. N., and A. O. Johnsen. 1982. *The History of Modern Whaling*. Berkeley: University of California Press.

US Department of Energy. n.d. "Map: Projected Growth of the Wind Industry from Now until 2050." Accessed May 8, 2019. https://www.energy.gov/maps/map-projected-growth-wind-industry-now-until-2050.

US Energy Information Administration (EIA). 2023a. "U.S. Imports of Crude Oil." US Department of Energy. Accessed January 8, 2024. https://www.eia.gov/dnav/pet/hist/LeafHandler.ashx?n=PET&s=MCRIMUS1&f=M.

US Energy Information Administration (EIA). 2023b. "U.S. Exports of Crude Oil." US Department of Energy. Accessed January 8, 2024. https://www.eia.gov/dnav/pet/hist/LeafHandler.ashx?n=PET&s=MCREXUS1&f=M.

US Energy Information Administration (EIA). 2023c. "How Much Gasoline Does the United States Consume?" US Department of Energy. Accessed July 6, 2024. https://www.eia.gov/tools/faqs/faq.php?id=23&t=10.

US Energy Information Administration (EIA). 2023d. "Nuclear Explained: Nuclear Power Plants." US Department of Energy. Accessed July 6, 2024. https://www.eia.gov/energyexplained/index.php?page=nuclear_power_plants.

US Energy Information Administration (EIA). 2024. "Henry Hub Natural Gas Spot Price Gas." US Department of Energy. Accessed July 6, 2024. https://www.eia.gov/dnav/ng/hist/rng whdm.htm.

US Energy Information Administration (EIA). n.d. "Electricity Generation." US Department of Energy. Accessed November 11, 2019. https://www.eia.gov/energyexplained/index.php?page=electricity_generating

Venturi. n.d. "Royal Dutch Shell Fascinating History of the World's Largest Anglo Dutch Company." Venturi Group. Accessed December 8, 2019 (page no longer available). https://www.venturi-group.com/royal-dutch-shell-a-fascinating-history-of-the-worlds-largest-anglo-dutch-company/.

Wells, B. A., and K. L. Wells. 2022. "First American Oil Well." American Oil and Gas Historical Society, August 21. Accessed October 23, 2022. https://aoghs.org/petroleum-pioneers/american-oil-history/.

Wood Mackenzie. n.d. "Power & Renewable Energy." Wood Mackenzie: Global Energy Research. Accessed December 23, 2019. https://www.woodmac.com/our-expertise/capabilities/power-and-renewables/.

World Bank. 2014. "Electricity Production from Hydroelectric Sources (% of Total)." World Bank, OECD/IEA, 2014. Accessed December 1, 2019. https://data.worldbank.org/indicator/eg.elc.hyro.zs.

World Nuclear Association. 2019. "Current and Future Generation Plans for New Reactors Worldwide." Accessed January 9, 2019. http://www.world-nuclear.org/information-library/current-and-future-generation/plans-for-new-reactors-worldwide.aspx.

Zachos, Elaina. 2018. "Why 2017 Was the Deadliest Year for Environmental Activists." *National Geographic*, July 23. https://www.nationalgeographic.com/environment/2018/07/environmental-defenders-death-report/.

CHAPTER 2

Current Impact and Linkages

OVERVIEW

This chapter focuses on the warfare that has been caused by the struggle over natural resources from the 1930s to the current day. It also discusses international geopolitical conflicts. Specifically, it covers wars fought over access to petroleum and natural gas. Some of these wars have been overtly about oil and gas, but most of them have been presented as wars about something else. Both international conflict and local civil wars will be covered. The four international wars discussed in detail are the Second World War, the First Gulf War, the Second Gulf War, and the Russo-Ukrainian War. The geopolitical conflicts that are focused on are the Arctic, Iran, Iraq, Saudi Arabia, Qatar, and Venezuela. These are places that have petroleum and gas reserves that draw many countries into conflict. The local internal conflicts that are explored are US fracking, Canadian Tar Sands, the Sudan Civil War, as well as conflicts in Nigeria and Angola.

Major Wars

By the end of the First World War, two things were clear: war was mechanized and petroleum oil (or oil for short) was the fuel of war. Oil provided gasoline for planes, ships, tanks, and automobiles of all types, lubrication for machinery, and some of the raw materials for explosives. Access to oil could easily decide the outcome of a war. Wars were fought over access to oil. Sometimes the battle for oil access was part of a larger war, as was the case in the Second World War. Other wars were fought primarily over oil; this was the case with the First Gulf War. Oil, or at the least the continued access to it, was central to almost every war fought after 1920. This chapter explores these modern wars as they connect to natural resources, focusing specifically on the Second World War, the two Gulf Wars, and the Russian-Ukrainian conflict.

The Second World War did not start out as a war over petroleum resources, but it did evolve into a war over access to oil. Japan attacked the US Navy at Pearl Harbor on December 7, 1941, because US president Franklin Delano Roosevelt had cut off oil exports to Japan.

Japan had invaded French Indochina in 1941. US oil had provided the fuel for the Japanese Empire's wars of expansion and genocide in Manchuria in 1931 and China in 1937. French Indochina was the final straw, and the United States (due to American treaties and relations with France), along with the UK and the Dutch, froze Japanese assets. This move made it impossible for Japan to purchase oil, cutting off over 80 percent of Japan's oil supply. This was in July, and by December the United States was at war.

The First Gulf War was fought over oil. Iraq invaded Kuwait based on the claim that Kuwait was drilling oil on Iraq's side of the border. Iraq also claimed that Kuwait was producing more oil than OPEC (Organization of Petroleum Exporting Countries) had allotted them. OPEC had been able to set the price of oil by controlling supply since the 1970s, however if a member "cheated," prices would drop. Iraq also owed billions to its southern neighbor Saudi Arabia, which also happened to have the world's largest proven oil reserves. The possibility that Iraq would have control of 65 percent of the world's proven oil reserves motivated the United States to start Operation Desert Shield to protect Saudi Arabia. Eventually, with a UN resolution and large worldwide coalition of forty countries. Originally, Operation Desert Shield protected Saudi Arabi and one of the world's largest oil reserves. Eventually, Operation Desert Storm invaded Kuwait and then Iraq. The coalition had the advantage of advanced weapons and was able to free Kuwait and destroy much of the Iraqi army in a war that lasted six weeks, from January 17 to February 28, 1991. The war did not continue into the heart of Iraq and left Saddam Hussein as the dictator of Iraq. The retreating Iraqis set most of the Kuwaiti oil fields on fire. The land mines surrounding these fires made putting out these fires difficult, and it was not until November 1991 that the last oil well was put out. Oil prices did not come down to pre-war cost until February 1992.

The war left Hussein in power in Iraq, but it did stabilize worldwide oil prices. Gas prices continued to remain low until about 2005. Leaving Hussein in power eventually led to the Second Gulf War that was initiated after the 9/11 terrorist attacks in the United States. Although the 9/11 attacks were not in any way connected to Iraq, they were used as an excuse to invade the country a second time and depose Saddam Hussein.

The Second Gulf War (also known as the Iraq War in the United States) was begun by the United States in 2003 on the false pretense that Saddam Hussein had stockpiled weapons of mass destruction (WMDs). There were also false claims of a connection with Al-Qaeda (the terrorist organization responsible for the 9/11 terrorist attacks). Secretary Colin Powell, despite personal reservation, testified to the United Nations that Iraq possessed WMDs (biological and chemical), and had a nuclear weapons program that was working on creating Iraq's first nuclear bomb. Both these assertions were proven to be false after the American invasion. The war was essentially an opportunistic war waged under false pretenses and approved in the atmosphere of a wounded and angry America. The first four years of the war killed between 150,000 and 900,000 Iraqi citizens and soldiers and fueled uprisings and civil wars that added to that death count. Total Iraqi deaths are estimated to be between 1.5 and 3.4 million people. The United States took apart the military and political institutions of Iraq and did not have a clear plan for replacing them. The rise of the Islamic State from the ashes of war-torn Iraq created a worldwide terror organization rooted in Iraq and Syria where none had existed before. This war, unlike the First Gulf

War, did not have a large coalition nor did it have United Nations approval. The long-term international effects are still being felt across the planet. The Powell Rule of "you break it you own it" was in effect as the United States spent an estimated $2 trillion on the war and the rebuild that continued until the end of 2021, when the US forces exited and transferred bases to the Iraqi military. Oil prices did not fall after this war. Despite the United States and its Kuwaiti and Saudi Arabian allies having control of the majority of the worldwide oil supply, oil prices continued to rise through the war until the Great Recession of 2008 brought a huge drop in prices.

The Second Gulf War originally cost the United States upward of $2 trillion without any perceived benefits except for more control over oil resources. It could be inferred that the control of that oil was deemed worthy of the investment. However, the US fracking boom has demonstrated that proven oil reserves are not as static as was once perceived. The United States has produced many times the annual amount of petroleum and natural gas than experts expected just ten years ago. This war has increased terrorism and led to the death of millions. While the historical judgment is still to come, it does not look that this war was successful by most demonstrable standards. It may be that this war was begun with a pre-fracking perspective that overemphasized the value of controlling oil reserves.

Russia originally invaded Ukraine in 2014 under the false premise of protecting ethnic Russians. Ukraine was going through a political revolution against the pro-Russian government of Victor Yanukovych, and Russia sent in army personnel without official Russian Army insignia to secure the Crimean Peninsula for military purposes and also to gain control over the underwater gas and oil fields located within Ukrainian waters. Russian personnel also came into the Eastern Ukrainian regions of Donetsk and Luhansk. Russia officially denied sending in what were obviously Russian troops (due to their Russian equipment and weapons) to secure these areas. Russia held an election in Crimea that was largely viewed as illegitimate by the international community, and then annexed the region despite worldwide outcries to return the Ukrainian territory. All three of these regions have large Russian populations; however, polling has indicated that the vast majority of the people in the Eastern regions prefer to remain part of Ukraine. (There is limited data on the Crimean Peninsula, due to Russia's complete control of the area, but some studies indicate local support for Russia.) Russia has been clear that it intends to capitalize on the opportunity to drill in the newly acquired water surrounding the Crimean Peninsula. Russia invaded the whole of Ukraine on February 24, 2022. Russia perceived Ukraine's desire to integrate with Europe as an existential threat. Russia bet that the world needed their oil, as the second-largest producer after the United States, and that the Europeans were too dependent on Russian gas to interfere or support Ukraine. While no country brought in their troops, twenty countries sent weapons of all kinds to support the Ukrainian military who fought off the Russian onslaught at a level that far exceeded international expectations. This gave allies time to send Ukraine needed supplies.

International Geopolitical Conflict

The countries and areas discussed here are places that bring many nations into conflict due to their resources, location, and ethnic and religious compositions.

The Arctic Ocean has less ice each year and has been increasing in temperature faster than any other area on the planet. This melting and warming creates opportunities for countries that have lands that abut the Arctic. It is also creating conflict as Russia, Denmark (Greenland), Canada, Norway, Sweden, Finland, and the United States vie for new resources. Territorial water claims can overlap, and Russia's military buildup in the Arctic is an ominous development for competing countries.

In the Middle East, Iran is a regional power that supports the Shia minority within the Muslim world. They also support rebels and terrorists who are attempting to arrest control from Sunni leaders. And, they have nuclear ambitions. This has created a situation in which international pressure has been exerted to control Iran's behavior. The main pressure point is limiting Iran's ability to export oil. The United States, under President Trump, had pulled out of a multinational agreement that was signed under President Obama. This tense situation has continued to build as Iran tries to flex its strength under the unilateral US embargo. The Biden administration has not been able to return to the former agreement.

Iraq, after the two gulf wars, continues to be a flash point of Sunni, Shia, and Kurdish conflicting interests. The Islamic State terrorist organization that grew out of this conflict continues to fund terrorism across the world. The United States, Iran, Syria, and what is left of the Islamic State continue to battle for control of the disparate regions within Iraq and Syria.

Saudi Arabia continues to be both a wealthy modern controller of the world's second-largest oil reserves and an old-fashioned dictatorship that gives women little power or voice. The country still regularly executes political opponents, members of the press, and those that fight for civil rights. It is a longtime ally of the United States with close ties to the Bush family as well as many other politically and financially powerful US elites. This dichotomous existence makes Saudi Arabia both a needed ally and a pariah that is not welcome in the Western world. It also has the second-largest proven oil reserves and exports more oil than any other country, so its importance in the geopolitical oil game cannot be overstated. Saudi Arabia has also heavily invested in the US stock market and US securities, with investments totaling over $200 billion.

Qatar is the number one exporter of liquefied natural gas (LNG). It also ranks ninth on the list of petroleum exporting countries. Qatar is ruled by an absolute monarchy that has some trappings of a constitutional monarchy, but the vast majority of all power sits with the Emir of Qatar. Despite a very weak form of government, Qatar has supported a free press through the Al-Jazeera news network. It has also been supportive of revolutions that occurred during the Arab Spring starting in 2011. This has led to conflict with Saudi Arabia, which has supported counterrevolutions that have put dictators and strongmen in power across the Middle East.

In South America, Venezuela has the world's largest proven reserves. Much of those reserves are in oil sands (also called tar sands). Venezuelan mismanagement has caused precipitous declines in production since 2003. The country with the world's largest oil reserves was importing oil from Iran in exchange for gold. Political and economic crises have destabilized the country and led to hyperinflation of up to 1,000,000 percent in 2018. Food shortages have caused the average Venezuelan to lose 19 pounds of weight on average as of 2019. US sanctions attached to socialist government policies and political crises have exacerbated and added to the situation. Sanctions in 2017, 2018, and 2019 have made life miserable for the people of Venezuela. Up to 6 million people or 20 percent of the population had fled Venezuela by 2023. The United States is seeking regime change with explicit plans to reinstate private oil companies. It is clear that the United States sees the crisis in Venezuela as an opportunity to get private access to nationalized oil reserves.

Internal Conflicts

Fracking, or hydraulic fracturing, is the process of injection of high-pressure liquids into shale beds and old oil wells to create tiny fissures in the rocks in order to release petroleum and natural gas. Horizontal fracking in the United States has been controversial. It has led to methane in municipal water supplies, methane coming out of the kitchen sinks led to flaming taps. Fracking has also caused hundreds of earthquakes in heavily fracked areas. On the other hand, fracking, particularly horizontal fracking, which drills sideways along a hydrocarbon seam while fracturing the rock, has helped the United States become an oil and gas powerhouse. The United States as of 2023 produced more oil and natural gas than any other country. Some states have embraced and profited from the fracking boom (Texas); others have banned all fracking (New York). Fracking has turned on its head the idea of peak oil and endlessly increasing oil prices. (Peak oil is the theory that the world has or will shortly reach the highest level of oil production. This would cause the cost of the remaining oil to continually rise and/or stay very expensive because demand would outstrip supply.) The benefit of the fracking boom is that it has the potential to keep oil and thus gasoline prices low. This benefit also has an extremely negative effect on the environment, for two reasons: Cheaper gas prices encourage more gasoline-powered vehicles. The process of fracking for oil tends to cause methane leaks and methane flaring. Methane is a greenhouse gas that is twenty times as powerful as CO_2. Methane flaring is the process of burning off methane. This methane is wasted, and additional CO_2 enters the atmosphere. Fracking has changed the oil and gas industries. However, continued affordable fossil fuels may lead the world along a path in which the effects of climate change cannot be undone.

Canadian oil sands (or tar sands) are thick bitumen heavy oil that is mixed with sand, clay, and water. The major deposit is in Alberta, and it is named the Athabasca oil sands. It holds 1.7 trillion barrels of oil, which is equivalent to all the proven reserves on the planet. It is estimated that less than 10 percent of that would be recoverable, due to oil prices and the expensive process of extracting oil. Canadian oil sands are located in the center of Canada underneath ancient boreal forests, on land that is sacred to the indigenous tribes that live in the area. It is expensive and more polluting than any other type of oil. The process of removing the oil from the sands and the final process of refining the oil take copious amounts of energy and release more CO_2 than any other petroleum. Environmental groups and many First Nations have fought to keep the oil sands in the ground. Moving the tar sands across the country or across the United States to refineries and/or shipping ports has proven problematic. Trains are expensive and dangerous due to spills. Pipelines also spill and are sometimes built through pristine and sacred places.

Protests across Canada and the United States have tried to stop the oil production and transportation. The XL Pipeline Protests across both countries may have influenced the Obama administration's decision to deny permits for the completion of the pipeline. This decision was reversed by the Trump administration and then reversed back by the Biden administration. The XL Pipeline aims to transport Canadian Oil Sands to Louisiana for refining and shipping. Canadian drillers and American refiners have a lot of money at stake. Local governments tend to be supportive, due to increased revenues. The Canadian Oil Sands have become a symbol of the battle between continued fossil fuel extraction at all costs and the movement to slow or stop climate change by moving away from fossil fuel use and keeping it in the ground.

Angola, its oil, and its civil war are extremely complicated cases. In 1975, the south African nation of Angola fought for independence from Portugal and won, but quickly fell into a civil war that was to last for twenty-seven years (ending in 2002). This war was funded by countries across the globe. There were three main fighting factions: one was communist in ideology and the other two were friendly to capitalist countries. However, each faction was supported by a major tribe. This was actually due to local tribes being forced into one nation by colonial or imperial powers. The tribes of Angola used the support of international actors to increase their chances of controlling the government of Angola. International actors such as the USSR, United States, Cuba, and South Africa framed it as a war of democracy over communism. It was a war over control of oil by two superpowers (the United States and Soviet Union). The people of Angola were caught in the cross fire.

Nigerian oil has been a source of conflict continually since it was discovered by a joint project of British Petroleum and Royal Dutch Shell in 1956. Oil operations in Nigeria have been plagued by thousands of spills and millions in bribes, but most of the people in the country see little direct benefit from the oil revenue. However, indirectly, the Nigerian oil industry affects the entire country as most government revenues come from oil sales. As of 2022, Nigeria produced more oil than any other African nation. It has between 16 and 37 billion barrels of reserves (depending on the source of the estimate). Nigeria produced almost 700 million barrels of oil in 2022, averaging 2.05 million bpd. It also has extensive gas reserves and exported 28,625,000,000 cubic meters, which brought in an additional $5.9 billion. Fossil fuels are intricately linked to the fortune of Nigeria and the over 200 million people who live there.

Sudan has been in a state of war for decades. Oil has helped fuel a war that is also linked to race and religion. Originally Chevron was the major player in Sudan but was eventually pushed out by deteriorating relations between Sudan and the United States. The northeast area of Sudan is populated by Arab Muslims. The West Darfur region is populated by African farmers that are mostly Muslim. Nomadic Arab Muslims and nomadic Arabized indigenous Africans that are Muslim also live in the area. These two nomadic groups have been recruited into paramilitary groups called *Janjaweed.* They are funded by Sudan using oil royalties and paid by the booty of war. Thousands of towns have been destroyed. Rape and murder have proliferated as well. Young children have been conscripted into war. The infamous Lost Boys are from this and the previous war in Sudan. Sudan has been in an almost constant state of war between the Christian Africans in the south and the Arab

Muslims of the north. This war would have ended decades ago, but the vast majority of the oil is in the south and all the refineries and shipping ports are in the north. The southern part of Sudan succeeded in becoming its own country: South Sudan. There are still lingering conflicts about borders and sharing oil royalties. These wars over oil and funded by oil profits have killed hundreds of thousands and displaced millions.

This chapter focuses on the conflicts over oil in the modern era from the 1930s until today. Some of these conflicts are overtly oil based. Some of these conflicts are motivated by oil, however nations claim a wide variety of other more palatable reasons in order to garner public support. All of these conflicts demonstrate the power of oil to drive wars, uprisings, and political movements. Oil was the defining resource of the twentieth century; the struggle to restructure human society without oil may define the twenty-first.

MODERN WAR

The Second World War

Buildup The First World War demonstrated that human warfare had changed forever. The more than 5,000 years of draft animal-powered warfare was over. The motorized division was replacing the cavalry. Sailing ships had long been replaced first by the steam ship and then, by the 1930s, the diesel-powered engine. Air forces were being created by all major militaries. Their combustion engines demanded copious amounts of fuel. The new modern army would run on oil.

As the Axis powers began to prepare for the Second World War, they were keenly aware that they did not have the necessary oil resources and that it would be imperative to secure permanent and safe sources of oil. Germany was able to secure the oil fields of Romania. Germany was able to successfully keep themselves supplied with both local oil sources and those in the many countries they conquered during the war. The Nazis also attempted to secure oil in Iran but were thwarted by the British. Italy had few fossil fuel resources, but they did appear to have a large army. Italy's lack of resources, both fossil fuels and the raw materials needed to build weapons, led to a number of unsuccessful campaigns during the war. Italy was in many ways a fossil fuel liability. Germany had to ship coal by train and supply Italy with oil. Italy controlled Libya, but oil was not produced there until well after the Second World War.

The islands that make up Japan have very little oil resources. Japan imported 90 percent of its oil (pre-war), with 80 percent of its oil imports coming from the United States. The Dutch East Indies (today's Indonesia) was the fourth-largest exporter of oil after the United States, the USSR, and Venezuela. In order to conquer the Dutch East Indies, and especially the island of Borneo where most of the oil wells were located, it would be necessary to first stop the US Navy from intervening. The US Navy was the only navy in the world that could compete with the Japanese Naval might. Japan decided that part of their strategy would be to first invade French Indochina (today's Vietnam) and use that as a base for their planned invasion in Borneo and the Philippines, among other targets.

The Japanese invasion of French Indochina was the final step in a series of events that led to the inevitable war with the Allies. Japan had attacked and occupied Manchuria since the 1931–2 war, which was seen as an act of conquest for the now growing Empire of Japan. Japan began a war against China in 1937. It was a bloody war and the Japanese murdered and raped civilians all along the way. The city of Nanjing experienced six weeks of murdering, raping, and looting that saw up to 300,000 killed in that one city alone. There had been some public pressure in the Western countries of Britain, France, Netherlands, and the United States to put some economic pressure on Japan to end the bloody war with China. Oil was the biggest lever that the West had since Japan was so dependent on oil from Western powers.

After Germany defeated France, the puppet government of an area now known as Vichy France granted Japan's repeated request to use French Indochina as a base for operations of the Empire of Japan. This proved to be one step too far for Britain and the United States. Sanctions began, and the United States banned all scrap metal imports to Japan and closed the Panama Canal to Japanese ships as well. In 1941, Japan moved into Southern Indochina and thus threatened British and Dutch territories in modern-day Indonesia. The United States froze Japanese assets in the United States and then shortly after that created a de facto oil embargo against Japan by denying Japan access to the US banking system.

The oil embargo was equivalent to an act of war with Japan because they had only 1–2 year's supply of oil. Japanese emperor Hirohito wrote that Japan went to war with the United States because of oil and they lost the war because of oil. Japan had long planned for this contingency and put that plan into action. Japan sent ambassadors to negotiate with the United States, all the while sending a flotilla of ships and aircraft carriers to attack the US Navy's main base at Pearl Harbor in Hawaii.

Conflict The bombing of Pearl Harbor was the worst attack on US soil in over a century. It propelled the United States into a world war that killed over 100 million people. It changed the course of events for the Empire of Japan, the United States, and the entire world. In the attack on Hawaii, over 2,400 US servicemen and civilians were killed and another 1,100 wounded. All 8 US battleships were damaged or destroyed along with 11 other ships, and 340 planes were damaged or destroyed. The US Navy was severely damaged. Japanese forces had completed two sorties that destroyed or damaged most of the ships at Pearl Harbor; curiously they did not attack nor damage the oil storage facilities. This proved pivotal.

The US Navy with its oil storage and fueling capacity undamaged was able to continue operations after the massive attack. Repair facilities that were also unscathed quickly repaired many of the lightly damaged ships. Admiral Nimitz, commander of the US Navy in 1941, estimated that destruction of the US Navy oil supplies would have prolonged the war another 1–2 years.

Battleships had been the most important class of ships in naval warfare until this point. However, the US Navy had no aircraft carriers at Pearl Harbor on December 7. This would also prove pivotal. The combination of an intact carrier fleet and an all-important fuel supply ready in Hawaii proved the combination that eventually led to US victory in the Pacific theater of the Second World War.

Battleships, which could be up to twice as heavy as aircraft carriers, proved to need too much oil. Many of these heavy ships were also slower than aircraft carriers and the smaller, faster destroyers and cruisers that accompanied them. Fuel requirements limited navies so much during the Second World War that many countries moved to nuclear-powered aircraft carriers and submarines after the war despite their significant costs and long-term risks.

Japan had taken control of oil fields in the Dutch East Indies but had difficulty transporting that oil. Of the 4 million barrels of oil a month produced in Japanese controlled areas, less than 1.4 million barrels ever reached the Japanese mainland. Japanese tanker fleets were insufficient, and US submarine attacks continued to destroy oil tankers throughout the war. US submarines along with air and surface ship support were able to sink 1,392 ships during the war (56 percent of those sinkings were by submarine). The inability of Japan to obtain the amount of oil needed hampered them especially from 1943 to the end of the war in 1945. In the final months of the war Japanese people were digging up pine tree roots and converting them to a very low-quality oil in order for the military to continue to function.

Continued access to oil by the United States and their allies helped lead them to victory. In the European theater of war, Germany's ability to create synthetic gas out of coal as well as their control of oil fields in Eastern Europe allowed them to continue in a two-fronted war much longer than they otherwise would have been able. In the end Germany had to change strategy in order to look for oil. Germany was close to victory over the USSR (Soviet Union), but lack of fuel stopped their progress. The Nazis were unable to complete their victory over the Soviet Union because they put considerable resources into working south through Stalingrad on their way to the Baku oil fields in the southern part of the Soviet Union. They never reached those oil fields and were eventually pushed back and out of the USSR. The lack of continued oil supplies led to the defeat of both the Nazis and the Japanese.

Continued Effects The ability of the Soviet Union and the United States to have continued oil supplies led to their respective rise as superpowers. Their access to oil was what allowed them to overcome two countries (Japan and Germany) that started the wars with vastly more advanced technology. (While the war with Japan ended with two nuclear bombs, Japan no longer had fuel to fly their planes and to move their ships.) These two empires that won the Second World War (the United States and the USSR) fought a forty-five-year cold war (without direct conflict) from 1945 to 1990 that included many hot wars between one or the other superpower and proxy nations that had financial and/or military support from the other superpower. Access to oil and, later, nuclear power defined this period, and their access to oil coincided with their respective rises to power. The continued access to oil defined many of the foreign policy decisions of both the United States and the USSR. The oil-rich countries of the Middle East became focal points for both empires.

Oil production in the United States originally peaked in 1970 and demand for foreign oil continued to affect the country. An oil embargo by Middle East nations in 1967 was absorbed by increased US oil production. OPEC began to flex its muscles by 1971 and was better able to control supply and thus set prices. In 1973 when the United States supported

Israel in war against the Arab-Muslim nations that surround it, the Arab nations banned exports to the United States and Great Britain, causing gasoline price spikes, shortages, and long lines at the pump as countries enforced rationing. This was a turning point for US foreign policy. The United States decided to focus much of its energy and time going forward on oil-producing nations.

The Soviet Union had an abundance of oil until a steady decline began in the late 1980s. This decline coincided with the breakup of the Soviet Union and the eventual fall of communism in all the former USSR republics, including Russia. Soviet oil production dropped by 30 percent between 1988 and 1992. The Soviet Union began its precipitous decline at the same time. Oil may be what had kept the Soviet Union and its satellite communist nations running (Reynolds 1998). The USSR's ability to provide other nations with cheap oil may be what kept communist governments in power for so long. The Soviet Union, which formed out of what was a mostly agriculturally based Russian Empire, did have the oil (and the coal) required for massive industrialization. It appears that the fall of communism may have had more to do with the fall of oil production than US foreign policy or the inefficiencies of a state-controlled economy.

First Gulf War (Desert Storm)

Buildup: Iraq and Saudi Oil The border disputes between Iraq, Kuwait, and Saudi Arabia had existed for centuries and involved kings, princes, the Ottoman Empire, and the British Empire. However, by the early 1920s, a French and an English diplomat agreed upon new borders that basically split control of oil production evenly between French- and British-controlled countries. That would lead to problems that continue to this day. Iran and Iraq were at war from 1980 to 1988. Kuwait supported their neighbor, Iraq. After the war, the Iraqi dictator Saddam Hussein demanded that Kuwait forgive Iraqi debts, but Kuwait refused. Saddam claimed that Kuwait was releasing more oil than was allowed under OPEC rules, thus depressing the price of oil. Iraq and Kuwait share the Rumaila oil field, and Iraq accused Kuwait of slant drilling the oil field and stealing Iraqi oil. Iraq demanded payment for the "stolen" oil. Hussein also reasserted the historical claim to Kuwait as part of its southern province. Kuwait refused to pay the $10 billion demanded for the oil pumped and sold from their shared oil field but offered to pay $500 million. Iraq invaded shortly after they received the offer to pay 1/20 of the demand.

It is important to note that Iraq in 1991 had the fourth-largest military in the world with nearly a million soldiers. Iraq had also received funding, support, and military equipment from the United States. Iran is the largest, richest, most powerful, Shia Muslim country and is always a regional threat to Saudi Arabia, the most powerful, richest, Sunni Muslim country in the Middle East. The United States had at first been neutral during the Iran–Iraq war, as it did not have close ties with either country. Iraq had been a Cold War ally of the Soviet Union. Despite this, the United States came to Iraq's aid because Iran had begun a push into Iraq, and the United States feared that the war would become regional and that Iran, if successful against Iraq, would continue into Kuwait, Saudi Arabia, and the surrounding countries.

After the war with Iran ended, Iraq engaged in genocide of the Kurds of Iraq (one of the largest ethnic groups in the world without their own country). Iraq was not targeting the entire estimated 7 million Kurds in the country but was focused on killing the Kurds in the oil-rich belt of land in northern Iraq. Between 100,000 and 180,000 men, women, and children were systematically murdered, simply for being Kurdish. (They are also Muslim.) Many villages were gassed with chemical weapons; the technology for creating these weapons was received from the United States according to the US Senate Riegle Report. Villages that were taken over with conventional weapons had all the men separated and executed while the women and children were abused and displaced. The lucky ones were forcibly moved to another area within Iraq or were able to flee across the border into neighboring Iran.

Iraq's army, fresh off an eight-year war and a one-year genocide, was a capable fighting force that gave the US military pause. Iraq's advanced army with US equipment had invaded Kuwait and began moving troops toward the Saudi border. The United States felt that they needed to bring the power of the US military and its allies to bear to soundly defeat the Iraqi military.

International reaction was swift and united behind Kuwait. The United States had already been building up forces in Saudi Arabia as part of their Desert Shield initiative in which over 500,000 US troops were placed in Saudi Arabia along with hundreds of planes and thousands of tanks as well as other military equipment. US secretary of state James Baker began what has been called the "tin cup tour." Baker went on a tour of countries asking for military and financial support. He was very successful; he put together an alliance of thirty-four countries that pledged troops and money. Billions of dollars came from the countries threatened: Saudi Arabia ($13 billion) and Kuwait ($13 billion). Billions more from US allies like Germany and Japan (who due to US rules after the Second World War could not send troops). With the huge coalition, almost a million troops, and plenty of money, the United States gave the Iraqis an ultimatum to unconditionally withdraw from Kuwait. Hussein responded by saying that all occupations in the Middle East should be dealt with together. They demanded that Israel withdraw from the occupied territories in the West Bank, the Gaza Strip, and the Golan Heights. This demand was a ploy aimed at uniting Arab leaders with Iraq. The United States refused to intertwine the issues. (The Palestinian-Israeli Conflict began when the British took over administration of the area from the Ottoman Empire in 1925. The attempt to partition Palestine in 1947 turned into a series of wars that were won by the Israelis and allowed them to expand control over all of Palestine. This did not relate in any way to the Iraqi invasion of Kuwait.)

The US military and its allies readied for war as the January 15, 1991, deadline came closer. The deadline came and passed without any reaction or response from Iraq and its invading forces. Operation Desert Shield became Operation Desert Storm, and the United States and its coalition began airstrikes on January 16, 1991.

Conflict Desert Storm, as the US military had code named the operation, was an apt name for the conflict. The actual conflict only lasted six weeks. The United States and its coalition partners began heavy bombardment of Iraqi positions with over 100,000 sorties (individual bombing flights) over forty-two days. The US coalition obtained air superiority quickly. Iraqi forces initially engaged coalition forces but significant losses in the first week of combat deterred further air-to-air combat. Thirty-six Iraqi planes, mostly MiGs and F1s, were shot down in aerial combat. Eighty-one helicopters and planes were destroyed on the ground. Iraqi pilots, in a surprise move, flew to Iran in the hope of someday retrieving the aircraft. Iran, not surprisingly, confiscated the aircraft and eventually incorporated them into the Iranian Air Force. Coalition planes also destroyed eleven Iraqi Naval ships that were attempting to flee Iraqi waters. At the end of the first week of fighting, the Iraqi Air Force and Navy were no longer a factor in the war.

Coalition forces quickly destroyed communication infrastructure and began the attempt to destroy Iraqi modified Scud missile sites. (These were Soviet Scud missiles that were modified by the Iraqis for extended range.) Iraq, in preparation of superior air power, placed Scud sites throughout the desert, making it difficult for coalition forces to locate and destroy them. The Scuds were used to attack Israel, Saudi Arabia, Bahrain, and Qatar. The vast majority of Scuds were sent to Israel and Saudi Arabia. Israel did not retaliate at the request of US secretary of defense Dick Cheney. The United States feared Israeli involvement would cause Arab allies to withdraw from the coalition or even to switch sides. The United States deployed Patriot missile interceptors to Israel and Saudi Arabia. Although the United States claimed they were effective, later research demonstrated that they might have only had minimal success. However, coalition success at destroying infrastructure throughout Iraq led to a fairly quick ceasefire. The coalition was able to destroy the vast majority of the electrical infrastructure in Iraq. At the war's end, Iraq had only 4 percent of the electrical grid in operation.

Iraqi forces invaded Saudi Arabia on January 29 and occupied the lightly defended town of Khafji. Coalition air power was able to destroy over 350 tanks and almost 150 armored personnel carriers in a 24-hour period. Many Iraqi soldiers simply abandoned their vehicles to avoid death. Qatari and Saudi forces, supported by US air power, were able to push back the Iraqis and move into Iraq in pursuit.

The coalition partners were engaged in a number of lopsided battles. The coalition's more advanced M1 Abrams, also called M1A1, built in the 1980s was much more effective than the T-72 and T-62 tanks that the Iraqi forces used; they were built in the 1970s and the 1960s, respectively. The largest tank battle was the Battle of Norfolk, located near the IPSA pipeline road. It was the second-largest tank battle in US military history. Iraqi forces lost over 850 tanks, either destroyed or captured. US forces lost 5 tanks. The British who were engaged on the side of coalition forces lost zero tanks and destroyed or captured 300. In total, 880 Iraqi armored vehicles and artillery pieces were destroyed or captured. This battle effectively ended the war in Southern Iraq.

In the north, the coalition encouraged Kurdish forces that had suffered greatly under Saddam Hussein and his Baathist Regime to rise up and rebel. They heeded the call and overtook Iraqi government forces in fourteen of the eighteen provinces in what they would call Kurdistan. The United States, however, decided to allow an Iraqi general's request for helicopter flights. The stated purpose was for travel because, after all the bombing, most of the roads, rails, and bridges were damaged or destroyed. However, the helicopters were instead used to attack the Kurds. Armed helicopter gunships easily overpowered Kurdish held areas leading to the deaths of upwards of 200,000 Kurds in north as well as Sunnis in the south. This sent millions of Kurds fleeing to Turkey and Iran. Thousands died along the border where Turkish and Iranian forces stopped them from entering. The Kurds are not generally welcome in these countries as the countries already struggle with large Kurdish populations yearning for freedom and independence. The coalition abandoned these peoples.

The coalition partners made the decision to not seek regime change. This allowed Iraqi leader Saddam Hussein to remain in power. Coalition forces created no-fly zones in the north and the south of Iraq (except for helicopters as discussed). Kuwaiti oil supplies and Saudi Arabian oil supplies were safe. Once Kuwaiti supplies were back online, oil prices were once again low. As long as Iraqi oil was flowing, it did not matter to the market if it was in the hands of the Kurds or the Iraqis. Texas oil firefighting legends like Red Adair, Boots and Coots, and Wild Well Control, along with Canadian venture Safety BOSS, which actually put out the most fire, came out to put the fires out in Iraq, and by November of the same year, all the oil well fires were extinguished. Oil markets were stabilized and most of the coalition partners packed up and went home.

Continued Effects The First Gulf War left many things unfinished. Saddam Hussein had an iron grip on Iraq. The Kurdish people suffered horribly under his rule, and many died. Hussein also attacked the Shia minority in the south of Iraq. Many people argued that the coalition did not go far enough. However, the UN resolution, which was the legal and moral justification for the war, was only intended to free Kuwait, and that was done. The United States kept a large military presence in the Persian Gulf. Tension between the United States and Iraq, as well as tension between Iran and Saudi Arabia, continued. Add that unease to the perennial tensions between Arab-Muslim states and Israel, and the area was a powder keg waiting to explode. The US military had pivoted from its Cold War stance that emphasized containing communism and confronting the Soviet Union to a focus on protecting US economic interests, primarily oil.

The unfinished business of the First Gulf War led to the Second Gulf War, also called the Iraq War in the United States. Members of US president George H. W. Bush's cabinet were also members of his son's cabinet eight years later, George W. Bush. Many members of George W. Bush's cabinet saw Iraq and Saddam Hussein as unfinished US business. As we will discuss next, the September 11 terrorist attacks on the United States in 2001 were used as the needed excuse for the Bush administration to go back and "finish the job."

Second Gulf War (Iraq War)

Buildup: Muslim Resentment to Western Imperialism, 9/11, and the War in Afghanistan The Muslim world, specifically the Arab-Muslim world, has been under the influence of Western powers since the 1800s. After the First World War ended in 1919, France and Great Britain sliced up the remains of the Ottoman Empire and created countries that in many cases were unnatural and not ethnically homogenous. They created a lot of tension in these countries by separating peoples with arbitrary borders. They also imposed rule over these areas, many times after promising home rule in exchange for help during the war. European powers used these countries for economic gain and did not allow local people to have the basic rights and freedoms that people of Europe and the United States had. After the Second World War, the United States became more involved in the Middle East, which encompasses mostly Arab-Muslim areas as well as Israel. Israel was formed in 1948, and the eventual backing of the United States was seen as a violation against all Arab-Muslims. The United States intervened in some countries even after they were able to shake off the chains of colonial rule. For example, the United States suppressed a democratic revolution in Iran and installed the Shah of Iran, who many saw as a puppet of Western Nations. These simmering tensions have been met with a movement to get away from Western culture, values, and norms and return to traditional Arab-Muslim traditions. This movement came out of Egypt and is called Wahhabism, named after its founder, Muhammad ibn Abd al-Wahhab (1703–1792), from Arabia.

This movement emphasized a return to traditional Muslim values. This included traditional roles for women: keeping women in the home unless accompanied by a male guardian, post-pubescent women covering themselves (sometimes including the face), and women not having jobs outside of childcare and homecare. The Western ideals of gender equality, as well as a secularism that emphasized the sexuality and freedom of women, was anathema to this return to classic Arab-Muslim rules. Wahhabism was limited to the Saudi Peninsula and connected to the House of Saud that ruled the Kingdom of Saudi Arabia. The money made from oil sales in the 1950s and the 1960s went to creating and funding universities that taught through a Wahhabist worldview. This began a spread of these ultra-conservative ideals that exploded in the 1970s across the Muslim world. Oil money allowed the Saudi version of Islam to be promoted across the Sunni world. (There are several different sects of Islam, but the two most dominant sects are Sunni and Shia at about 89 percent and 10 percent, respectively.) The Saudi version of Islam also did not promote democracy and civil rights, as Saudi Arabia is led by a monarchy that they believe was chosen/blessed by God—a Muslim version of the Christian kings' divine right to rule.

The center of this movement was located in Saudi Arabia. There were many newly wealthy Saudi princes that supported this movement. One of those wealthy Saudi royals was Osama bin Laden. He was a founding member and leader of Al-Qaeda. Al-Qaeda's stated goal is to drive Western power and influence from the Muslim world, especially the Arab-Muslim world. The United States has the most powerful military in the world, possibly the most powerful of all time. Unable to fight the United States in a traditional war, al-Qaeda engaged the US military in asymmetrical warfare, also called terrorism. Early successful attacks included a bombing of the US warship USS Cole on October 12, 2000, as well as US embassy bombings in Tanzania and Kenya.

On September 11, 2001, members of al-Qaeda hijacked four US commercial planes. One plane was flown into the Pentagon (the administrative center of all US military operations, located in Washington, DC); two planes flew into the Twin Towers of the World Trade Center (the physical and symbolic center of the international financial system dominated by Western powers in general and the United States in particular), and the last plane crashed into a field in Pennsylvania. Passengers were able to contact friends and relatives using plane phones and cell phones. Once they realized they were on a suicide mission, they attacked the hijackers en masse and were able to break into the cockpit. The plane crashed in the struggle for control. Investigations demonstrate that the hijackers were planning on flying the plane into the Capitol Building that houses the US Senate and the US House of Representatives. Overall, the attacks killed 2,977 people not counting the 19 terrorists who also died.

Of the nineteen terrorists who carried out the 9/11 attacks, fifteen were from Saudi Arabia. The funding came from Saudi Arabian oil profits. The attack was organized and planned by Saudi Arabian nationals that led al-Qaeda. The United States was a staunch ally of Saudi Arabia and did not want to disrupt the oil flowing from the most prolific producer of oil in the world. Al-Qaeda had previously fought against the Soviet Union in Afghanistan with funding and weapons from the administrations of Ronald Reagan and George H. W. Bush. They had strong ties with the Taliban leadership of Afghanistan, who helped put them into power. The Taliban shared al-Qaeda's hate and distrust of Western powers. Al-Qaeda training centers were in Afghanistan, and these provided a target for the US military that would not affect world oil supplies. The Taliban was openly opposed to what they would call Western Imperialism and American Imperialism in specific. The United States gave the Taliban government of Afghanistan an ultimatum to allow the US military access to al-Qaeda training bases. The Taliban predictably refused, and the United States came in full force with the intention of obtaining revenge for the thousands of American lives lost in the September 11 attacks.

Afghanistan is the place where the great empires of history come to die, or at least to end their string of conquests. It is sometimes called the Graveyard of Empires: Ancient Persians, Ancient Indians, Alexander the Great, the Scythians, the giant Buddha building Kushans, the Huns, the Arab conquerors that had devastated the countries of the Mediterranean, the British Empire, and more recently the Soviet Union all conquered or tried to conquer Afghanistan. None of them were able to hold on to it. A classic example is the Arabs who after unparalleled military success sent 20,000 soldiers into Afghanistan and returned with 5,000; it took 200 years to spread Islam across what was then Buddhist-influenced Afghanistan. Afghanistan is home to some of the steepest and most jagged mountains in the world. The centuries of wave after wave of attacks have created cities that look like fortresses, and people that have accepted warfare as a way of life. The United States attacked in 2001 and remained there until May 2021. They had a strong hold in the capital (Kabul), but a tenuous hold on other parts of the country. The Taliban held many parts outside of the major cities. (The eventual US withdrawal led to the Taliban swiftly taking back control of the entire nation.)

The administration of US president George W. Bush then turned its attention to the unfinished business of Saddam Hussein and Iraq. The United States claimed that Iraq had broken one of the key tenants of the agreements that ended the First Gulf War. It claimed that Iraq was creating weapons of mass destruction (WMDs) in producing chemical weapons as well as attempting to build long-range ballistic missiles. The United States brought the evidence of these claims to the United Nations. The United Nations was not impressed and did not pass a resolution authorizing force. The United States put together a group they called the coalition of the willing. It was only a small fraction of the members of the original coalition. Countries such as Spain, the United Kingdom, Australia, and Poland supported the war, but only the last three sent troops. The war in Iraq unfolded in many surprising ways. As Secretary of Defense Donald Rumsfeld said, "There are the known knowns, the known unknowns, and unknown unknowns." Those unknown unknowns are what came back to bite the United States and the world.

Conflict It is important to understand that Iraq is not a homogeneous society. The people of Iraq are made up of many different ethnic and religious groups. Iraq has Sunni Muslim and Shia Muslim populations that make up the majority of the population. Most of the Shias and Sunnis are part of the population that has adopted Arab culture as well. (Arab-Muslims conquered Iraq in seventh century CE.) Most Iraqis are more genetically related to Iranians and Syrians than Arabs. The Kurds in the north adopted Islam, but not Arab ways, and in that way see themselves as culturally distinct people. A number of people remained Christian as well; these are the Assyrian people of Iraq (also called Chaldo-Assyrians). There are also people who have a monotheistic religion that is not Abrahamic. They believe that there are seven holy beings or angels that rule and protect the world. They have faced persecution from Muslims and Christians alike. These are the Yazidis. Islamic State of Iraq and Syria (ISIS) heavily targeted them. There are also 1.5 million Afro-Iraqis. These people are of mixed African and Iraqi descent. They are the result of 1,000 years of importing African slaves. Unlike slavery in the Americas, slavery was not passed down to the next generation. Children of slaves could marry and own land.

The official war in Iraq that consisted of the US invasion, code named Operation Iraqi Freedom, lasted only six weeks—from March 19, 2003, to April 30, 2003. US Armed Forces estimate that 9,000 Iraqi combatants died along with 4,000 civilian casualties. In this initial phase of the war, 122 US soldiers died.

Operation Viking Hammer was carried out by US Special Forces and Kurdish Peshmerga forces against Ansar al-Islam in the north of Iraq. This operation neutralized a terrorist group and freed up Kurdish forces to attack Iraqi forces from the north. US forces and allies came in from Kuwait into the south of Iraq. Bombing of military positions across Iraq led to widespread desertion among Iraqi troops. High-level officers had been contacted by CIA and other US agencies ahead of time asking for surrender. Many officers did take advantage of the opportunity to surrender. Iraq had been dealing with years of sanctions, and the Iraqi military was undertrained and ill equipped.

Despite this situation, there were areas of intense resistance to the US invasion. The Battle of Karbala was the only Iraqi victory of the war. The United States deployed thirty-one Apache helicopters, but the Republican Guard (highly trained and skilled Iraqi force) was well hidden in an urban area and ready for the attack. All of the helicopters were hit (15–20 times on average), but only one was shot down and another crashed. The Battle of Baghdad was also an intense and hard-fought multi-day battle. Iraqi Army losses are estimated at 2,320. The battle took nine days. The US military lost thirty-four soldiers and seventeen vehicles along with two tanks and an A-10 plane. Iraqi soldiers vigorously defended a number of other locations, including Basra, Najaf, and Al Faw.

Most Iraqis were glad to be rid of Saddam Hussein, but they were unaware of the dangers that would linger in a post-Hussein Iraq. The invasion toppled the Hussein regime, but it unleashed decades of sectarian violence.

Continued Effects The iron grip of Saddam Hussein forcibly kept the peace, through brutal tactics, in a country that should not logically exist. The Kurdish people in the North fought for their own country and freedom from the series of oppressive regimes in Iraq, Iran, Turkey, and Syria. The Shia-dominated southeast chafed under the regime as well. Throughout the country of Iraq, Islamic extremists struggled under the secular Baathists, who hoped for a pan-Arab nation-state that superseded sectarian religious concerns. When the United States disbanded the Iraqi military, they released millions of soldiers and leaders without direction. Many of these ex-military personnel found a place in the sectarian and religious uprising that began across Iraq. Violence erupted across the country.

Shia or Shiite uprisings started almost immediately after the United States declared victory in Iraq over Saddam Hussein. The largest and most successful occurred in 2004. Shia cleric Muqtada al-Sadr led the movement from Baghdad, and it spread across the Shia-dominated south of Iraq led by a group calling themselves the Mahdi Army. Sunni uprising began at this time as well, led by Baath Party loyalists and Sunni mujahideen. The cities of Fallujah and Ramadi were at one time or another completely controlled by rebellion forces. The United States was able to put down these uprisings fairly quickly, but they did lead to power-sharing agreements that further weakened the central Iraqi government the United States was trying to install. These regional power-sharing agreements, a weak central government, and a weak military led to a power vacuum that allowed for the explosive growth of ISIS ten years later.

The Kurds have been US allies against Saddam Hussein and against ISIS. The United States has rewarded the Kurds with de facto autonomy for Kurdistan in Iraq. Kurdistan sits on large oil reserves: between 4 and 45 billion barrels. Almost every major oil company operates in Kurdistan. It is speculated that Kurdish military assistance and an openness to foreign oil companies has kept Iraqi Kurdistan operating as a de facto country. The United States appreciates the blood and oil that the Kurds have given.

ISIS or ISIL, which stands for the Islamic State in Iraq and Syria or the Islamic State in Iraq and Levant (the area stretching from Lebanon and Israel on the Mediterranean to where the Tigris and Euphrates rivers meets the Persian Gulf in Iraq) was a religious movement that held similarities to the Wahhabism movement out of Saudi Arabia. It was in fact a splinter al-Qaeda group. It began in 2006 but did not start to demonstrate its power and strength until January 2014, when it took control of Fallujah, Iraq. It embraces the idea that there should be strict adherence to Sharia law and that religion and government should be combined. A big difference in this movement is that there was no tolerance for those who are not strict adherents of Sunni Islam and follow the Salafi movement, which is a movement that came out of Egypt in the nineteenth century as a counter to Western Imperialism. ISIS members would massacre non-Muslims and Muslims who were not Sunni as well. The Yazidi people suffered especially, with thousands killed and over ten thousand kidnapped, raped, and tortured. Yazidi women were treated as war booty and personal sex slaves, traded among soldiers. ISIS was able, with authoritarian efficiency, to tax every transaction that happened in the territory it controlled. Public executions were a regular experience in towns and cities under its control. At the height of its power, ISIS controlled an area as large as the entire country of Jordan, roughly 34,000 square miles. It controlled large swaths of land in both Iraq and Syria as well as a large area in Libya. ISIS was also an internet savvy international terrorism organization, coordinating hundreds of terrorist attacks across the planet, resulting in over 1,400 civilian deaths.

The rise of ISIS created a huge coalition of unlikely bedfellows. Russia and the United States, as well as al-Qaeda and Hezbollah, fought against ISIS. The ground fighting was mostly done by Kurdish Peshmerga forces. (Peshmerga means those that face death.) Over 11,000 Kurds died fighting and defeating ISIS. Coalition forces sent over 25,000 airstrikes in support of Kurdish Peshmerga and Iraqi forces. It took three years for the coalition to rid the area of ISIS. In 2019, President Trump abandoned Kurdish forces and civilians in Northern Syria and allowed Turkish forces to invade the area.

The fight against ISIS coincided with the revolution in Syria. This war includes Russia, Iran, the United States, Syrian Kurds, Iraqi Kurds, Syrian Freedom Fighters, Syrian Religious Extremists, Hezbollah, al-Qaeda, as well as the Syrian Army and ISIS. The geopolitical importance of this region hinges on its location near countries with vast amounts of oil. Syria has oil enough to account for 25 percent of government revenue, but less than 1 percent of worldwide supply. The fallout from the Second Gulf War continues to reverberate across the Middle East and continues to ensnare almost all major world powers.

Russo-Ukrainian War

The Buildup for the First Invasion Russia (or at least the Russian dictator Vladimir Putin) views Ukraine as part of its sphere of influence and works diligently and deliberately to influence the political and economic landscape of Ukraine. This has been true since Ukraine declared itself independent of the USSR in 1991. Russia made up the vast majority of USSR, and 29 percent of Ukrainians speak Russian as a native language. There are two reasons for that. Russian control of Ukraine led to Russian becoming the language of power in Ukraine, especially in the cities. There are also areas in the Ukraine that have been either settled by Russians (over the 80 years of Soviet rule and 120 more of Russian Empire rule before that) or that have had large portions of native Russian speakers due to the proximity to Russia. The northeastern area of Ukraine has a larger portion of Russians than the rest of the country. The isolated peninsula of Crimea has the largest percent of Russian-speaking people. This is due to 200 years of Russia basing its southern fleet out of Crimea. Fifty-eight percent of Crimea people self-identify as Russian. Russia pipes natural gas to Europe through Ukraine and pipes oil through Ukraine to ports on the Black Sea. Russia also had its eye on the natural gas resources in the Black Sea. The Crimean Peninsula is nearly an island in the Black Sea, connected by a slim isthmus of land to Ukraine, and that location gives it control of most of the known natural gas in the sea. This is part of why Russia has been looking for a reason to regain control of the area (Cohen 2019).

Conflict in the Initial Invasion Protests in Kyiv (also spelled Kiev), the capital of Ukraine, erupted after corruption allegations surfaced about pro-Russian president Viktor Yanukovych in 2014. Yanukovych has a history of corrupt election victories that were annulled by the Ukrainian Supreme Court. Russia has repeatedly interfered in the elections of Ukraine.

Ukraine sits directly between Europe and Russia. It borders six European countries on its Western side and is almost surrounded by Russia on its eastern side. Russia has been interfering in Ukrainian elections since at least 2002. Russia views Ukrainian ties, both economically and militarily, to Europe as a direct threat to Russia and its interests. There are millions of Russians in Ukraine. There is a long history of Russian rule of Ukraine, Russian colonization of the Ukraine, as well as periods of forced repression of the Ukrainian language in favor of Russian, however that has left Russia with deep contacts and a lot of support among some of the people who live in Ukraine and identify as Russian. There is also deep resentment from those that feel that Russia has imposed its language and culture on Ukraine over the years. Russia had financially supported Yanukovych from the beginning of his career. Anytime elections did not go Yanukovych's way, Russia would threaten to cut off gas supplies to Ukraine. Russia used its military and economic might to bully Ukraine out of economic and military alliances with European countries. Russia's biggest fear is that Ukraine would sign on with NATO (North Atlantic Treaty Organization). NATO includes most European countries as well as the United States. This would make the annexation of Crimea or other land grabs in Ukraine impossible without starting a new world war.

There were allegations of voter fraud after Yanukovych "won" the presidency in 2010. Yanukovych bragged about his corrupt ways and how they enriched him personally, according to other foreign leaders. He appeared to have almost unlimited funds, presumably made available by Russia, and is accused in court of bribing top Ukrainian leaders and officials with $2 billion worth of bribes. His enormous mansion on the edge of Kyiv had millions spent on light fixtures alone. The estate is valued at over $100 million. Yanukovych's government salaries over the years amounted to an average of less than $2,000 a month.

The pressure on President Yanukovych began after he abruptly canceled plans to begin an EU (European Union) association which could lead to eventual European Union membership. This abrupt turnaround was seen as directly connected to Russian influence and control of Yanukovych. Tens of thousands of protesters filled the streets in what came to be known as the Maidan Revolution. The protests began in November 2013, and by February 2014 they had turned violent with seventy-seven protesters shot by government forces. This blatant murder of civilians increased the pressure, and Yanukovych fled to Russia with the protection of Russian troops that picked him up in Kyiv. The opposition took over and Russia could see what was coming next: Ukraine would move toward EU membership and join NATO as well. Vladimir Putin ordered the annexation of Crimea by military force. Russian troops and weapons started pouring over the border with Ukraine. These soldiers did not wear their official Russian insignia but were obviously Russian soldiers with Russian weapons and they spoke Russian. Russia's official line was that it took over the Crimea to protect Russians who lived there and wanted to be a part of Russia. The Russian soldiers in the eastern part of Ukraine are referred to as Ukrainian separatists. It is likely that there are some Ukrainians who may want to fight to become part of Russia, but data from surveys have demonstrated that there is little support to join Russia in Eastern Ukraine and there is some support in Crimea. Crimean data is suspect as Russia had already taken control of Crimea at the time of its collection, and people may have been afraid to voice dissent.

Russia conducted a referendum in Crimea and claimed that 90 percent of the voting residents wanted to join Russia. International condemnation and sanctions have not swayed Russia, and the annexation of Crimea remains popular with the Russian people. The Russian-Ukrainian war in Eastern Ukraine continued at a stalemate with Russian troops controlling the Donetsk and Luhansk regions of Eastern Ukraine until February 2022. Russia then began a full-scale invasion with the intent to conquer Kyiv and kill the elected leaders there. Thousands have died in the fighting. Ukraine is dependent on military aid from Western countries in order to contain and limit the loss of land to the Russian invasion.

Russia's Second Invasion When Vladimir Putin decided to move from a slow and steady campaign to gain control of the oil and gas assets of Eastern Ukraine to a full-scale invasion of the entire country, he was counting on two things: One, that Ukrainian defenses would be quickly overwhelmed by the superior numbers of Russian invaders; and two, that the European Union, due to its dependence on Russian gas for electricity and home heating, would not be willing to support Ukraine. However, the Ukrainian military, after eight years of fighting Russian forces in the Donbas region, were more of a match for the Russians than anyone imagined. And, despite dependence on Russian gas exports, many of the nation-states that make up the EU were willing to support Ukraine with arms and money.

The Donbas-Dnieper region of Ukraine produces 90 percent of Ukrainian natural gas. It also has a large ethnic Russian population. These two facts are why Putin has supplied and supported a separatist movement in Eastern Ukraine. When the democracies of the EU, North America, and across the globe did not stand in solidarity with Ukrainians as Russia illegally annexed the Crimean Peninsula, they signaled to Putin that there would be very little consequences for Russian aggression in Ukraine. The former Soviet Union was one of the two superpowers of the post-Second World War era. It could be assumed that Putin planned to hold the same sham election he held in Crimea in other parts of Ukraine. He could then declare those areas independent, but in reality he could then integrate them into the Russian state. This would once again have put Ukraine into the "Russian Empire." It would have increased Russian power and clout as a legitimate world power. Currently, Russia is regarded as the third most powerful country, well behind China and the United States.

Russia has invested much blood and money into this war. Russia has seen an average of 800 casualties a day in the summer of 2023. It has not achieved its major goals of Ukrainian regime change at minimum and the reincorporation of Ukraine into a new Russian Empire. The Russian invasion has also had unintended consequences. It has shifted the EU toward American natural gas and accelerated the transition toward renewable energy in the EU. Sanctions on Russian oil exports have forced Russia to pivot toward China, and to a lesser degree India, to find markets for its oil. In this new world order, Russia finds itself at the mercy of China. Without Chinese oil sales Russia would be cut off from the funds it needs to wage its unprovoked war of murder and destruction of the Ukrainian people.

Unless Russian fortunes change, the Russian invasion of Ukraine will only further marginalize and diminish the prestige and power of Russia. Military weaponry pouring into Ukraine is turning Ukraine into a regional military powerhouse. The question that remains is will Russia in desperation use nuclear weapons against an emboldened Ukrainian military. It is also unknown what Russia would settle for. It is likely that Russia would not accept returning Crimea, as it sees the area as the needed location base for its Black Sea-based navy. Russia might settle for recognition of the "break away" provinces. This would allow it to have control of the vast majority of Ukraine's natural gas proven reserves. Will the West pressure Ukraine to cede territory in order to end the war? That is yet to be seen.

Continued Effects Russia has gained political power and stolen natural resources. Ukraine is in a much weaker position than they were before the invasion and illegal annexation of Crimea. Russia has made it impossible for Ukraine to join NATO. It would be extremely unlikely that NATO would allow Ukraine to join while they are at war with Russia. The Western nations want to deter Russia from invading other countries, but they do not relish the idea of engaging in a protracted land war with Russia. The fear of that kind of war escalating to a nuclear war is clear.

Russia has become a political pariah to almost all world leaders. Former president of the United States, Donald Trump, is one of the few Western leaders to support Russia. The president's affinity for Vladimir Putin and Russia is befuddling to most foreign policy experts. It remains unclear why Trump appeared to defer to Putin. China has not condemned Russia because they hope to retake Taiwan someday, by force if necessary. China has also taken advantage of the situation by stocking up on discounted Russian oil.

President Trump abruptly pulled US troops away from the Syrian-Turkish border in 2018. This is an area that is mostly Kurdish. The Kurds have been staunch US allies against the regime of Saddam Hussein in both Gulf Wars and in the 2014–15 battle against ISIS. It is unclear why Trump would be willing to abandon an ally in lieu of a traditional enemy of the United States. What is clear is Russia is attempting to become the new power broker in the Middle East. With so much oil at stake, it is not surprising that Russia would like to be viewed as the country that keeps the peace in the Middle East.

Depending on how quickly electric cars are adopted, control of the Middle East could be either a long-term liability or a long-term asset. The natural gas resources that Russia stole from Ukraine will continue to be viable into the foreseeable future. Transitioning away from coal has involved both renewables and natural gas plants. While renewables are growing at a faster pace than natural gas, natural gas makes up 22 percent of all energy production. Renewables account for 30 percent, which includes all sources but primarily hydro, wind, and solar. Russia is likely to see fossil fuel profits from its politically unpopular invasion and annexation of Crimea. If Russia can continue to control Eastern Ukraine, Ukrainian gas will continue to enrich Putin and kleptocratic associates. The United States with her allies can hold Russia accountable and apply international pressure, but China, Iran, and India are supporting Russia by buying oil, and Iran is supplying weapons.

CURRENT GEOPOLITICAL CONFLICT

Arctic

The Arctic Ocean was once a place only suited for polar bears, seals, fish, and traditional hunters and herders. Only the hardiest of humans had adapted to survive, including the

Inuit, Sami, and the Chukchi. The over forty ethnic groups that have survived there for millennia have learned how to harness nature's bounty and use it to protect themselves from nature's wrath. Covered by thick, decades-old ice and visited by storms twice the size of hurricanes (up to 200,00 km^2), it was a death trap for European and American explorers and empires that dreamed of a northern passage between Eurasia and North America. Global warming has changed that. While the global warming caused by climate change has warmed the planet by an average of 1.9° F, parts of the Arctic Ocean have warmed over 7° F. Warmer waters mean less ice and more open ocean.

Russia and the United States fight with Canada, Denmark (Greenland), Norway, Sweden, Iceland, and Finland for control of the resource-rich waters. The once inaccessible Arctic waters are becoming more navigable every year. Sea ice is becoming thinner and younger. Some ships can now cross the Arctic Ocean without an icebreaker making a path. The first large freighter, the *Nordic Orion*, crossed the Northwest Passage (the waters North of North America) in 2013 and the first cruise ship made the passage in 2016 with 1,700 passengers aboard. Animal and plant species are migrating from the Pacific to the Atlantic. In 2011, scientists recorded a plankton species not recorded in the Atlantic for 800,000 years. Oil and gas that has been inaccessible is now available for extraction. The race is on to claim and drill these resources.

Russia has the biggest head start. Russia also has more coastlines facing the Arctic Ocean than anyone else. It has had military operations in the Arctic from before the Cold War and, unlike the United States, it continued to have a large military presence long after the Cold War ended. In a world in which many times might makes right, Russia is able to enforce its competing claims in Arctic waters. Russia has threatened to use force if necessary to secure its ability to drill for oil and gas in areas claimed by Denmark (Greenland), Sweden, Finland, and Norway. Russian claims also butt up against US claims as Alaska as Alaskan coastal waters run up against the coastal waters of Siberia (the easternmost province of Russia). For all its bellicose posturing, Russia has only been able to put one oil rig into operation as of 2023, and that rig is not a deep-sea oil rig. It has also drilled a few wells for oil and gas in the Kara Sea area, which is at the edge of the Arctic Ocean. While Vladimir Putin claims that there will be dozens of oil platforms in operation by 2025, Russia does not have the support ships needed for each platform. Each rig requires fifteen additional support vessels. The one operational rig is using most of the vessels that are built to survive Arctic conditions (Kireeva 2019). Russia, which had built most of these types of vessels in the past, is not currently building any of these ships. The physical conditions of the Arctic also require specialized covered rigging platforms to shield workers from the cold and harsh conditions. It is extremely expensive to build these rigs. These technical and logistical problems are exacerbated by the Russian bureaucracy, which requires companies to "prove" resources by putting in two test sites before starting operations. Each region is twice the size of Belgium (Kireeva 2019). Current geopolitical sanctions imposed against Russia for their unprovoked invasion of Ukraine also create a less-than-hospitable gas and oil export market. Given the fact that oil demand is likely to plateau by 2030 and decrease thereafter, it is an open question whether Russia or any other nation will be able to capitalize on Arctic resources before falling demand will make expensive Arctic drilling economically unfeasible.

The United Nations Convention on the Law of the Sea (UNCLOS) has been signed by every Arctic Ocean nation except the United States. Its purpose is to weigh competing claims to the fossil fuel and mineral rights under the Arctic Ocean. These claims do not affect the 200 nautical miles that are internationally recognized as the territory of each country. Denmark, Russia, and Canada all claim that their continental shelves extend far enough to claim the mineral rights under the North Pole. The UNCLOS has not yet ruled on these competing claims, but Russia has indicated they would be willing to fight for their claim if it is not approved by UNCLOS even though they are a signatory on the agreement. These competing claims for rights to fossil fuels will depend on the long-term markets for these fuels and if the expenses are worth the extraction.

The United States has not set up an oil rig outside of its territorial waters, but it does have several rigs in the Arctic Ocean, north of Alaska. The United States appeared to have neglected the Arctic in many ways since the end of the Cold War. However, it does patrol the Arctic with nuclear submarines. There is an off-again, on-again debate over drilling in the Arctic National Wildlife Refuge, but otherwise the United States appeared to de-prioritize the Arctic. That is, until 2019 when Secretary of State Mike Pompeo, who appeared at the biannual meeting of the Arctic Council, stated that the United States was reengaging in the Arctic for trade and oil and would be backing that up with an increased military presence. Despite the rhetoric, not much has changed to date. The United States hopes for peace but has witnessed some bad actors (Russia and China) that do not appear to play by the rules. The United States stands ready to aggressively defend its rights in the area to both resources and trade routes. Climate change will completely alter the Arctic, and if humanity continues on a path that is fossil fuel driven, the Arctic could become a battleground for the remaining fossil fuels. Scientists estimate that the Arctic Ocean and the surrounding land masses may hold upward of 33 percent of all remaining undiscovered oil and gas. The "Pompeo doctrine," as it has been called, may be the United States' first verbal salvo in a war of words that may one day become a war in earnest.

Iran

Iran was one of the original oil-producing countries in the world. Iran's history with oil stretches back to 1901 when it gave concession rights to William D'Arcy. More importantly, D'Arcy discovered oil in 1908. The Anglo-Persian Oil Company or APOC was born in 1909. (Iran was called Persia at the time.) In 1914, a young Winston Churchill realized the importance of oil in warfare, and the British government bought a majority share in APOC, changing it to Anglo-Iranian Oil Company (AIOC) in 1935. Britain and the United States managed to keep control of the oil in Iran; this included a coup d'état led by the CIA and British intelligence agencies in 1953. Widespread resentment of US and British meddling and the miniscule amounts of oil revenue that actually got to Iranian citizens helped fuel the Islamic Revolution of 1978–9. US hostages were held for over a year and US-Iranian relations have been on the rocks ever since. Iran continues to produce over four million bpd. Despite US attempts to limit the power and influence of Iran, Iran continues to be a regional powerhouse and the de facto leader of the worldwide Shia Muslim community.

Iran's revolution gave religious clerics, specifically Ayatollah Khomeini, complete power over the country. Iran had been considered modern and moderate, but the revolution changed that. Iran has supported and supports Shia groups that the world considers terrorist organizations. Groups such as Hezbollah in Lebanon received arms and funding. While these actions angered many Western countries, little action was taken due to the importance of Iran's oil on the oil market. A steep decline in Iranian production would likely spike oil prices globally.

Iran went to war with Iraq. The United States tacitly supported Saddam Hussein, the dictator of Iraq, in that war due to the US dislike of the overtly religious and anti-US government that rules Iran. However, President Ronald Reagan and his administration arranged for illegal arms sales to Iran in order to fund the Contras, a conservative paramilitary group operating in Nicaragua. This affair, called the Iran-Contra affair, led to disgrace for the Reagan administration as well as jail time for some members of the administration. Reagan's repeated phrase "I do not recall" during the hearings have led many to wonder if the president was already showing signs of Alzheimer's disease and possible dementia by the end of his second term. The Iran-Contra affair was an illegal scheme to directly go against the will of the people, expressed through Congress. Possibly due to the fact that the Iran-Contra affair came to light at the very end of Reagan's second term as president, it did not lead to his impeachment. It did demonstrate to the lengths an administration was willing to go to foment change from a socialist government to a pro-capitalistic, anti-socialist mold. The Contra revolution was not militarily successful, but the Contras had some later success politically. Iran continued to be generally left alone after the war with Iraq ended in a costly stalemate for both sides. That changed when Iran began the process of creating nuclear weapons.

Iran started enriching uranium for the stated purpose of creating fuel for a nuclear plant. This weapons-grade uranium could also be used to create nuclear weapons. The United States and her allies began putting sanctions on Iran. These sanctions had the desired effect of bringing Iran to the negotiating table. In 2015, the Obama administration struck a deal with Iran and five other nations that arranged to end uranium enrichment and end Western sanctions against Iran. However, the Trump administration viewed the Iranian deal as too lenient on Iran and unilaterally pulled out of the agreement without consulting with the other cosignatories. The United States then put an embargo on Iranian oil. Iran resumed its uranium-enrichment program with hopes to have functional nuclear weapons.

The growing US oil shale/fracking boom has helped drive down oil prices. It has given the United States the ability to absorb a decrease in oil from either OPEC or an individual country. The United States has used its power of banking and financial markets to force other countries to stop buying Iranian oil. This has greatly decreased the amount of oil Iran is able to sell. This is a huge financial blow to Iran. It has not happened all at once, however, because many countries were given temporary waivers to continue to import Iranian oil until 2019 or 2020. By 2020, the full force of the US-led embargo on Iranian oil came into effect. The Trump administration had hoped the embargo would convince Iran to come back to the bargaining table. The hope was that pressure could be used to force Iran to stop supporting terrorist organizations and to remove their militias from other parts of the Middle East such as Iraq and Syria. This approach failed.

It is likely that Iran is fast tracking its attempts to have a nuclear weapon. If they are successful, it will likely create a Middle East arms race. Saudi Arabia has stated that if Iran obtains nuclear weapons, they would need to obtain nuclear weapons as well. Saudi Arabia is the de facto leader of the Sunni Muslim majority and sees Iran as a direct threat to itself and Sunni Islam.

Iran once produced over 5 percent of the world's oil but that share of oil continued to decline temporarily. Eventually, Iran found other avenues to export oil. Iran's oil exports had dropped to 1.5 million bpd, but by 2023 they were up to 2.5 million bpd. Of course, an oil embargo is difficult to enforce because Iran is motivated to move oil and will sell at a discount. The amount of money to be made in this situation is high, and many people will try to smuggle oil by truck, train, and ship to other locations where it can be mixed with the oil of other countries and hidden from US intelligence. A nuclear missile may give Iran additional leverage in negotiations. It will also further destabilize the Middle East.

Iranian oil will continue to fund Shia militias as well as politically aligned organizations that are designated by other countries as terrorist organizations. Oil will continue to fund a religious Iranian government that stifles free speech and enforces Sharia or religious law that represses secular people as well as those that have other religious beliefs or those who do not believe in any religion. The rise of the religious government of Iran is a direct reaction to British and American imperialism in Iran. Suppressing the will of Iranian people and their secular leadership to reinstall the Shah of Iran led to the Iranian Revolution. It is likely that if Iran was without oil, the United States and Britain would not have intervened. The Iranian leader at the time of the coup d'état was planning on nationalizing Iranian oil, and the United States and Britain did not want to let that happen. The current situation in Iran is another example of the long-term effects of foreign meddling in the affairs of oil-producing nations. What remains unclear is how Iran will transform, as oil becomes a smaller and smaller part of the world's energy mix.

Iraq

The Gulf War and the Second Gulf War/Iraq War explain why Iraq and Iraqi oil have been in the middle of Middle East wars. These wars explain why Iraq continues to a complicated place with many competing peoples and countries vying for influence and access to oil. Iraq is split between Shia and Sunni Muslims. Wars and skirmishes inside of Iraq are at times proxy wars between Sunni and Shia powerbrokers. The plight of the Kurds (the largest ethnic group of people without their own state) plays out in Iraq. The Kurds' de facto state within Iraq gives hope to Kurds in Iran, Turkey, Syria, and across the planet for a permanent homeland. The independence of Kurds in Iraq causes fear and alarm across Syria, Iran, and Iraq. Leaders in those countries are always fearful that Kurds in their countries will try to join forces with Iraqi Kurds and carve out part of their country for a new Kurdistan. Iraqi oil continues to flow out of Iraq and Iraqi Kurdistan. The US invasion has destroyed the national monopoly on Iraqi oil and allowed all the major international players into the competition to pump more oil out of Iraq. The people of Iraq are once again in a situation in which the oil money does not make it to the common people; many are still without basic services, poor, and unemployed. Protests in 2019 demonstrate the frustration felt by Sunni, Shia, and everyone else. Young people especially are frustrated that decades of wars and rebuilding has only enriched corrupt politicians and foreign oil companies. They feel that promises that the United States and her allies gave have never come true. All they want is the ability to work and support themselves and their families, but they still struggle with the very basics of life. This war-torn nation has been blessed and cursed by oil, but for the common person, they have mostly been cursed.

Iraq was the first petrol nation. During the First World War, a French diplomat, François Georges Picot, and a British aristocrat, Sir Mark Sykes, sat down with a map of the Ottoman Empire and split up the area between regions controlled by the French (Syria and Lebanon) and those controlled by the British (Iraq and Palestine). The British eventually, after some haggling with France and a guarantee that they would share the oil wealth, were able to have control of the entire area that makes up modern-day Iraq. The Sunnis, the Shias, and the Kurds were bound together by political lines that paid no attention to historical, religious, or ethnic boundaries. The people of Iraq did not accept this. Gathering tribes of Sunni and Shiite people together, 100,000 men gathered to fight against the British. The British brought in 70,000 troops (mostly Indian) to suppress this rebellion. The British also brought in air power and started bombing civilians and soldiers alike from March 1920 until April 1921. Winston Churchill advocated using mustard gas against the people of Iraq but, fortunately, it was not feasible for the British air force at that time. After 8,500 Iraqis were killed, the rebellion was eventually suppressed. The British, French, Dutch, and Americans signed a deal guaranteeing an even share of the revenues for the oil companies of each nation. Once the oil began flowing from Kirkuk, they built the world's longest pipeline across Iraq and Syria with a split so that oil would flow to French-controlled Lebanon and British-controlled Palestine.

Iraq was built by combining three provinces or Ottoman vilayets named Mosul, Baghdad, and Basra after the capital city of each province. The British put a leader from Saudi Arabia on the throne of Iraq. King Faisal was the son of Husayn ibn Ali, who had taken over the eastern half of present-day Saudi Arabia including the holy cities of Mecca and Medina and created the Hejaz or Hashemite Kingdom. Faisal was a true Arab hero. He had led the attacks against the Ottoman Empire and worked with the British to drive the Ottomans out of Arab lands. He worked to create a pan-Arab movement that brought together the various religious sects of Islam with the hopes of creating an Arab country that stretched across the Fertile Crescent and into modern-day Saudi Arabia. The Arabs were betrayed by the British-French partition of the Ottoman Empire. They had fought believing they were fighting for independence. Faisal went to Paris to fight for an independent Arab state in the area including present-day Syria and Lebanon, but the French were determined to have influence over the people and the oil in the region. His father Ali was so upset by the French-English partition of the Arab world that he would refuse to sign on as British protectorate and was later defeated by ibn Saud, who eventually created Saudi Arabia. King Faisal, who was chased out of Syria by a people incensed over Faisal's inability to negotiate a peace without European power over them, would rule under the British. He and the Iraqi people chafed under the British rulers. He did not accept the throne until the people of the newly formed Iraq voted. Over 96 percent of the people accepted him as their ruler. In fighting for their freedom, they had exchanged one overlord for the other. However, the new overlords were mostly only concerned with oil.

As the roots of the Second World War began to take form in Europe, King Faisal was working to be sure that the British kept their word and allowed Iraq to have their freedom. Iraq was granted freedom as promised in 1932. However, the process of a departing British allowed the small Assyrian Christian population to be massacred. The British had hired Assyrians to fight with them during their years of rule and paid them higher wages than Arabs. This was probably due to the fact that the British and the Assyrians were both Christians. Resentment and anger toward the Assyrians were at an all-time high at this time. What started as acts of theft, violence, and raiding quickly became an act of genocide. Tribes of Kurds, Yazidis (monotheists that are not directly connected to the Abrahamic religions of Judaism, Christianity, and Islam), as well as other Sunni and Shia tribes in the areas, looted the towns and drove out the Assyrians that had lived for centuries across the Nineveh Plain. Women and children fled, and Assyrian men were rounded up and shot by military and paramilitary groups. They had been promised safety in a town called Simele and that was true for a couple of days. The Assyrians, some who were ex-military, disarmed and depended on the Iraqi army for protection. On August 11, 1933, the commanding Iraqi officer Ismael Abbawi Tohalla ordered the Assyrians to disperse and/or return to their homes. The soldiers began to fire on the unarmed defenseless Assyrians. He ordered soldiers to not target women. This unleashed a methodical massacre. Men, women, and children were killed. All the men were systematically murdered. Women were raped and forced to march naked in front of military officers. Children were run over by military vehicles. Bayonets pierced pregnant women and children (Shimun and Patriarch 2010). Over 6,000 Assyrians were killed. Over 20,000 Assyrian refugees eventually resettled in French-controlled Syria. Unfortunately, this massacre was one of the few things that three larger factions of Iraqis could all agree on. The new Iraqi nationalism did not see the Assyrian Christians as part of that new nation. As this did not affect Britain's ability to access Iraqi oil, they did not intervene in a genocide that the British helped create by singling out the Assyrians and treating them as superior to the Muslims, due to their shared Christian religion.

The British would not Intervene in Iraq again until Iraqi leaders worked with the Nazis to expel British forces from Iraq during the Second World War. The new leaders of Iraq dethroned the British-placed king of Iraq, Faisal II (the grandson of Faisal I). They worked with the Nazis because "the enemy of my enemy is my friend." The Nazis were only interested in oil, and Iraq had enough to support all the needs of the Third Reich. Fortunately, the British and the Allies of the British, were able to win with a small scrappy force and a Palestinian Arab who was able to call in a phantom column of tanks that spread fear and panic to the ruling junta of Iraq. The king was put back in place, until 1958 when he was executed by a democratic revolution.

Saudi Arabia

The US government, while publicly supporting freedom, democracy, and human rights, has a long history of supporting brutal dictators and repressive monarchies if they are willing to help them economically, especially if that country has oil to export. Saudi Arabia has a long history of human rights abuses by a monarchy that is absolute. Women did not even get the right to drive a car until 2018. Public beheading of those with differing religious views than the Sunni Muslim leaders was and continues to be common. Women have fewer rights than male children. These second-class citizens cannot leave the house without a male escort. This creates situations where women can be living in slavery-like conditions with their treatment completely dependent on the arbitrary desires of the male guardian (father, husband, son, or uncle) who controls their every movement.

Mohammed bin Salam bin Abudulaziz al Saud, also known as MBS, was appointed crown prince in June 2017, displacing his cousin Muhammad bin Nayef. His father, King Salman, made the decision to remove Nayef from all positions in 2017 and appoint his son as heir to the throne. MBS is spearheading a dichotomous approach to governing. He is allowing women to do things they have never been allowed to do in the past, such as attending some sporting events, driving, working in public spaces, and having a greater level of autonomy outside of the traditional guardianship system. However, at the same time he is also jailing and torturing women rights activists in Saudi Arabia. He has jailed and murdered journalists. He rules as an absolute autocrat that continually reminds his citizens that any freedoms or rights are only granted by the grace of the monarchy and can be retracted or rejected at any time.

Saudi Arabia assassinated Saudi journalist Jamaal Khashoggi, who had US residency, because he was critical of Saudi leadership. US intelligence sources identify Mohammed bin Salaam as ordering the assassination. Khashoggi entered a Saudi embassy in Turkey where he was drugged, murdered, dismembered, and his body parts burned in an outdoor oven of the Saudi consulate's home in Istanbul, Turkey. This caused many countries to halt arms sales to Saudi Arabia. The US Congress passed three resolutions to do so as well, but President Trump vetoed them all. Trump believed that the lucrative weapons contracts were more important than human rights. While this case is particularly egregious, it is "par for the course" for Saudi leaders. Saudi leaders have enjoyed close relationships with US presidents since Franklin Delano Roosevelt first met King Abdul Aziz Ibn Saud during the Second World War.

The US relationship with Saudi Ar"Iia'is complicated. Saudi Arabia has the ability to produce more oil than any other country on earth. They also have the ability to slow production and raise prices. Oil effects governments across the planet; good relations with Saudi Arabia give the United States some leverage with Saudi Arabia. Selling weapons to Saudi Arabia increases that leverage. Keeping Saudi Arabia dependent on US weapons and thus US replacement parts means that Saudi Arabia will usually try to not run afoul of US policy. Saudi Arabia was willing to increase oil exports at the request of the United States in 2018. The United States had put oil sanctions on both Venezuela and Iran, curtailing global oil supplies. The United States also needs allies in the Middle East. It is important that the United States has countries willing to host their bases and to use their airspace. The United States currently has over forty-five bases in the Middle East with five bases they are currently using in one form or another in Saudi Arabia. This relationship and Saudi oil is why Iraq's invasion of Kuwait and threats to Saudi Arabia in 1990 were so important to the United States and the world.

Saudi Arabia is the birthplace of Islam. The two of the three holiest cities according to Islamic tradition are Mecca and Medina, located inside of Saudi Arabia. (The third holy city is Jerusalem, which is located in Israel and holy to Christians and Jews as well.) Mecca is the holiest city of Islam, and every Muslim is expected to complete the hajj, a trip to Mecca, in his or her lifetime. The stories from the Koran (Islam's holy book) are mostly located within the borders of modern-day Saudi Arabia. For these reasons Saudi Arabia will always be the religious center of the Muslim world.

Saudi Arabia's oil wealth has also allowed it to be an economic leader in the Muslim world. When there are conflicts that pit Sunnis against Shias, Saudi Arabia has the ability to financially support the Sunni position. (It is important to note that the Sunni/Shia split comes from a disagreement on who was the true Muslim leader after the death of the prophet Muhammad.) In this way, the oil money of Saudi Arabia is used to fuel one side of a war or conflict, while the oil money of Iran is used to fuel the other side.

Saudi Arabia continues to be an important player internationally and locally. The power it wields is based in both its historically religious importance and the fact that it was the largest oil producer on the planet until surpassed by the United States in 2018. Saudi Arabia has taken tentative steps to diversify its economy. It is not known when peak oil consumption will occur. It will mark a time that dwindling demand will lower oil prices. Conversely, lowering demand will decrease investment in new oil discoveries, thus decreasing supply. This will push prices up temporarily until supply and demand balance. Saudi Arabia will be well positioned in these moments to capitalize. Its oil fields are well developed, and the supply appears to be plentiful into the foreseeable future. If Saudi Arabia can successfully transition to solar energy and a diversified economy that incorporates tourism and manufacturing, it may be able to weather the transition. However, if it stakes its future on endless demand for oil, there will be a rocky transition as internal and external stakeholders fight for a slice of an annually diminishing pie.

Qatar

Qatar is the most fossil fuel-dependent country in the world. The entire Qatari economy is driven by oil and gas, which make up 85 percent of all exports. Most other exports are based on petroleum or natural gas as well. Qatar is a small peninsula that is connected only to Saudi Arabia, its much larger neighbor to the South. Qatar's population was estimated to be less than 16,000 in 1949 when exports resumed after the Second World War. The population was made up of mostly nomadic Bedouin herders. As of 2024, the population is over 2.7 million people: 313,000 Qatari people and 2.4 million immigrant workers, mostly men. Men outnumber women (3:1) in Qatar; many men work in Qatar and send money to their families back home. The majority of these workers come from the countries that make up the Indian sub-continent: India, Pakistan, Sri Lanka, Bangladesh, Nepal, and Bhutan. Qatar has the highest per capita GDP of any country on earth at over US$128,000 per person. However, workers have few rights and can be subject to abuse. In 2017, the Qatari government set a minimum wage of the equivalent of US$195 per month. That is an annual salary of $2,340. This demonstrates that the vast oil and natural gas wealth is dispersed unevenly with Qataris, and that well-connected Qataris reap the vast majority of that wealth. Foreign investment is allowed, but all business based in Qatari must by 51 percent Qatari-owned.

Qatar is a Muslim nation that follows Sharia law (strict laws taken from the Quran that include stonings and lashings for adultery, consumption of alcohol, and apostasy). While a small number of lashings are carried out each year, people are rarely stoned to death, despite it being a legal punishment. Some of the older traditional ways of Qatar stand in direct contrast to the modernization of the country that has been happening for seventy years. The wealth from oil and gas exports has created a modern high-tech city in Doha, the capital and home to over 70 percent of the population. Immigrant workers provide services for wealthy Qataris who are able to afford the luxury of a private workforce in the home. This situation creates a very small owner class and large workforce that live in conditions that are well below poverty. This entire system is beholden to the natural resources of Qatar and the waters that surround the nation.

Qatar produces more natural gas than any other nation on the planet. It produces 14.5 percent of total world natural gas imports. These are valued at US$40.5 billion. This is from one of the smallest countries in the world. Qatar has a total landmass of 4,416 miles, which is 20 percent smaller than the US state of Connecticut. Despite being the third smallest country in Asia (only behind Singapore and Bahrain), its wealth gives it a lot of power. Qatar has been supportive of the Muslim Brotherhood after they rose to prominence during and after the Arab Spring. (Qatar also funds Hamas, which is considered a terrorist organization by most countries.) Other autocratic governments see the Muslim Brotherhood as a direct threat to their ability to rule and have labeled them a terrorist organization. The Qatari monarchy does not see them as a threat to their ability to rule and continues to fund them. This has caused a rift between Saudi Arabia and Qatar. Saudi Arabia has cut off relations with Qatar and cut off all imports to them as well. Other Sunni countries joined them in an attempt to isolate the country and force it to change its position on the Muslim Brotherhood. It has not been effective, and Qatar uses its abundant wealth to import food and goods from many other countries. The majority of its food imports come from Australia. Saudi Arabia is currently threatening to dig a canal across the border, effectively turning Qatar into an island, but has yet to do it. The battle for political power in the region continues and is fueled by natural gas and oil sales.

Qatar may have to diversify its economy over the next twenty years. Changes in oil and natural gas demand could have intense effects on the Qatari economy. Because Qatar is more dependent on natural gas than oil sales, it is likely to have a longer time before it has to look for new sources of income.

Venezuela

Venezuela has transitioned from a functioning democracy to a de facto dictatorship. Elections that were considered a sham by many countries in 2018 have contributed to the crises. The socialist leaders of Venezuela have been singled out by the United States. Crippling sanctions were implemented in 2017 and tightened in both 2018 and 2019. While Venezuela has the largest oil reserves in the world, much of that oil is in oil sands (also called tar sands), which is more difficult to extract. The difficulties in their oil industry started when Venezuelan leader Hugo Chávez fired the majority of state oil workers in 2003 and replaced them with political allies. Venezuela also expropriated the equipment and property of private oil companies in 2007 that had been in Venezuela recovering tar sand oil. These two events led to a precipitous drop in oil production. It is interesting that the United States is allied with many absolute monarchies that suppress human rights such as Saudi Arabia, but it is intent on regime change in Venezuela.

Venezuelans elected Hugo Chávez in 1999, and he quickly began the process of turning a democracy complete with checks and balances into a one-man dictatorship. Chávez called a constitutional assembly that weakened the legislature and stacked the courts in his favor. The people, despite low voter turnout, approved the new constitution. Chávez was now able to run for two consecutive six-year terms instead of one five-year term, and he controlled all three branches of government.

Chávez at first was looking to create a democratic socialist system in the model of the countries of Northern Europe. As he received push back from some centrist forces in Venezuela and from business leaders, he started to move the country to a more control and command central government socialist model. He rewrote the deals with the international oil companies and gave the state oil company Petróleos de Venezuela, S.A. (PDVSA) 51 percent control in all ventures. As mentioned, during an oil strike in 2002–3, Chávez fired all 19,000 striking workers and replaced them with political partisans and ex-military members that were loyal to the Chávez regime. This move caused the PDVSA to lose the majority of its experience and intelligence and caused a deep drop in annual production, despite new discoveries. Venezuela has the most proven oil reserves of any country on the planet. Chávez eventually nationalized all oil infrastructures, manufacturing, and refining companies that would not agree with the new terms in 2007. ExxonMobil and ConocoPhillips are still fighting to be reimbursed for their losses (Rapier 2017). Despite these self-inflicted wounds, Venezuela was able to temporarily increase production; however, without the major oil companies investing billions in new oil discoveries and continuing to invest in critical upgrades to oil infrastructure, the country has witnessed a steady decline of oil production. This was further exacerbated by the shale oil fracking boom in the United States that pushed oil prices lower in 2014. Even when oil prices began to rise, Venezuela was not able to produce the amount of oil needed to fund its economy.

Chávez died in 2013 and was replaced by his handpicked successor, Nicolás Maduro. Maduro won his first election by 1 percent. In 2015, he lost control of the legislature. He used the lame duck National Assembly (congress) to grant him the power to rule by fiat. The new National Assembly pushed for a recall but was foiled by Maduro and a court stacked with party loyalists. In 2019, after a sham reelection, the assembly elected an interim president, Juan Guaidó, who was recognized by most countries in the West. Most countries run by dictators (Russia, China, Cuba, Iran, and Turkey) still support Maduro. Venezuela's economy is in free fall. There is not enough food to feed people. Maduro has amplified this problem by making it illegal to bring food in from other countries. He has also blocked food aid from the United States and other countries. People are starving to death, and he has executed over 9,000 people for "resistance to authority" and other political crimes. There are also armed paramilitary groups loyal to Maduro that carry out political murders across the country. Millions of people have left the country out of fear and hunger.

Chávez and Maduro used oil royalties to pay for social programs for the country's poor. Venezuela had huge disparities between the poor and the middle class. It also had a wealthy class that controlled a lot of the country's resources. It could be argued that the United Socialist Party, which was originally called the Fifth Republic Movement, bribed the poor to vote for them through social programs. It could also be argued that years of economic disparity stretching back to colonial times created a situation in which the poorer classes were fed up. Oil was the means to create equality in the country, but it was mismanaged. If a country uses its oil resources like Norway does to directly benefit the people and they continue to reinvest in oil infrastructure and discovery, it can benefit the people indefinitely. Chávez's decision to fire all of the country's oil workers coupled with nationalization driving out foreign investment crippled Venezuela's oil production. Maduro's obsession with staying in power even when public opinion has turned against him has turned Venezuela into an international pariah. His choice to turn Venezuela into a de facto dictatorship has been met with crippling sanctions from the United States and others. Lower oil prices and the refusal to accept food aid is leading to starvation in a once extremely prosperous country. People with the resources and ability to leave are exiting the country by the millions. Even if oil prices return to their highs of over $100 a barrel, it is unlikely that Venezuela will be able to regain its prosperous economy and functioning democracy any time soon. Venezuela's fall from a thriving democracy to a failed petrostate with autocratic rule demonstrates the dangers that are inherent when an economic system is too dependent on one natural resource for economic security. It also demonstrates that all democracies are inherently fragile, and an elected leader can destroy the checks and balances of a democratic system to become a dictator. As of 2023, 8 million people have fled poverty and starvation in Venezuela.

LOCAL CONFLICT

US Fracking

Fracking (or hydraulic fracturing) has been a divisive issue that has brought about an unexpected oil boom in numerous places across the United States and has also caused small earthquakes and polluted local groundwater supplies. It has divided local environmentalists and other locals hoping to cash in on the fracking boom. What is undisputed is that fracking has propelled the United States to the number one producer of oil from 2019 to the time of writing this book. The fracking boom has also kept worldwide oil prices lower than they otherwise would be. This affects many countries, but it especially affects those countries that depend on oil revenue for the majority of their funding. OPEC has been unable to counter the fracking boom completely. OPEC+, which includes Russia, has been able to steady the price of petroleum by agreeing on deep cuts in production. They have not yet achieved their goal to return oil prices to over $100 per barrel but will likely obtain that goal soon. It is yet to be seen if US fracking is a blip in oil history or if it will continue to produce record amounts of oil across the country. Fracking is spreading across the globe, and its ability to produce oil from fields that were thought to be dead or in decline is likely to vastly increase the amount of recoverable oil across the planet. The fracking process's ability to keep oil prices close to one-third of their historic highs has the unintended consequences of prolonging worldwide dependence on oil and the use of combustion engines. That may be countered with deep OPEC+ cuts, led by Saudi Arabia. (Long-term oil demand may decline at a quicker rate if OPEC+ continues to find ways to increase prices and drive people toward electric vehicles.)

Fracking is the process of using horizontal drilling combined with the injection of water, water plus chemicals, acids to dissolve rock formation, gasses, and/or sand. Most of the time a combination of these injectables are used. By cracking or fracturing the rock formations that contain hydrocarbons and injecting substances, oil companies are able to extract oil that was long thought of as unattainable or unrecoverable.

Many oil-producing formations occur underneath local water tables. Most cities and towns obtain their water from local wells that tap these resources. Fracking has, in some cases, caused natural gas to become mixed with local water supplies. This has created situations in which people were able to actually light their water on fire. Of course, the water itself is not on fire, but it can appear so as the gas mixed with the water ignites. The Environmental Protection Agency (EPA) has found that fracking comes with significant risk to local water resources (Burden et al. 2016). Oil and gas drilling companies are not required to report what chemicals are used in their fracking process. They claim that this information is proprietary, and they do not release it so that other companies do not use their methods. Many people believe they will not release the ingredients used in their process because they are dangerous chemicals to both humans and other animals. Reports to the US Congress have noted over 750 different chemicals that are injected into rock formations. Some of these chemicals have leached into local water supplies, according to studies undertaken by the EPA. These studies indicate that fracking chemicals can be a problem for local communities and their water supplies. There are places that already have naturally occurring methane in their water sources. The EPA was able to identify the chemical signature of the various methane sources to determine whether the methane was from a driller or from a naturally occurring source. The National Institute of Health has determined that early exposure to fracking chemicals can harm the immune system in mice. In order to protect public health, a number of states have put either temporary or permanent bans on fracking.

The process of fracking causes rock formations to break apart. This process can cause earthquakes. In Oklahoma, fracking has caused thousands of these small earthquakes. While most earthquakes have been relatively small, there have been fracking earthquakes up to 5.7 magnitude. These types of earthquakes can cause significant damage to homes and other buildings. The US Geological Survey (USGS) states that the injection of wastewater during the fracking process is more likely to cause larger earthquakes than the fracking process itself. A number of states affected by fracking quakes have set up monitoring and created organizations to assess the causes and solutions, but only Arkansas has actually banned fracking due to earthquakes (and only in a specific area of the state).

Some US states have taken steps to limit or outright ban fracking due to concerns over fresh water, chemical pollution and contamination, and earthquakes. Only Vermont and

Maryland have laws permanently banning fracking. New York State has a fracking ban as well, but it could be reversed, as it was a decision by the executive branch of the state government. Other states have placed limits as well, but they do not outright ban fracking. Overall, the money that is being made through fracking is too lucrative for most states to ban the process. This has also allowed many private landowners to profit by leasing their land. Environmental and public safety groups continue to fight for limiting or outright banning fracking. Currently, the process of fracking is being used in most of the states, and it has been extremely effective at increasing access to oil and natural gas. Its long-term effects are yet to be known.

Canadian Tar Sands/Athabasca Oil Sands

Underneath the boreal forest that stretches across the Athabascan lands located in the center of the Canadian province of Alberta are 1.7 trillion barrels of the thickest and dirtiest oil in the world. At 2020 prices, only about 170 billion barrels were economically feasible to recover. (As prices have risen in 2022 and 2023, more of these oil sands may eventually be recovered.) This oil still makes Canada the holder of the third largest oil reserves on the planet. The problem is that it is both expensive and destructive to recover this oil. It is also located underneath a pristine forest and is thousands of miles away from any sort of port. The "easier" oil is oil that can be obtained by clearing the forest and strip mining the tar sands underneath. The tar sands must then be processed. It is important to note that the tar sands are also called bitumen or asphalt. It is the heavy tar-like substance that is excellent for sealing a canoe or paving a road. The process of converting bitumen to usable oil requires massive amounts of energy. Huge earthmovers load 400-ton trucks with tar sands. The tar sands are steamed and spun in a water slurry in which sand sinks to the bottom and impure bitumen froth rises to the top. The froth is deaerated and diluted with naphtha so that solids and impurities can be removed. The bitumen is then put through inclined plate settlers and centrifuges that both work to further clean it up. After all of this, the bitumen is transported thousands of miles by truck, train, and pipeline to refineries that will convert it into oil, gas, and other petroleum products. All of the energy inputs that are required to turn tar sands into fuel make this oil the most carbon intensive of any oil on the planet. Despite these difficulties, Canadian tar sands produce 3.2 million barrels a day (Canadian Centre for Energy Information 2023). Newer in-situ techniques, which separate the sand out of bitumen, are allowing greater production at lower costs. The greater difficulty is balancing Canada's desire to reduce its carbon footprint and its desire to profit off of its vast resources while oil is still used for transportation.

The political pushback has been intense in Canada and in the United States. The issue of leaving dirty or carbon-intensive oil in the ground spilled into the public sphere with the Keystone XL pipeline protests starting in 2012. This is not to be confused with the Dakota Access pipeline protests in 2016. The Dakota Access pipeline stays within the United States and does not cross into Canada as the Keystone XL does. These protests helped ignite a larger debate over the logic of using the tar oils at all, highlighting the local environmental risks to wildlife and habitat destruction. The global environmental risks in the creation of additional CO_2 emissions also became center stage. Both pipeline projects were put on hold by the administration of President Barack Obama. President Trump reinstated them both. The Dakota Access Pipeline projecl is completed. (This pipeline does not transport Canadian oil sands. It only transports US oil from the Bakken Formation in North Dakota.) The Keystone XL project was canceled by President Biden, and the builder of the pipeline TC Energy is in the process of removing the completed parts of the pipeline.

The Athabasca Oil Sands have also been a political football in Canada. The Conservative Party of Canada has always supported it, but the Liberal Party has had a more tortured relationship. The progressive wing of the party does not support the oil sands, but the leader of the Liberal Party, Justin Trudeau, has supported it for economic reasons. While there has been pushback and protest, the policy has not changed, and Canada continues to grow its production of petroleum from tar sands.

Angola

Oil has been used by the peoples of Angola for millennia. Natural oil seeps and asphalt deposits in the area around Libongos provided abundant material for caulking and sealing. So much so that the Portuguese began exporting it back to Portugal and to their colonies in modern-day Brazil when they arrived in the area in the late 1700s. Oil exploration began in earnest in 1915, as the worldwide oil and automobile boom began. It took until 1955 for a viable well to be found. However, one of the largest deposits turned out to be offshore. In 1968, an American oil company named Cabinda Gulf Oil Company (later bought by Chevron) discovered oil in an area called Mangola, Angola. This area still produces 500,000 bpd. These discoveries played a role in Angola civil war and involved many international actors. A French company, Elf Petroleum, made the final major oil discovery in 1996. This deep water well hit oil at 1,300 m, and the discovery led to a black "gold rush." Today, there are multiple major oil companies producing over a million barrels per day. It is important to understand how all this oil shaped the twenty-seven-year civil war in Angola.

The Angolan Civil War began in 1975 immediately after the various tribes of Angola had united to successfully win a war of independence from Portugal. Countries funded this war across the globe. The MPLA (The People's Movement for the Liberation of Angola) was a communist-Marxist revolutionary group that first came to power. Its base of support was the Ambundu people. They were supported by the USSR, Cuba, and many fellow African countries. The MPLA was also funded by the offshore oil resources that they controlled. The National Liberation Front of Angola (FNLA) drew its support from the Bakongo people who worked to restore the Kongo Empire that ruled Southeastern Africa during 1390–1914 CE. Finally, the National Union for the Total Independence of Angola (UNITA) was founded by an ex-leader of the FNLA and supported by the Ovimbundu people. While MPLA was funded by both Soviets and oil sales, FNLA and UNITA were funded by diamond sales (sometimes called blood diamonds) and the United States, South Africa, Saudi Arabia, and Kuwait. Angola, during their civil war, was alternatively the largest or the second-largest exporter of oil in Africa. The reason this civil war, which was based on ethnic groups competing for control, received international intervention and support was because oil was involved. The United States, South Africa, Saudi Arabia, and Kuwait have only one interest in common: access to oil resources. Saudi Arabia and Kuwait are ruled by absolute monarchies and South Africa at that time had an apartheid government that did not allow the Black Africans to vote or have civil rights. This was not a war for democracy over communism. This was indeed another oil war. Angola produced 1.16 million bpd as of 2022, second only to Nigeria in African oil production. Deep-water oil production is just beginning, and Nigeria looks as though it will produce even more oil in the future. With 9 billion barrels of proven oil reserves and 11 trillion cubic feet of natural gas reserves, oil and gas will be pivotal in Angola's future.

Nigeria

As stated in the introduction to this chapter, Nigeria produces more oil than any other African nation at 1.18 million bpd Oil in Nigeria was originally discovered by the British energy giant BP in a joint project with the Dutch giant Shell in 1956. Nigerian lands and waterways have been plagued by thousands of spills caused by both people trying to pirate oil and by corporate disregard for safety and environmental protections. The people of Nigeria do not see the economic effects in their daily lives, but they do know that the money is used to fund the government. Unfortunately, the government is notoriously corrupt, and millions of bribes have been paid to officials in Nigeria by international players in the oil industry. The most recent case ended up in US and EU courts. Shell and Italian oil company Eni S.p.A. were in court over a $1.3 billion payment that was supposed to go to the Nigerian government but instead went to Malabu Oil and Gas, a front company for Nigerian oil and gas minister Dan Etete. There is a long history of those that oppose Shell being persecuted by the government. The most famous case was the 1995 execution of the Ogoni Nine, who were against Shell's oil extraction due to its destruction of the Niger Delta.

This conflict has raged from the early 1990s and still simmers today. As companies and local tribes fought for oil wealth, a local arms war ensued. Tribal violence over who has the rights to certain lucrative oil areas spills over into piracy of oil and kidnapping of oil workers both foreign and domestic. The military has moved to crackdown on the kidnapping, but they have been as brutal and lawless as the militant groups. Government amnesty from 2009 to 2016 has brought over 15,000 militants in for retraining. Violence has decreased, but the horrendous destructive oil pollution that started the movement continues. The lull of violence has allowed for the tripling of oil exports, but without addressing the deeper issues of deep poverty, environmental destruction, and a corrupt system of bribery that highly rewards a few while ignoring most. Violence continues at a slower pace with deaths of military members guarding oil platforms and pipelines as well as military reprisals targeting men, women, and children in the villages that they believe the militants come from. The oil and gas of Nigeria has brought about economic advancements, but it has also brought death and environmental destruction. It has also created a group of billionaire Nigerians. It could be argued that it is better that at least some of the money is going to Nigerians, but most Nigerians would argue that the money would be better used fighting poverty and increasing educational opportunities across Nigeria.

Sudan

The conquering British originally set up the borders of modern-day Sudan in the mid-1800s as a territory that held many distinctive tribes and peoples. Roughly speaking, the North was made up of Arab or Arabized people that were mostly followers of Islam. The South was made up of a mix of sub-Saharan African tribes that were mostly Christian as well as many people who followed traditional or animist religious practices. The Western Darfur region has a mix of many peoples: nomadic Muslim Arabic people, nomadic Muslim African people, and town-dwelling Africans that are a mix of Muslims, Christians, and animists. When Sudan won independence from Britain, a civil war ensued. If oil was not involved the war in Sudan could have ended quickly with a separation between the North and the South of Sudan. Instead, this war lingered on for decades with oil money continuing to fund both sides.

The majority of the oil wells are located in present-day South Sudan. The refineries and the oil ports are located in the North side of Sudan. This is why Northern Sudan (modern-day Sudan) would not let the South (modern-day South Sudan) to secede. The South is landlocked with no access to the ocean, so the South was also dependent on the North for transporting and selling their oil. The warfare in the areas in which opposing peoples with different religions lived in close proximity was particularly fierce and inhumane. Villages were destroyed, and women were taken as booty of war. The adult men were executed. Young girls were taken as slaves. Young men either fled or were forced into becoming child soldiers fighting for the people who raped and murdered their mothers and murdered their fathers. Some of the young men who fled and were on their own for years as they tried to find their way to safety became known as the Lost Boys. The Lost Boys walked hundreds of miles in search of safety and refuge. Some of them made it to refugee camps. The lucky ones were able to immigrate to countries that were stable and had education and financial opportunities. Those that did make it became well-known in the United States and other Western countries. However, most Sudanese who were involved in these wars were killed, hurt (physically and mentally), or displaced. Millions have died and millions more have been displaced either internally or externally and will never again live in their hometown. Villages were destroyed. Minority populations were expelled from their lands without a path to ever regaining their place in the world.

Oil had a big part to play in the wars in Sudan, as did the arbitrary borders created by colonial powers without regard for the peoples that lived in these areas. Religion was also a factor in the long drawn out civil wars and genocides in Sudan. Oil was the reason that these wars have been able to perpetuate. Most civil wars financially exhaust a country and at a certain point the financially exhausted sides must come to a peace. The oil in Sudan allowed fighting to continue until present day.

CONCLUSION

Oil wars have existed since oil was first used for lighting at the beginning of the twentieth century. Oil has been a major driver for the majority of wars since the 1930s. Even the wars that did not begin about oil became about oil. The modern machinery that is needed to be successful in war demands oil. Without oil the machinery will not run. The country that runs out of oil will lose the war. A country with oil can still lose a war, but a military without oil cannot win a war. The twentieth century has been defined by wars that were indirectly or directly affected by oil.

The twenty-first century has already had its share of war over fossil fuels. It appears that the world's dependence on oil may be waning. If that is true, then the wars of the future may be less frequent and more easily avoided. If the world will one day be powered by renewables, then wind and sun may be what powers nations and armies alike. Let us hope that each country has enough wind and sun that they will not feel forced to fight for the renewable resources of another country.

FURTHER READING AND REFERENCES

Burden, S., M. M. Fleming, J. Frithsen, L. Hills, K. Klewicki, and C. D. Knightes. 2016. *Hydraulic Fracturing for Oil and Gas: Impacts from the Hydraulic Fracturing Water Cycle on Drinking Water Resources in the United States*. Washington, DC: US Environmental Protection Agency.

Canadian Centre for Energy Information (CCEI). 2023. *Energy Fact Book 2023–2024*. Accessed July 12, 2024. https://energy-information.canada.ca/sites/default/files/2023-10/energy-factbook-2023-2024-section6.pdf.

Cohen, A. 2019. "As Russia Closes in on Crimea's Energy Resources, What Is Next for Ukraine." *Forbes*, February 28. Accessed July 12, 2024. https://www.forbes.com/sites/arielcohen/2019/02/28/as-russia-closes-in-oncrimeas-energy-resources-what-is-next-for-ukraine.

Izuaka, M. 2023. "Nigeria Regains Position as Africa's Biggest Oil Producer." *Premium Times*, June 16. Accessed July 12, 2024. https://www.premiumtimesng.com/business/604891-nigeria-regains-position-as-africas-largest-oil-producer.html.

Kireeva, A. 2019. "For All of Russia's Talk about Oil Drilling in the Arctic, Most Arctic Oil Will Likely Go Untouched." *Bellona*. Org. April 24, p. 2019. Accessed July 12, 2024. https://bellona.org/news/arctic/2019-04-for-all-of-russias-talk-about-oil-drilling-in-the-arctic-most-arctic-oil-will-likely-go-untouched.

Leahy, S. 2019. "This Is the World's Most Destructive Oil Operation—and It's Growing." *National Geographic*, April 11. https://www.nationalgeographic.com/environment/2019/04/alberta-canadas-tar-sands-is-growing-but-indigenous-people-fight-back/.

Rapier, R. 2017. "How Venezuela Ruined Its Oil Industry." *Forbes*, May 7. https://www.forbes.com/sites/rrapier/2017/05/07/how-venezuela-ruined-its-oil-industry/.

Reynolds, D. B. 1998. "Soviet Economic Decline: Did an Oil Crisis Cause the Transition in the Soviet Union?," *Journal of Energy and Development* 24, no. 1: 65–82. http://www.jstor.org/stable/24808741.

Ritchie, H., H. Roser, and R. Rosado. 2022. "Energy." *OurWorldInData.org*. https://ourworldindata.org/energy.

Shimun, M. E., and C. Patriarch. 2010. *The Assyrian Tragedy*. Bloomington, IN: Xlibris Corporation.

CHAPTER 3

Responses

INTRODUCTION

Conflict over energy resources has defined human civilizations for millennia. The driving force behind these conflicts is the fact that resources are finite. However, recently this has changed. (Yet, financial conflicts remain.) The potential renewable energy that is available is currently much larger than all the needs of humanity. The question then remains: What is the conflict? The conflict is not made up of the traditional players. Empires and/or nations are not in conflict over limited megafauna, wood, coal, oil, or natural gas. (The exception may be Russian attempts to steal the vast resources of Ukraine through conquest.) Countries and multinational corporations are making trillions of dollars on fossil fuels. Many democracies are compromised by fossil fuel industries that donate tens of millions of dollars each election cycle to both individual candidates and organizations that support fossil fuel interests. This problem is particularly bad in the United States, which has produced more climate changing emissions than any other country. Four companies, ExxonMobil, Koch Industries, Chevron, and Marathon Petroleum, spent $94 million in the 2018 US election cycle alone (Parafiniuk and Smith 2019). This type of money makes it extremely difficult for the needs of the people to trump the needs of deep-pocketed companies.

This is just one particular example of why it is so difficult for countries to transition away from fossil fuels at the speed in which it is necessary in order to avoid the worst effects of climate change. Entrenched economic interests will fight to create a playing field that favors their industries. The fossil fuel industry received $5.2 trillion in government subsidies in 2017 (Coady et al. 2019). Worldwide subsidies for renewable energy are estimated to be at $150 billion. A 3,466 percent advantage is far from an even playing field. In a time when fossil fuels' huge advantage should be reversed, an even playing field would be enough to see an explosion of renewable energy.

This is not only a problem for democracies. Dictators across the world depend on oil exports to prop up their brutal regimes. Most countries in the Middle East have petro-dictatorships. Not all dictatorships depend on oil, but the ones that do tend to have the strongest hold on their populations. Iran and Saudi Arabia are the most well-known, but smaller countries like Kuwait, Bahrain, Oman, Qatar, the United Arab Emirates, and Yemen have petro-dictatorships. These countries also have an interest in keeping the status quo. Petro-dictatorships also exist in Venezuela, Russia, and Angola. Sudan and Algeria are fighting to overthrow their petro-dictatorships. Oil is quick money that is easily controlled by a government. That is why it has been used to repress the will of the people across democracies and dictatorships alike.

Despite the advantages of the fossil fuel industries, the world has made some strides in progressing toward renewable energy and decreasing its output of greenhouse gases (GHGs). The Paris Climate Accord of 2015 was a major step in this direction. The first binding agreement on climate change tackled limits on GHGs and funding for poorer developing nations. It is frequently criticized for not being enough, but it is a step in the right direction. Not all countries are living up to its pledges, and one very important country, the United States, had withdrawn from the agreement under President Trump but rejoined under President Biden. Overall, the agreement has helped slow the increase in GHG emissions, and it lays the groundwork for future decreases that could, potentially, slow the effects of climate change. The conflict over energy resources has changed. It is now a political and economic conflict over how governments support renewable and fossil fuel energy systems. Currently, the world is not moving fast enough to keep global warming to the stated goal of the Paris Climate Agreement of 1.5 °C. The battle to increase the speed of the energy transitions continues. This battle will decide exactly how humans alter the world through climate change.

INTERNATIONAL RESPONSES

The international community has been partly successful and unsuccessful at addressing issues that affect global common-pool resources. (These are resources such as the atmosphere and the oceans but can also include Antarctica and the Arctic. Future common resources may be the moon, the planets, and asteroids.) The first global treaty was signed in 1979 (Unies 2014). It was the Convention on Long-Range Transboundary Air Pollution. It was the first of eighteen major United Nations treaties. (For more information on these treaties see the United Nations Treaty Collection.) We will highlight some of the most consequential ones in this chapter. The global community came together to ban the aerosols that were creating a hole in the ozone layer that protects the planet from harmful solar radiation by designing, signing, and enforcing a treaty that limited and eventually banned the substances that were causing the problem. In the Arctic, countries are jockeying for position in order to extract fossil fuel resources. In Chapter 2, we covered the competing claims that may send countries into direct conflict. Here international cooperation does not appear to be winning the day. In Antarctica, where conditions are harsher and potential resources are less known, the Antarctic Treaty appears to be holding and countries are respecting it as neutral territory. The Paris Climate Accord is another example of countries working together to make big changes. Conversely, cities and states throughout the United States have stated their intent to do even more than required by the Paris Climate Accord.

So, it is yet to be seen how well the agreement will actually do at moving the world away from high GHG emissions and toward a more sustainable future. Currently, worldwide carbon emissions are the highest they have ever been, but so is international investment in green energy projects. The IEA estimates that worldwide clean energy investment in 2024 will top 2 trillion, compared to 1 trillion of investments in fossil fuels extraction and transportation. Just five years ago spending on green energy was even with fossil fuels (IEA 2023).

Ozone Model

Chlorofluorocarbons (CFCs) used in industry and in personal products enter the atmosphere and break apart the ozone layer, which contains O_3 (three oxygen atoms connected by molecular bonds). This thin layer of ozone is a natural shield; it blocks out harmful radiation from the sun in the form of medium-frequency ultraviolet light rays. This ultraviolet radiation can cause cancer, suppress the immune system, and damage DNA (that is, cause genetic damage). In 1976, it was realized that CFCs were depleting the ozone layer. The true scope of the damage was realized in 1985 when holes in the ozone layer were discovered at the North and South poles. National bans on aerosol sprays that used CFCs were put into place in the United States, Norway, and Canada in 1978, but these bans allowed for CFC use in many other products and processes.

The twenty governments of the largest emitters of the world signed the Vienna Convention, which created a framework for negotiations of what would become the Montreal Protocol. The Montreal Protocol was signed in 1987 and was eventually ratified by every member country of the United Nations—all 197 of them. DuPont, the largest producer of CFCs, had the most to lose and fought this agreement. Using the Alliance for Responsible CFC Policy as well as its CEO, DuPont argued against the science and claimed there was no connection between ozone layer depletion and the use of CFCs. DuPont later supported the Montreal Protocol when they realized that they could produce a profitable replacement product (Maxwell and Briscoe 1997).

The Montreal Protocol is often lifted up as one of the most successful international environmental agreements. It was designed, signed, and implemented relatively quickly. It has had its intended effect. The ozone layer has begun the natural repair process, and estimates indicate that it will be completely healed by the middle of the twenty-first century (EPA 2019). Some argue that the success of this treaty was due to how few corporations were affected and due to how small a part of the global economy was affected. Climate change, on the other hand, affects all of the largest economies in the world. It also affects the wealthiest companies in the world and some of the most profitable. The ozone hole was also something that people could easily understand as an immediate threat, because it could cause cancer and genetic damage. DuPont did not have the resources or time to mount a counter science campaign the same way that the fossil fuel industry has been able to do. However, the Montreal Protocol does remain a shining example of international cooperation succeeding to address a common problem.

Antarctic Treaty System

The original Antarctic Treaty was written in 1959 and eventually signed by twelve countries by 1961. The agreement banned all military options across the entire continent, turning it into a scientific preserve and guaranteeing freedom for scientific investigation. The original signers were the United States, the Soviet Union, the United Kingdom, Japan, Chile, Argentina, South Africa, Australia, New Zealand, Belgium, France, and Norway. Before the treaty, there had been a number of military skirmishes over which a country could claim Antarctica as its own. These skirmishes were small in the number of military personnel that were involved. (The United States did bring 4,700 soldiers to Antarctica in 1946, but that was just an extreme cold training exercise.) The Antarctic Treaty put an end to all military conflict on the continent. It may be that countries were more willing to agree to this because it was unlikely that any country could extract any resources from Antarctica with the 1959 technology, due to the harsh climate. Antarctica is not only extremely cold but also prone to long intense storms. It is surrounded by treacherous waters that are only passable at certain times of year. The United States had originally planned to do nuclear testing on the continent, but opposition from Argentina, Chile, and the Soviet Union was enough to persuade the United States to drop its proposal. This makes the Antarctic Treaty the first arms control agreement of the Cold War, appropriately about the coldest place on earth.

The Antarctic Treaty has been augmented by many other agreements. These agreements combined with the Antarctic Treaty have come to be known as the Antarctic Treaty System (ATS). ATS has added forty-two signatories and now includes fifty-six countries. It protects the flora, fauna, and environment of Antarctica. Waste disposal, which at one time had no regulations, is strictly limited, and most waste is removed from the continent. The ATS has been a bright example of successful international cooperation and diplomacy that makes it quite unique in the annals of human political interaction. Rarely have countries been able to come to an agreement that protects not only the animals in an area but the entire environment of that area. It gives hope that the nations of the world will be able to come to agreements on the shared atmosphere and oceans. It is increasingly imperative that the nations of the world work together to address their shared needs for clean air, clean water, and clean land. They must combine efforts to combat the effects of climate change while they change behaviors that have cumulatively caused and continue to contribute to climate change. The ATS provides hope that they can.

Climate Change Agreements

United Nations Framework Convention on Climate Change The United Nations Framework Convention on Climate Change (UNFCCC) was drafted in May 1992 and was open for signing at the Rio de Janeiro Earth Summit in June of the same year. It entered into force on March 21, 1994. There were 197 countries that were part of the drafting process, and it has been signed by 165 of those parties. It was established to limit worldwide greenhouse gas emissions in order to prevent anthropogenic (caused by human activity) warming of the planet. The framework was non-binding and had no enforcement mechanisms. This means that while the agreement was a positive first step it did not in actuality limit greenhouse gas emissions. GHGs have increased by over 50 percent since the agreement was signed.

Kyoto Protocol The Kyoto Protocol was an extension of the UNFCCC. It was signed in Kyoto in 1992 and went into force in 2005. The agreement was binding for almost every country on the planet, but not all countries were expected to decrease their GHG emissions. The United States was a signatory on the protocol, but the US Senate never ratified it. Because of this the Kyoto Protocol was never binding for the United States. Canada also pulled out of the protocol when it became clear that they would not meet their targets and the government was prioritizing the mining of the Alberta Oil Sands over any GHG reductions. Its overall goal was to reduce CO_2 emissions by 4.7 percent by 2012, and that was achieved.

The Kyoto Protocol, if followed, continued, and strengthened, could have created a framework in which the world was able to deal with the growing problems caused by climate change. Unfortunately, the United States and other countries undermined the protocol, and it was not able to continue to drive declines in CO_2 emissions. Despite this setback, the Kyoto Agreement was partially successful. The fifty-five original members decreased CO_2 emissions by over 11 percent. However, the majority of the decrease came from the collapse of the Soviet Union and the collapse of heavy industries in Russia and Ukraine. Western European countries have built off this success and continue on the path toward de-carbonization, albeit at a slower pace than the UN recommends.

Paris Climate Agreement The Paris Climate Agreement was the first time that climate change was addressed in an international treaty. The treaty stipulates that countries must set goals and report those goals back to the United Nations. Each year the countries must set a goal for even lower carbon emissions than the year before. Each country sets its goal independently. The overall goal was for the combined efforts to limit global warming to 2°C. That is unlikely given the goals that were set and the fact that most countries are not achieving their goals. Research in 2020 demonstrated that at current rates warming would not be kept under 2°C (Lewis et al. 2019). Other research demonstrates that due to reinforcing feedback loops, it is likely that the earth will warm between 3°C and 4°C (Schwalan, Glendon, and Duffy 2020), unless more extreme measures are taken.

One of the biggest criticisms of the Paris Agreement is that there are no mechanisms for enforcement. If a country does not achieve its goals or does not set lower targets, they will not be punished. There is a general consensus that only an international tax on carbon or some other carbon market would create a situation in which CO_2 emissions would actually be reduced quickly enough to limit global warming to between 1.5 and 2 degrees Celsius.

The United States had, under President Trump's leadership, withdrawn from the Paris Agreement, although according to the rules and regulations of the treaty, the United States was not eligible to exit the treaty until after the 2020 presidential election. President Biden re-entered the treaty. At the time of his withdrawal from the treaty, President Trump had already taken steps to deregulate pollution of many types. He and his administration made it easier for companies across industries to create and release more CO_2 than during the administration of his predecessor, President Obama. He rolled back over 200 regulations that protected clean air, water, and soil (Popovich and Pierre-Louis 2020). The repeal of the Clean Power Plan will have some of the largest effects of any of the rollbacks. It was a plan to limit CO_2 emissions from coal and gas plants. Rules to limit leaks of methane were also scrapped. The Trump administration repealed California's right to create stricter rules for tailpipe emissions in an attempt to combat smog in the state. Combined, these rollbacks have led to a reversal of the declines in CO_2 emissions that had been trending downward since 2008. The Biden administration has been able to reverse the majority of these rollbacks, but there are many reversals that are still in process. In 2018, US CO_2 emissions rose by 3.1 percent (EPA 2020). It appears that market forces driving down the use of coal in power plants combined with shutdowns in response to the Covid-19 pandemic caused economic stagnation and decreased CO_2 emission in 2019 and 2020, but 2021 and 2022 were predicted to have increased emissions (EIA 2021). While CO_2 emissions did increase in 2021 and 2022, they are still below the high-water mark set in 2007 and below pre-pandemic emissions. It appears the United States will continue to decrease emissions as renewable energy and electric vehicles continue to increase.

Overall, the Paris Climate Agreement is a step in the right direction, but its overall success or failure will depend on if it is built upon. If the signatories come back and sign an agreement that has a mechanism to compel nations to continually decrease their CO_2 output, then the Paris Agreement will be seen as a true success. However, if this agreement is the final one, it will not be enough to help the people on earth avoid the worst effects of human-caused global warming.

TECHNOLOGICAL RESPONSES

All of the technological responses covered here will focus on the present and the future. Chapter 1 covers the history of each one of these forms of renewable energy. Please refer to that chapter for information about the beginnings of these renewable sources of energy. The technological responses here cover almost all known forms of renewable energy. They all have incredible potential. As of 2023, hydropower produced the most renewable energy worldwide, but solar and wind are growing at exponential rates, whereas hydropower is a mature, reliable resource that shows little current growth. The future will likely be powered by renewable energy, but there is still debate about which renewables will be most important. If the current trends in decreasing prices for both wind and solar continue, it is likely that the future will be powered by growth in wind and solar with battery and pumped hydro storage (EPA 2024).

Renewable Energy (Growth and Potential)

Hydropower Hydropower is the original renewable energy. It is the oldest renewable energy source, and at one time it provided more energy than all other renewables sources combined. The difficulty with hydropower is threefold. One, most of the low-hanging fruit has been picked (affordable and accessible locations for putting in large-scale hydropower dams), so any future large hydropower project would require either extraordinary resources or would be in a location far from large population centers. Two, the resources required to create a hydro powered dam are so great that private corporations or investors rarely fund large hydropower projects. (They do fund many small projects.) Finally, the environmental cost of building dams is extremely high, from the flooding of natural areas and farmland to the loss of spawning grounds for salmon, bass, sturgeon, eels, and other euryhaline fish (fish that can live in both freshwater and saltwater). Despite these difficulties, there are opportunities to grow hydropower while respecting the environment and creating paths for fish and eels to pass the dams unharmed.

In the United States alone there are 101 gigawatts (GW) of hydropower capacity. The National Hydropower Association (O'Connor 2016) estimates that there is the potential for another 50 GW of hydropower in the United States that could be harnessed by 2050 (O'Connor et al. 2016). Hydropower has the advantage over solar and wind in that it is almost always available. By building a dam and monitoring water levels, plant managers can be sure to have consistent electricity production. Overall production will vary from year to year depending on rainfall. The ability to have continual production makes hydropower the perfect renewable partner for solar and wind power. Hydropower can ramp up quickly and can produce more energy at peak load times. The process of holding back water and releasing it uses hydropower as a natural battery that requires little technology and no rare resources. The more hydropower that is available on a grid, the more solar and wind power can be used. Hydropower can act both as a base load and as a peaker plant (a plant, usually natural gas powered, that can quickly ramp up to match user's needs). These advantages of hydropower are limited to areas that are suitable to build traditional hydropowered dams. There is also another possibility for areas that do not have the needed geography for traditional hydropower: small or micro hydropower.

Many small or micro hydropower plants can be built as run-of-river (a type of hydropower that does not require building a dam and does not cause the environmental damage that traditional hydropower dams do) or built on canals. These smaller hydropower projects can be implemented almost anywhere that has running surface water. Another important hydropower potential is on existing dams. Only 3 percent of existing dams have hydropower capacity. According to the US Department of Energy (DOE), there is 4.8 GW of potential power from these dams. Upgrading current hydropower plants can add an additional 6.3 GW. Another 1.7 GW is possible using streams and small rivers. The biggest possible source of hydropower is through a process called pumped storage.

Pumped storage is the process of using solar or wind energy to pump water up to a storage facility, lake, or pond to be used later to create hydropower. This is how solar and wind power can be saved for later use. The DOE estimates that there is 35.5 GW of pumped storage capacity in the United States by upgrading current facilities and creating new ones (O'Connor 2016). The widespread use of pumped storage will depend on the price and availability of battery storage. Battery storage is the biggest competitor of pumped storage. It is likely that in areas that already have hydropower or that have affordable pumped storage possibility, batteries will not be the first option. In drier areas, batteries are likely to be the bridge to a more renewable grid.

Hydropower continues to hold promise across the planet. If it is done right, it can produce the energy needed without the GHGs that come with fossil fuel power plants. If it is implemented recklessly, it will come with other costs that can also be destructive. Most of the largest and easiest sites for large-scale hydropower plants have already been built. While there may still be little growth left for the traditional, sizable hydropower stations built on major rivers, the biggest room for growth lies in pumped hydropower. Pumped hydro energy storage (using existing dams) may be the key to unlocking solar and wind potential. This ready-built infrastructure could provide an alternative to batteries that is cleaner, greener, and more efficient. Hopefully, hydropower will continue to grow in a sustainable manner that assists in the worldwide transition to renewable energy.

Solar Power Solar energy is the fastest growing form of renewable energy. The price for solar panels decreases (or decays) at an exponential rate. Solar panel power cost $77 per watt produced in 1977, but in 2020 that price reduced to potentially less than $0.10 per watt produced or over 770 times less (77,000 percent less). The adoption of solar energy started off slow but has moved quickly since 2010. In many markets, annual growth of over 50 percent was the norm. That changed in the United States in 2017, when the Trump administration put a 30 percent tariff on imported solar panels. This led to a loss of 62,000 jobs and a first-time decrease in annual solar installations, as the two US manufacturers this tariff was meant to protect both filed for bankruptcy by 2019. The biggest impediment to the continued growth of solar power is political. Economically, in most regions, there are no reason to stop increasing the amount of solar energy used for electricity production. The continually decreasing prices of battery storage are adding to the incentive to increase the use of solar energy. Entrenched fossil fuel interests that are heavily subsidized by federal, state/provincial, and local governments have pushed for the rollback of subsidies for solar power both commercial and residential. They have also pushed for laws and tariffs that make solar more expensive and difficult to install. Despite these attempts, solar power has continued its meteoric rise to becoming a major international player in the energy field. By 2023, solar was producing more energy than ever. The Trump administration was only able to slow down the rate of growth. It could not stop the growth of solar, as it is one of the most affordable forms of energy.

There are many ways of capturing solar power. There are the prehistoric methods of using solar energy for drying foods, for creating salt from ocean water, and for warming homes. Designing a home to be warmed by the sun is called *passive solar heating*. It does not require additional equipment or advanced technology. It merely requires thoughtful design. Most simply, a house must be designed to receive and use the power of the sun when homes are the coldest. Homes are designed to allow the maximum amount of sun to enter the home when the sun is at its lowest point in winter. In summer, when it is warmer, a passive solar home will shade the home from the heat of the sun by using building design and the strategic use of deciduous trees. In this way homes, even in cold climates, are able to receive most (and in very sunny locales all) of their heating from the sun. Additionally, passive solar homes design floors and walls to absorb heat during the day and release that heat at night. Dark-colored stone, adobe, and other dense materials work best for this purpose. Ventilation systems are incorporated to move warm air from solar heating spaces to rooms that logistically may be difficult to heat with passive solar design, such as bathrooms. These ancient techniques, aided by advances in technology that modify a system to have an exact temperature, allow people to use the sun to their advantage. Unfortunately, most homes that are built with two priorities: style and cost-effectiveness. It is regrettable that most homes built in sunny states do not take advantage of this free solar energy and that many houses actually work against passive solar and allow more sun in during the summer months than in the winter months. It would be in the best interests of the planet to create government incentives or regulations that encourage passive solar design in new buildings. Passive solar heating is some of the low-hanging fruit of the solar world. As humanity continues to wrestle with how to effectively address climate change, passive solar power may have a large part to play.

Solar water heaters are another ancient technique. The Romans used them to heat water for their public baths. The concept is simple: expose water to the sun and then have a system to move it to where it is needed. Solar water heaters became popular in sunny parts of the United States in the early 1900s. The problem of course was that they depended on the weather, so they did not work on extremely cloudy days or at nighttime. It was also difficult to regulate temperature. The invention and widespread dissemination of modern water heaters in the 1940s caused the demise of solar water heaters. However, with the environmental movement of the 1960s and 1970s as well as the energy crises, solar water heaters grew in popularity once again. President Jimmy Carter placed solar water heaters on top of the White House and created government subsidies for solar water heaters. This helped spur a solar water heater boom in the United States. They are still used all over the world; however their popularity has waned with advances in photovoltaic systems that convert sunshine into electricity. (US policy also stopped supporting solar water heaters.) These systems can also be used to heat water by powering an electric water heater. Solar water heaters are still used all over the planet as they are effective and cost-efficient.

Photovoltaic systems or solar panel systems are what most people are talking about when they refer to solar power. Photovoltaic systems are the systems that have decreased in price to the point that they are the cheapest energy available (given sunny conditions). They can be installed on homes, businesses, or put in huge arrays by governments or utilities. They can create electricity even when it is cloudy, although not as much as when it is sunny. Photovoltaic systems are able to produce electricity without turning a turbine like other forms of electrical generation (wind, hydro, gas, coal, nuclear, and biofuel). Solar panels have photovoltaic cells inside them that create an environment in which photons (particles of light from the sun) knock free electrons from atoms. This creates an electric current.

China leads the world in solar capacity installed with three times as much solar power installed as the second country on the list, the United States, as of 2024. However, if the European Union is counted as one entity, they produce twice as much as the United States or about two-thirds of China's production. China, as a communist country, can decide how much solar will be installed in the country. China is a hybrid economy, part command and control socialism and part capitalist. China can heavily subsidize solar manufacturing as well as demand that state-run and private energy providers install solar. This mixed-methods approach has proven successful as China's solar industry continues its exponential growth. The United States has spurred growth in the solar industry by providing tax incentives for individuals and companies to purchase and install solar panels. The Inflation Reduction Act of 2022 has been the most recent set of tax incentives to boost green energy, and it has been tremendously successful. Many countries have done the same. Italy at one time was offering 100 percent reimbursement for installing panels on a house, but when more conservative political parties took power, they completely eliminated the subsidy. These political swings can prove difficult for the continued growth of solar. If the various political parties in Italy, or anywhere else, could agree on a long-term subsidy that is consistent, then solar power can continue its exponential rise. Solar power will continue to grow with or without subsidy, but all businesses thrive better in an environment that is both stable and predictable. One thing that has been predictable in the solar industry is Swanson's law, which states that solar prices drop by 20 percent every time the total shipments of solar panels doubles. Solar panels will continue to be more affordable, and they will also continue to be more efficient.

Solar panels are converting sunlight into electricity at an average rate that has increased from 15 percent to over 22 percent. Efficiency has been increasing, and recent developments have demonstrated efficiency rates of over 27.3 percent using perovskite (Ambrose 2020a). Once these more efficient solar panels become available and affordable for consumers, people will be able to produce the same amount of power in less space. As new solar panels replace older ones, power will increase dramatically. Solar panels continue to be less expensive and more efficient. As batteries continue to be more affordable each year, more individuals and utilities will use that combination to create energy when the sun shines and store it, so energy can be proved when the sun does not shine.

The exponential growth of solar power cannot be overstated. It is clear that solar energy will continue its growth and coal will continue its decline. The lower prices of solar are already bringing solar to developing nations across the globe that do not have energy infrastructure.

Overall, the total capacity of photovoltaics has increased worldwide an average of 28.67 percent a year from 2014 to 2019. If total capacity is 633 GW at the end of 2019, and growth continues at a conservative estimate of 28 percent a year then the world will be producing 9,565 GW of solar power from photovoltaics by 2030 (SEIA n.d.). It is possible that solar will be the most abundant form of electricity production. Coal had for many years been the leader and it has recently been replaced by natural gas. As many countries are attempting to meet their goals through the Paris Climate Agreement, they will turn to solar as a continually less expensive energy resource. Further advances in solar and battery technology are likely to accelerate the speed of this transition. As solar is affordable, portable, and can be as small as a single panel, it is likely to be the energy of choice for people in developing countries that lack the necessary infrastructure for centralized electric power plants and the grid system that is required to put electricity in many rural homes.

In the United States, the growth of solar saw some setbacks during 2017–19, but has rebounded in 2020–3 and is on pace to smash all previous records for installation. The low cost of solar combined with the de-carbonation goals of many cities, counties, and states has led to this explosion in installations. Barring a breakthrough technology that makes current forms of creating electricity obsolete, solar power is poised to continue its historic growth in the United States. China and India will also continue to grow their solar portfolios.

Concentrated solar power (CSP) plants concentrate the power of the sun using mirrors. Mirrors are either aimed at a central tower or at a pipe going through a parabolic trough. The tower, with the concentrated solar power, heats water or molten salt, and that heated liquid is used to create steam that turns a turbine, just like almost every other form of power except photovoltaics (Ho and Iverson 2014). Concentrated solar power that uses molten salt has the advantage of storing heat in the molten salt that can last for hours. This allows CSP plants to create electricity in the dark during times that electricity is in greatest demand. Electrical demand starts out strong in the morning and dips as people leave for work. At 4:00 p.m. in the United States and most other countries, demand increases as people return home; at the same time, solar power is beginning to decline and will stop sooner or later depending on the time of year. This situation is sometimes referred to as the duck curve or the California duck curve because California has been ahead of most of the world in implementation of solar power. Basically at the time that the most power is needed, after work in the evening, solar power begins to decrease and then stops completely.

A CSP system using molten salt can continue to produce the amount of energy required. This around-the-clock energy appeared to fix the nighttime problem of solar. However, the constant lowering cost of both photovoltaics and battery storage may make CSP plants financially untenable, comparably. CSP can be used in a variety of industries, such as desalination and advanced oil recovery (Weinstein et al. 2015). There are currently 6,450 MW worth of CSP plants installed worldwide. The largest plant is the Ivanhoe power plant in California. This represents 729 GWh a year or about 1 percent of the total solar power in the world. It appears that unless there are unforeseen decreases in the cost of CSP plants, they will continue to operate as a niche industry similar in scope and size to geothermal plants. Some CSP plants use natural gas to start up their operations, so that increases the amount of CO_2 in the atmosphere. An average plant still produces 4–5 times as much energy (with solar) as would a natural gas plant using similar amounts of fuel (gas alone).

CSP has the ability to store energy relatively inexpensively, but it still has the same drawback as all other solar types do. They cannot produce energy when the sun does not shine. CSP plants can flatten the duck curve but they cannot fix the everyday reliability problem. The future of solar will likely be in conjunction with pumped hydropower, natural hydropower, and geothermal wherever these resources are available.

Agrivoltaic systems are solar systems that work with agriculture and/or animal husbandry. Solar photovoltaic systems have been demonstrated to increase crop yields, to decrease water usage, and to increase moisture retention in the soil. Agrivoltaic systems benefits run both ways (NREL 2019). Solar panels are also more productive due to the cooling effect of having plants underneath them. The evaporative effect of moisture leaving the soil and the plants kept solar panels 9 degrees cooler on average (Barron-Gafford et al. 2019). Keeping solar panels cooler increases their electricity production. This synergy can provide financial benefits as well due to more crop production, less water costs, and increased profits from electricity production.

There are three main barriers for agrivoltaic systems. The most important one is that most farmers and ranchers are not aware of agrivoltaic systems or their affordability and possible benefits. New technology takes time to spread in the beginning, but once the information is available and adoption rates hit a tipping point then adoption of the technology explodes. An example of this phenomenon is wind power in Texas. In 2005, wind power was miniscule but, with government support, public opinion began to embrace wind and the financial benefits it can provide. Person-to-person outreach is an important part of this process. The second biggest barrier is infrastructure. Remote, rural areas need electrical grid infrastructure in order to financially benefit from producing electricity. The final barrier is support, both government support and private business. Farmers and ranchers are busy people just trying to keep their heads above water, financially. They need government and electric company regulations that ease the installation and integration of solar panels. Loans must be made easy to obtain, and businesses should be encouraged to work in rural areas that are best suited to agrivoltaic systems. With proper support, agrivoltaic systems will be able to provide energy and economic stimulus to the sunnier agricultural areas across the planet.

New solar technologies have demonstrated that solar can be smaller, thinner, and more efficient. These new technologies can be as small as printable thin-film solar cells. This technology was ahead of the curve but the steep drop in photovoltaic prices due to subsidized Chinese manufacturers quickly closed their financial advantage.

Photovoltaic paint is another technology that would have the advantage of being able to be placed anywhere at minimal cost. Currently, solar paint is still in the experimental phase. It may be more efficient than regular photovoltaics, but it is not commercially available. Currently, there are three forms: hydrogen solar paint that can split water into hydrogen and oxygen, quantum dots (nanoscale semiconductors), and perovskite solar paint (Gautam n.d.). Photovoltaic paint may develop into a major part of the solar industry, or it may become a niche product. People may use it to paint cars, bikes, and backpacks, giving them a mobile energy source whenever and wherever the sun shines.

There are grander proposals for creating solar outside of the atmosphere. Space-based solar power (SBSP) is currently being pursued by Japan, China, and Russia. An SBSP station could be placed in an orbit in which it is always in the sun to create maximum power. It could then transport that power using a microwave transmitter or a laser emitter. This would require extremely large receiving antennae that would cover large swaths of land. SBSP does have the potential to produce almost unlimited amounts of electricity. The vast expenses and the refinement of the technology are the biggest barriers stopping the growth of SBSP for now. SBSP could possibly play a large role in some countries as they strive to power their societies without carbon.

Overall, solar power is growing faster than any other form of energy. As energy storage continues to advance and prices decline for both solar and storage, solar is likely to become the largest producer of electricity on the planet. If the estimates of Ramez Naam are correct solar will be cheaper than all fossil fuel produced electricity, in all locations, by the year by 2035 at the latest.

Wind Power There is enough wind power potential to power the electrical needs of the entire planet eighteen times over. Wind power continues to grow at exponential rates and has been the leader of non-hydropowered renewable energy, surpassed by solar power in 2020. Wind power is the process of turning a turbine that produces electricity using the wind. Wind has been used for millennia to move water and grind grain among other things. Today, countries across the planet are attempting to meet the goals of the Paris Climate Accord by putting in wind turbines onshore and offshore. The economics of onshore wind power has made financial sense since the early 2000s. Offshore wind is more expensive and more complicated, but it has the potential to tap into resources that can power entire countries. The European Union has been the leader in wind since the early 2000s with Germany and Spain leading the way. China and the United States continue to increase their wind energy production as well. China once again has demonstrated its ability to move quicker than any country in the European Union. By the end of 2020, there was 651 GW of wind capacity installed worldwide (GWEC 2019). China has experienced a number of years in which it has increased wind energy capacity by over 20,000 MW per year. This has led to China's complete dominance in the field. The country installed 211,000 MW at the end of 2018 and 278,000 MW at the beginning of 2023, and it expects to install 437,000 MW by 2028. This will average over 25,000 additional megawatt per year. The United States installed 107,000 MW in 2020 and 144,000 MW by the end of 2022. The United States has no major offshore wind operations (just two small pilot projects off the East Coast), but it does have a lot of plans for offshore wind. States on the east coast have committed to over 30,000 MW of projects. Combined with continued interest in onshore wind in the central United States and possible offshore projects on the west coast, the country is poised to continue to be one of the largest, if not the largest, per capita wind producers in the world. However, when the European Union is considered as a whole, it can be seen as a leader in wind energy production. Most European countries enjoy wide public support for wind energy. Over 80 percent of people in most European countries approve of increasing wind power capacity. The EU's wind capacity continues to grow at a rapid pace, especially in the area of offshore wind farms.

Overall, worldwide wind power capacity was at 651 GW installed by the end of 2019, and 906 GW installed in the middle of 2023. Wind is another energy source that is growing exponentially. Worldwide wind power installation increased 19 percent year over year in 2019 and increased by 20 percent in 2020 (Ambrose 2020b). The increased use of wind power may increase the rate of expansion, barring geopolitical and economic barriers that may arise. In the United States, installations continued to break records in 2020, even amid the Covid-19 crisis. The United States installed 1,800 MW of new wind capacity to bring the total to 107,443. There are over 24,000 MW under construction in the United States. Overall, worldwide wind power capacity was well over 1,000 GW as of 2024.

Onshore wind has been booming since the 1990s, in places that one might expect (Northern Europe) and in areas that appear less likely (Texas and Oklahoma). China and the United States have been leading the growth of wind power, but some European countries continue to lead the way in offshore wind projects. Great Britain has more installed offshore wind capacity than any other country, over 14 GW as of 2023. Offshore wind has great potential because many countries have territories where turbines could be located and new technology is bringing down the cost of offshore wind power. Floating wind turbines have expanded the possibilities for offshore wind, allowing for wind turbines to be placed in deeper waters than ever before. Offshore wind costs decreased by 32 percent in 2018 and continued to decrease through 2019 and 2020. In many places, offshore wind can compete with other forms of energy production without subsidies.

In the United States wind power surpassed hydropower as of 2020, and in all likelihood will continue to grow due to low costs and environmental benefits. The entire world is moving toward renewables and wind is likely to be a very big part of that movement. In the United States, wind energy provides 9.2 percent of all energy. It is likely that wind power will continue its decline in price and will be the energy of choice for people with wind resources across the globe.

Geothermal Energy Geothermal energy is a contender for the title of most underused renewable energy source. This is because it is expensive to drill the wells necessary to access this energy in most places. It can cost between $20 and $30 million to drill test wells at a site. There is always a chance that despite excellent research and good technology, the wells could turn out to be a bust. This is what has been the biggest obstacle to renewable and continuous geothermal energy becoming a major player in the renewable energy field.

Geothermal energy provides about 1 percent of the world's electric power. It is created by either tapping into natural areas in which water is turned into steam by the heat of the earth or by creating steam by pouring water into a well that is then converted to steam by the heat of the earth. It is available 24 hours a day and 365 days a year. This gives geothermal energy a huge advantage over both solar and wind power.

Some countries are addressing the problem of high start-up costs without guarantees of success by either drilling themselves or paying private companies to drill for them. Some of these countries are looking to tap into the potential of supercritical fluid, which is water that is heated to over 400°C. This fluid is somewhere between a liquid and a gas. It is difficult to control and can be much hotter than 400°C. The heat and pressure of this fluid can destroy equipment. It also has the potential to be among the most energy-dense energy sources on earth. The United States, Japan, Iceland, Italy, Russia, and Mexico have all been investigating and experimenting with supercritical fluid. While it is unknown if and when this form of geothermal every will be harnessed, there are three forms of geothermal energy that the world is already using: dry steam that comes up directly in the form of steam, flash steam that comes up as hot water under pressure (above 182°C) and becomes steam as it nears the surface, and binary cycle plants that use the heated water to warm a secondary fluid that has a lower boiling point. All three of these types of plants are in operation, but flash steam plants are the most common.

The future of geothermal is uncertain but the potential to produce enough energy for the entire planet is proven. The expertise to drill the wells needed for geothermal energy is widespread. It is the same knowledge used to drill for oil and gas. The collapse of the oil markets during Covid-19 in 2020 put many drillers out of business. This collapse created an opportunity to drill for less. The price to hire a team to drill is controlled in part by supply and demand. A lower demand for new oil wells temporarily created a lower price to drill for geothermal energy.

In the United States, the Western United States has plentiful resources of geothermal energy. Every state west of the Great Plains has access to geothermal heat that is close to

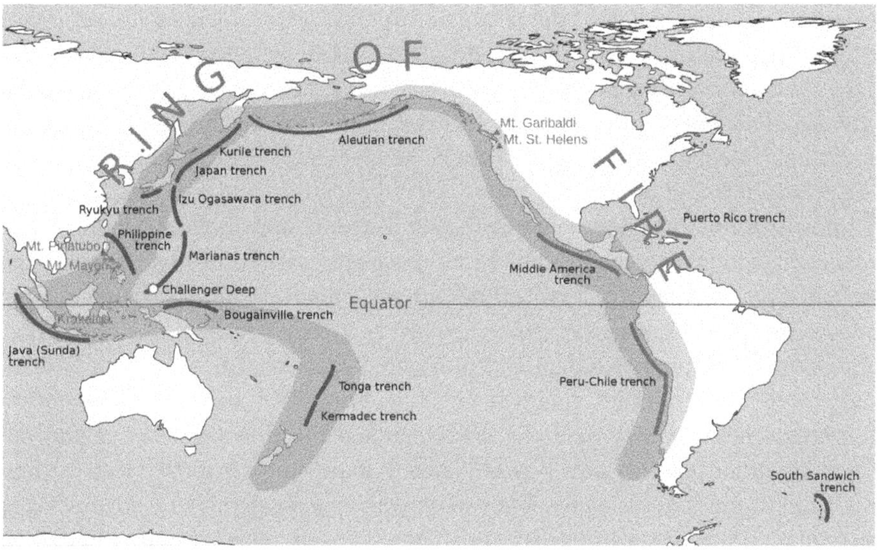

FIGURE 3.1 Pacific Ocean Ring of Fire. *Source*: Wikipedia, https://en.m.wikipedia.org/wiki/File:Pacific_Ring_of_Fire.svg.

the surface. All states west of the Rocky Mountains have hot water that is near the surface. With proper investment, the western half of the United States should be able to have enough geothermal energy to augment wind, solar, and hydropower. The constant nature of geothermal energy helps fill the gap that can be created by the intermittency of both solar and wind.

The United States is the international leader in geothermal energy with 3,676 MW capacity, but there are many other countries that are using geothermal energy to create electricity as well. Indonesia has 2,133 MW capacity, the Philippines has 1,918 MW, and New Zealand has 1,005 MW (Richter 2020). These countries are located in what geologists call the Ring of Fire. This ring surrounds the Pacific Ocean and is the result of plate tectonics. Areas where one tectonic plate goes under or over another create intense heat nearer the surface. The down side is increased volcanoes and earthquakes. Countries abutting the Ring of Fire have the opportunity to tap into geothermal resources that far exceed most other countries (see Figure 3.1). However, there are resources outside of the Ring of Fire. Turkey, Italy, Mexico, Kenya, and Iceland all create significant amounts of electricity using their natural geothermal resources. The future use of geothermal resources will be pivotal in how quickly the world can transit to renewable energy.

Wave and Tidal Energy Wave and tidal energy have limitless potential but are stymied for two reasons: tidal energy plants are very expensive to build, and wave technology is still developing and more expensive than wind and solar. In Chapter 1, we covered some of the major tidal plants in France and Korea; there are also smaller plants in China, Canada, and Russia (EIA 2022). It is important to note that the oldest plant, La Rance in France, is still in operation forty-seven years after it was built in 1966. The longevity of this renewable resource makes it the perfect option for governments who can look at the whole life of the plant and are not driven to make immediate profits. That is the reason most large hydropower plants are government owned. Governments will also be able to boost the fledgling wave power industry by investing in wave, wind, and sun combination power stations. Wave and tidal power has the advantage of being constant, as waves and tides are predictable and dependable.

Tidal power is not available everywhere, but it can be very effective in areas that have dependable high tides. Great Britain, France, Russia, and South Korea have the most potential tidal power; however, new technology is expanding where tidal power plants can be located. Tidal power can be disturbing to a local ecosystem if the barrage system is used. A barrage system is similar to a dam and is usually built across an estuary or bay. Most of the tidal power plants built are this type of plant. Fishermen and environmentalists both dislike barrages because they alter the landscape and the flow of water in and out of a bay. There are other types of tidal power plants that are less disruptive. Tidal turbines are basically wind turbines in the water. They work well in areas that have a strong predictable tide, and they do not disturb local sea life. There are a number of experimental stations that are vetting the effectiveness of tidal turbines. There are tidal turbines in the East River next to Manhattan in New York and in Grand Passage in Nova Scotia, a province in Eastern Canada just North of the United States. Grand Passage is part of the Bay of Fundy, which has one of the world's largest tidal ranges, up to 16 m (the height of a four-story building) between low and high tide. The experimental turbines there are already generating power. It is hoped that they can be scaled up and that there will be less controversy, as they do not appear to harm fish or change the tide. It is likely that these tidal turbines will advance in the same ways that wind turbines have. They will be able to produce more electricity with less flow as the technology improves. Tidal power looks to be an important niche energy source that can be used by communities that have good tidal resources. This will include both coastal and island communities. Sustainable Marine Energy and Verdant Power are a couple of the companies that are trying to turn this hope into a reality in the near future.

Wave energy is abundant. It has been estimated that waves have enough power that, if harnessed, would create more electricity than the entire world needs. As mentioned in Chapter 1, the harshness of the marine environment coupled with new and unproven technology has stopped wave power from living up to its potential. That paradigm is changing. Technology has matured and has been proven. Companies have connected wave power with wind and solar power to create offshore renewable energy power plants that have stable and continuous power. German wave energy company Sinn Power is putting together the first offshore wind, solar, and wave power plant. It will be located off the island of Crete near Heraklion, Greece. Seabased is the name of a Swedish company that has been improving their wave energy parks through small experimental joint projects with governments in Northern Europe, and after fourteen years of smaller demonstration parks, they have been hired to produce 100 MW of energy off the coast of Ghana (Harris 2018). This will be the first large-scale wave plant. If this plant proves successful, it is likely that wave energy plants will become common across the planet. Forty percent of the human population lives within 100 km of the coast. Wave power may prove to be the best option for this large slice of the human population.

Hydrogen Hydrogen fuel cells may be the future of transportation energy. They take the chemical potential of oxygen (O_2) and hydrogen (H_2) and convert it to heat, electricity, and water vapor. The process creates no greenhouse gases and uses no fossil fuels. Oxygen is easily captured from the atmosphere, but hydrogen, while a very common element, is not easily found in its pure state. Hydrogen can be separated from water (H_2O) through electrolysis, which requires electricity. For hydrogen to be truly carbon-free it must be harvested using carbon-free electricity. There are a number of methods to make "green" hydrogen (Dincer 2012). To do this one only needs access to water and wind or sun or hydro or any other renewable energy source. Currently, most hydrogen is harvested in a process that uses natural gas. Some areas have too much wind or solar capacity at certain times or seasons. Creating the necessary infrastructure for hydrogen harvesting could save this energy from being wasted and could turn remote outposts into modern-day hydrogen mining operations (splitting hydrogen off from H_2O or other less green sources such as methane). New advances in electrochemical water splitting technologies promise to bring down the cost of electrolysis (Chatti et al. 2019). There are many places all over the planet that are sunny and/or windy but where there are few people. Areas like this are perfect for hydrogen mining because there will be very little NIMBY (not in my backyard) pushback.

Hydrogen does have many critics. Hydrogen mining is energy-intensive when compared to powering an electric car directly from renewables, for example. Hydrogen requires new infrastructure that many argue is unnecessary. It is argued that advances in battery storage will make hydrogen an unnecessary intermediate step. The counterargument is that hydrogen can be stored and used to power anything including semi-trucks, large ships, trains, and airplanes. Current technologies make it difficult or impossible to power some things using battery power alone; this is especially true with aviation. Hydrogen can also be used as a type of jet fuel that is carbon-free. So despite some critiques of hydrogen, it does have some advantages that may help it succeed as an energy source of the future.

There are pros and cons to hydrogen, and the feasibility of a future hydrogen-driven economy will depend heavily on governments if they are willing to invest in creating the infrastructure necessary to encourage business investment in hydrogen. EV (electric vehicle) technology has at least a ten-year head start on fuel-cell (hydrogen) technology. Hydrogen may be part of a carbon-free economy. Hydrogen can be used directly as a fuel in liquid form (extremely cold), cold/cryo compressed, or a compressed gas form (energy.gov 2020). Hydrogen can also be used to create electricity directly through fuel cells. This versatility may allow hydrogen to supplant fossil fuels in areas in which a charged battery may not be feasible. As with many types of renewable energy, hydrogen's future is tied to local, national, and international commitment to end fossil fuel use and attempts to create a carbon-free economy.

Energy Storage Energy storage capacity will be absolutely necessary for any transition to renewable energy. Energy storage can take many forms. The most common form of energy storage is the one most people use every day: batteries. Batteries are used to power laptop computers, cellphones, and electric vehicles. Batteries are charged by electricity, storing it in the form of chemical energy and then eventually converting that back into electricity. Some batteries are built for one-time use. From a carbon-free energy perspective, these kinds of batteries are extremely wasteful because it takes a lot of energy and resources to create a battery made of metals and chemicals that will power something for a relatively short period of time. Rechargeable batteries on the other hand are able to be charged thousands of times before they wear out. New EV batteries can be used for well over a million miles. Batteries come in many different shapes, sizes, and types. They can be lithium-, lead-, or sodium-based. They can be liquid or solid. They can be used in almost any device or vehicle that needs energy, but they are not the only way to store energy.

Any type of energy storage has some loss of energy in the process of storing the energy, so it is preferred to use energy as it is produced. When this is not possible, batteries, pumped hydro (94 percent of all electricity storage), compressed air, flywheels, as well as raised weights can be used to store energy (EPA 2020). Pumped water storage is the process of taking water that has come through a hydropower plant and pumping it back uphill behind the dam, where it can be used to create electricity again. Raising weights is a similar idea. Electricity produced at times when it is not needed is used to raise heavy weights into the air. The weights are released when electricity is required, and they turn turbines, creating that needed electricity at the right time. These are the major types of energy storage systems that are being used, but they are not the only types. As the world continues to transition to a more renewable energy system, new techniques will continue to push the boundaries of how energy can be stored. Advances in battery efficiency will continue to decrease the dependency on fossil fuels as sources of portable energy.

Nuclear Energy

Nuclear energy creates no GHGs and is touted by many as the path away from fossil fuels. Nuclear energy also creates radioactive waste that is dangerous for 10,000 years. People are afraid of nuclear energy due to the potential for accidents that can kill any humans nearby, as well as poisoning air, water, and soil. As discussed in Chapter 1, the nuclear accidents at Three Mile Island in the United States, Chernobyl in the former USSR, and especially the most recent one in Fukushima, Japan, have made many countries question the safety and efficacy of nuclear energy. Germany has closed plants, many other countries have pledged to phase out nuclear power, but other countries are moving ahead with new nuclear power plants. The majority of new plants are Generation II and III nuclear reactor power plants. This technology, while safer than Generation I reactors that were mostly experimental and demonstration reactors, is not as safe as some of the newer Generation IV nuclear reactor technologies. Most Generation II and III reactors use pressurized water to cool the reactor. They are called Pressurized Water Reactors (PWRs). There are two types of PWRs: light water reactors that use normal water to cool the reactor and moderate the speed of neutrons and heavy water reactors that have more of the hydrogen isotope deuterium. Over 90 percent of the reactors that are generating electricity are light water reactors. Both types of PWRs operate with water at extremely high pressure (15–20 times normal pressure) (World Nuclear Association 2020).

Generation IV designs have the advantage of creating hundreds of times more power with the same amount of fuel. They can be made safer with "walk away" safety features.

They can also produce less and safer nuclear waste. Nuclear waste from Generation IV reactors only needs to be stored for a couple hundred years instead of tens of thousands of years.

One of the new Generation IV technologies is the Molten Salt Reactor (MSR). In an MSR, nuclear reactions are used to melt salt mixtures at high temperatures. They are not under high pressure and have "walk away" safety features. Most PWRs require constant electricity in order to keep the plant cooled, but an MSR has a cooled plug that melts if the plant loses electricity. The molten salt mixture would drain under the reactor to cool and end the nuclear reaction. The biggest problem with this technology is that there have only been demonstration reactors and no country has yet built a full-scale plant, despite the fact that it has been demonstrated technology since the late 1960s.

Another set of Generation IV reactor technologies are called fast reactors. These reactors have the advantage of being able to consume much more of the fuel than the currently dominant light-water reactors. The ability to use up more of their fuel (up to 99 percent more) solves the problem of nuclear waste created by most nuclear reactors in use today. The Russian BN-800 is the only operating Generation IV fast reactor in the world (Nuclear News 2020). (It is also cooled using liquid sodium, but it is not considered an MSR.) It was created as part of a joint US–USSR project to start destroying weapons-grade plutonium in 1983. The United States pulled out of the project in 2016 and has not provided any plutonium. Both the United States and USSR/Russia have reduced the amount of nuclear weapons they actively maintain. Decommissioned nuclear weapons create weapons-grade nuclear materials that can be used for nefarious purposes if they get into the wrong hands. The idea of using weapons-grade radioactive materials to create carbon-free energy has a "swords into plowshares" ring and is still embraced by many people, organizations, and countries. However, the rest of the world is not close to having nuclear reactors that destroy this type of material while producing electricity.

Overall, nuclear energy has the potential to help the world create electricity without contributing to climate change and global warming. Expenses and public fears appear likely to derail most nuclear plants in the near future. An organized international effort to tackle climate change has the potential to change that, but currently the world does not have the political will to engage in that work. It is yet to be seen if the leaders of the world will be willing to use nuclear energy as a way to create GHG-free energy or to even create plants specifically to pull carbon out of the atmosphere.

CARBON STORAGE

Power Plant Capture and Storage

The successful and affordable capture and storage of CO_2 has been the perennial hope for those that are attempting to find ways to continue to use abundant fossil fuel resources, especially coal. The hope for so-called clean coal is that on-site carbon capture and storage would be able to transform coal from a CO_2-intensive and polluting energy source into an energy source that is neither of those things. Canada and the United States attempted this. Most coal smoke is regulated. The smoke must go through a process called scrubbing, which removes some of the particulate matter and also removes some of the more dangerous chemicals inherent in coal smoke, such as mercury. While it is not expensive to remove large particulate matter, it becomes increasingly expensive to remove smaller particulates and many of the chemical components. It is also incredibly expensive to remove CO_2. While politicians have promoted (and a few continue to promote) the idea of clean coal, it has not come to fruition. The process is difficult and can create additional environmental problems that may need to be remedied (Eldardiry and Habib 2018). It had the potential to become economically feasible in the 1990s when the price of renewable energy was prohibitively expensive. As the cost of renewable power, particularly solar and wind, continues its precipitous decline it becomes even more unlikely that clean coal would make economic sense.

The process of CO_2 capture is fairly straightforward. A coal burning plant runs the smoke and fumes from the burning of coal through one of a number of capture processes, including chemical looping, membrane gas separation, adsorption, absorption, and a newer technique that runs the emissions through organic soda lime. These various techniques remove the carbon from the exhaust stream and store it separately. At that point the CO_2 is usually compressed and moved to a storage facility. The CO_2 can be stored underground in old oil or gas wells. It can also be directly injected into rock formations that geologists determine will contain the CO_2 and, in some cases, react with the it to create permanent chemical compounds that are unlikely to change. CO_2 also has both industrial and commercial uses. It can be used to extract oil and gas in the hydraulic fracturing (fracking) process. It is added to all kinds of beverages, most commonly soda pop such as Coke and Pepsi products. None of these uses create enough income to come close to covering the costs of the carbon-capture systems. This is one of the main reasons that carbon-capture coal-fired power plants are not economically feasible. If carbon were priced through a market mechanism, it may become economically feasible to add carbon-capture mechanisms to existing coal plants. However, it is more likely that building more renewable energy capacity will be more affordable and practical than retrofitting old coal-powered power plants. As long as natural gas remains an affordable alternative, there is unlikely any situation in which coal would be the preferred energy source.

Sometimes, however, coal is the de facto method for electricity production, and in that case preference is a moot point. There are countries that are currently building coal plants, and these countries will want to recover their sunk investment costs. In these situations, coal carbon capture will be necessary. While it is a difficult and expensive process, it has been proven to be effective.

There have been some advances in turning pollution from power plants, refineries, and steel manufacturing plants into ethanol. This process may be viable if it can be proven economically feasible or if nations put a price on carbon. Companies like LanzaJet are currently operating a number of sites that are producing hundreds of thousands of tons of ethanol from what would have been pollution. This will continue to be important with manufacturing processes such as steel and cement that are necessary for modern society but produce a lot of CO_2 and other pollutants.

Direct Air Capture

CO_2 captured directly from the air affordably and effectively is the Holy Grail of climate change solutions, and like the mythical Holy Grail it remains far out of reach. CO_2 packs a big punch for something that takes up an extremely small percentage of the atmosphere. As of 2023, atmospheric CO_2 levels are at 418 parts per million. CO_2 concentrations are both too much for the planet to absorb in the land, air, and water, and simultaneously too little to be easily removed from the atmosphere. The processes of direct air capture (DAC) or direct carbon capture and storage (DACCS) tend to be expensive. Most processes require heating air, moving air using a fan, and or a chemical process to actually remove the CO_2 from the air. The process of condensing and storing CO_2 requires energy as well. Successful DACCS requires an energy source that is carbon free such as renewable energy or nuclear. It requires large infrastructure to produce electricity and large infrastructure to remove a meaningful amount of carbon from the atmosphere. It is within human capabilities, but under current economic incentives it would require a worldwide "moonshot" mentality to scale up DACCS to a level that would significantly lower atmospheric concentrations of CO_2. There are a couple of new technologies that give hope to the idea that there is financially viable, scalable, technology that can be implemented on a timescale that makes sense for climate mitigation. Climeworks is a Swiss start-up that began a project in Iceland using a process that uses little heat (usually low-grade waste heat) and reusable filters (up to a point) and is currently operating at a very small scale (Landgraf 2019). Carbon Engineering is a company out of Canada that uses a process in which CO_2 is pulled in through an air contactor and then captured as a carbonate salt. The salt pellets are then heated in a calciner (furnace), and the CO_2 is released, concentrated, and stored. The current cost for this is \$100 per ton of CO_2. Unfortunately, most governments have not calculated the true cost of CO_2, and CO_2 trades for less than \$10 a ton on many markets. The success of direct air capture of CO_2 will depend on the ability of governments to create markets that actually price CO_2 according to the potential atmospheric damage it can cause. Once this has been implemented, companies will have financial incentive to draw out of the atmosphere the same amount they put in. For example, if a government charges \$200 for each ton put into the atmosphere, companies would save \$100 a ton by pulling that CO_2 back out of the atmosphere.

Mineral and Geological Storage

After CO_2 is captured, there are various methods for storage. There are business and industrial uses for CO_2, but they make a small fraction of the amount of CO_2 that needs to be removed from the atmosphere. As long as coal plants continue to operate, CO_2 will also need to be "scrubbed" out of emissions from coal-fired plants. There has been much research into the subject of how to store CO_2, and we will look at two major methods here: structural trapping, which can lead to residual trapping and solubility trapping, and mineral trapping (Snæbjörnsdóttir et al. 2020). Structural trapping involves pumping CO_2 underground below the surface of impermeable cap rock. The CO_2 will be unable to surface, but it will continue to react underground. In this position, it can interact in a number of ways, but the two most likely ways are residual trapping and solubility trapping. Residual trapping is when CO_2 becomes trapped inside small pores in the surrounding rock. Solubility trapping is when CO_2 dissolves into groundwater. The downside of structural trapping is that some CO_2 can eventually work its way out if the water it has dissolved into is brought to the surface. Warmer water will release CO_2. The second form of carbon storage we mentioned is mineral trapping. Mineral trapping is an even more permanent method of storage than structural trapping. Mineral trapping involves injecting CO_2 into rock that reacts with the CO_2 to form stable carbonate minerals. These new "rocks" are unlikely to release carbon on a human timescale.

Research has demonstrated the effectiveness of these methods, and they are already being implemented. The most successful carbon storage projects are connected to power plants, gas processing plants, and synthetic fuel plants. These plants capture millions of tons of CO_2 per annum (year). The shorthand for this is Mtpa. Capturing many Mtpa is easy enough to do at a power plant where emissions are concentrated. If the world is going to achieve its goal of limiting global warming to 1.5°C–2°C, then it will be imperative that all gas, coal, wood, and oil power plants are required to remove CO_2 before release into the atmosphere. This must be the first step, but there should eventually be a complete transition away from CO_2 producing plants. Finally, there must be massive investment in air capture through the various methods we will discuss.

Natural Carbon Storage Options

Biochar/Terra Preta Deep in the Amazon rainforest are areas of what is called *terra preta*, which is Portuguese for dark soil. This soil is much more fertile than the surrounding soils, sometimes called *terra comun* or common soil. It was created over the last two millennia by the native peoples that lived in the Amazon River Basin. *Terra preta* is made up of a combination of charcoal, pottery shards, and plant and fish refuse. *Terra preta* retains higher mineral content as well as much higher levels of microbial life that makes those minerals available to plants. Once *terra preta* has been established, it becomes self-sustaining. It can increase at a rate of up to 1 cm a year and has been measured to depths of 2 m.

The most important part of *terra preta* is the inclusion of charcoal, specifically a type called biochar. Biochar is created by burning biomass in an oxygen-free environment (pyrolysis). Carbon is preserved in the process, and it is carbon negative (Glaser et al. 2009). Biochar can be used to store carbon that will then remain stable in the soil for 2,000 years. The addition of biochar quickly changes the amount of carbon stored in the soil (Farrell et al. 2013). Biochar has the ability to store carbon and improve agricultural practices. Researchers in Brazil have grown crops on biochar for over forty years and have not depleted the soil. This suggests that biochar can not only store carbon but also decrease the need for oil-based fertilizers. This can also help achieve the goals of the Paris Accords of limiting emissions, reducing CO_2, and keeping the planet from increasing temperatures past 2°C.

The problem with biochar is that it is a completely different system than farmers across the planet are currently using. Biochar would need to be buried under the soil along with compost. It would be part of a low-till or no-till farming regimen. It would take large financial incentives by governments to make these types of changes to agricultural systems possible. Farmers tend to live on the very edge of economic solvency and are unlikely to risk a change without guarantees from governments that they will be supported through the transition. Throughout the world, there are overgrown forests that need to be thinned in order to decrease the likelihood of catastrophic forest fires. The biomass (in this case small trees) can be used to create biochar. This biochar will naturally and inexpensively store carbon in the earth while increasing agricultural yields and decreasing the need to buy expensive and sometimes polluting fertilizers. The success of what some call "the black agricultural revolution" will depend on the help and support of governments and NGOs to lessen the risks for farmers (Hawken 2017).

It is important to note that the Amazon region is experiencing unprecedented deforestation. The creation of biochar would only have a positive effect on overall CO_2 concentration if it is created using forest products from places like the Western United States, where thinning operations are often stymied by the inability of companies to find markets for small-diameter trees. The Amazon itself would be better used as deep carbon storage. Protecting the Amazon from deforestation due to farming and cattle grazing will need to continue to be a priority. However, the biochar techniques discovered in the Amazon do offer an alternative to fossil fuel-intensive farming.

Carbon Farming: Agricultural Techniques for Storing Carbon There are multiple ways that farmers can adapt their techniques and practices to store carbon. Currently, most mainstream farming practices add carbon into the atmosphere. This is due to the fact that intensive tilling releases carbon that is in the soil and that most farmers and fossil fuel fertilizers add carbon to the atmosphere as well. Runoff from farms carries nitrogen rich fertilizer to rivers, lakes, and oceans, where algal blooms that absorb all the oxygen can create toxic dead zones free of oxygen. Importantly, these farming practices produce mass amounts of food. Farmers have been increasing yields for decades. What many farmers do not understand is that they can continue to have excellent yields while including measures to reduce carbon and increase biodiversity around their fields. As farming has moved away from small subsistence farming to industrial farming, the natural carbon cycles have been broken. In the past farmers used their animals' manure to fertilize their fields. They fed their animals plant remains that humans could not digest. In this system there was little waste, and carbon was balanced. Some older techniques actually sequester carbon. The transition to carbon sequestration from carbon-intensive farming will require a number of big and small changes to farming, but these changes have been demonstrated to work for both the farmer and the environment.

The techniques covered here are ranked by the United States Department of Agriculture (USDA). We will discuss each method in brief. For future reference and updated information see NRCS (n.d.) in the Bibliography section or visit www.carboncycling.org, where much more information is available.

- *Mulching/Compost Application* is simply the process of using plant life and creating cover (mulching) or creating fertilizer (composting). This process is effective in two ways: it decreases the amount of unused plant materials that end up in local dumps and create the potent GHG methane, and it limits the amount of oil-based fertilizer that is needed by farmers.

- *Residue and Tillage Management, No Till, Strip Till, and Direct Seed* are methods that aim to not disrupt the soil and allow microbial fungi to thrive. These methods allow the soil to retain more carbon and moisture as well.
- *Anaerobic Digesters* are composters that operate without oxygen and can compost almost anything that breaks down, meat and dairy products included. They create compost and methane gas that can be collected and used as fuel.
- *Multistory Cropping* is an old technique; this method has been used for millennia. Coffee is a traditional multi-story crop. It is grown under the shade of fruit trees or other commercially useful trees. In some cases there can be three or four layers of agriculture: taller trees, understory trees, bushes, and crops in the ground. This method does not work with all crops and in all regions, but it can be very successful in tropical and sunny desert regions. It can be used with coffee, tropical fruit trees, tomatoes, and peppers as well as many others.
- *Windbreak and/or Shelterbelt Establishments* are beneficial for a number of reasons. The first is that by planting a line or belt of trees, farmers protect themselves from water- and wind-caused erosion. In the process, they store carbon in the form of trees. These shelterbelts also provide a place for wildlife to live. They can also create important connecting pathways for wildlife, particularly large predators that need an extensive range.
- *Silvopasture Establishment* is the process of creating forests that are also effective grazing areas. This can benefit the rancher if the trees have commercial uses and planting trees creates shade for animals as well as sequestering carbon.
- *Tree/Shrub Establishment* can be thought of as simply switching crops. Trees and shrubs sequester carbon in themselves and the undisturbed soil that surrounds them. There are many crops that grow from trees and there are many perennial shrubs (bushes) that produce fruit. The difficulty in this process is that a farmer may need financial assistance in the years it takes to transform a farm from annual production to perennial production.
- *Riparian Restoration and Riparian Forest Buffers* are two separate types of carbon sequestration, but they are intertwined. Riparian areas are areas around water, most commonly rivers and streams. Trees stabilize the shorelines and reduce erosion thus protecting farmland from flooding. Riparian areas are vital to wildlife of all types: fish, birds, mammals, reptiles, and amphibians alike. Restoring riparian areas and creating forest buffers benefits wildlife, stores carbon in plants, trees, and the soil, and it protects farmlands.

These are a few of the many important ways that farming can be transformed from carbon-intensive to carbon-sequestering. There are new technologies that can help create carbon-sequestering farms, but there are many more techniques that have long existed that have been forgotten in the rush to switch to industrial agriculture. Governments have encouraged the switch to industrial agriculture in many countries, but there has been a counter movement as well. Many European countries have moved to subsidize all types of farming, and this has preserved many traditional methods. The United States has subsidized staple crops, especially corn and soy. This has pushed out traditional integrative small farms and replaced them with large industrial monoculture farms. This transition can be reversed by a change in public policy, in the US Farm Bill. If the United States and other countries create new incentives that promote carbon sequestration and traditional farming methods, then farms can once again become a natural part of the carbon cycle and part of the solution to climate change.

Reforestation, Afforestation, and Stopping Deforestation The first and most important step countries can take in this area is to stop deforestation. For example, US policies have created a system in which trees are valued as a product and harvested sustainably. The United States has the same amount of forest as it had in 1900. (However, the original settlement of European farmers to the east coast decreased total forest coverage by about 25 percent.) Forests are increasing in volume as large trees are worth more money. Most private landowners allow trees to grow significantly before harvest. Another incentive that may be used in the future would be a carbon pricing system that places value on standing trees.

The second thing that can be done is reforestation. This is a fairly straightforward process of planting trees in areas that have been clear-cut for short-term agriculture, wood harvesting, and/or cattle grazing. However, success is not always as simple as planting a tree. Forests may not grow back for a variety of reasons. Climate change can make an area too hot and/or dry for a sapling to succeed. Intense forest fires can be so hot that they sterilize the soil, killing the soil microbiome necessary for many trees to grow. There are many areas in the Western United States that have never been reforested after a forest fire. Clearing a forest can also change the weather. In tropical climates, rainforests release moisture that becomes clouds and that moisture is returned in the form of rain. Restarting this cycle can be extremely difficult. Despite these difficulties, many reforestation programs have been extremely successful. South Korea had been almost completely denuded of trees by decades of Japanese occupation and years of war with North Korea. Today, South Korea is a shining example of how to reforest. Over two-thirds of the country is covered in forests, and forests continue to expand (Kim and Zsuffa 1994).

The third thing that can be done is afforestation. Afforestation is different from reforestation in that it is planting a forest where one did not exist. This is done in areas where that land is no longer useful or profitable for agriculture. Trees can be grown for fuel, fruit, or logging. If managed sustainably they can sequester tons of carbon. Some afforestation efforts are not commercial in nature; they are grown for the enjoyment of the community and the positive effects on the area and the planet. They can also be used to stop the process of desertification.

Humans have deforested large expanses of the planet, sometimes whole countries. Ireland, for example, was at one time covered by forests. Human settlement over the millennia has turned forests into farms and pastures. This process has played out across the planet. Anywhere humans bring agriculture and animal husbandry, the forests disappear. Some areas were and are still kept clear of trees by hunters who set deliberate fire to increase rangeland over forest. Rangeland produces more game than forests, so keeping an area clear of trees increases the amount of game available for hunters.

Humans require a lot of land to grow food both for human consumption and to feed cattle that become meat or that produce dairy products. However, there are many abandoned farmlands that have depleted soils and that no longer support productive agricultural endeavors. These areas are prime locations for reforestation projects. Trees effectively sequester carbon. The bigger the tree, the more carbon it can sequester. Planting forests involves planting many very small trees. These trees take at least ten years before they are sequestering large amounts of carbon. Reforestation can have a huge impact as a climate change mitigation strategy. Large swaths of land can be used to create these carbon-sequestering forests. However, some countries have found that by using thousands if not millions of smaller areas they can reforest more than they would have with a few large swaths of land. Individual countries have started "1 billion new trees" campaigns, and an international effort to plant 1 trillion trees has the support of almost every country on earth.

As mentioned, the biggest problem to adding more trees to the planet is that people must first stop cutting down forests. This is especially important because large old-growth forests sequester the most CO_2. The two biggest culprits in the loss of forests are cattle and palm oil plantations. In the areas around the Amazon, local cattle ranchers use slash-and-burn techniques to create more rangeland. Estimates of land loss are between 75 and 80 million acres per year in the Amazon alone. The problem in Southeast Asia is that the forest is being replaced with palm oil plantations. While palm oil trees are trees, they replace tropical jungles and sometimes peat bogs that sequester large quantities of carbon. It is estimated that conversion of each hectare to a palm plantation releases 9–20 tons of carbon per year (Carlson and Curran 2013). (The development of palm plantations also leads to loss of biodiversity, loss of species, and loss of forest services for local people.) Palm oil is inexpensive and in demand, so it is difficult to stop the growth of palm oil plantations. The situation is another example of where a robust carbon market could drive market behavior. If palm oil producers had to pay for the CO_2 that is released during the creation of a plantation, they would be encouraged to avoid using forests and peat bogs in favor of using grazing lands or unproductive farmlands instead. Converting pasturelands into palm oil plantations can actually decrease CO_2 emissions and possibly sequester CO_2, especially if it is replacing cattle grazing.

Protecting existing forests, reforestation, and afforestation offer an opportunity for countries to increase their biodiversity while they also fight against climate change and global warming. By creating or preserving both large forested areas and smaller areas connecting forests, people as well as the many species that call forests home benefit. National and local policies are the tools that can make that happen, but individual actions can also create more forest lands. An international carbon market can also be a valuable tool. Countries such as the United States and South Korea have shown that there are many ways to support a robust forest that benefits everyone.

Blue Carbon "Blue carbon" refers to coastal and marine ecosystems that capture and store carbon. These systems include mangrove forests, underwater kelp forests, underwater seagrass meadows, and tidal marshes. Together, they store massive amounts of carbon. The ocean is estimated to have absorbed up to 40 percent of the CO_2 released by human activity. Blue carbon is also responsible for creating stable coastlines that protect against coastal flooding as well as storm surges. Without these blue carbon protections, rising oceans will be more destructive, causing more damage to human habitation and displacing millions of people. One way that countries are looking to address the effects of global warming and climate change and to create more resilient coastal communities is to restore these forms of blue carbon.

As North Atlantic hurricanes and tropical cyclones increase in intensity, duration, rainfall, and frequency, blue carbon and especially coastal mangrove forests offer a natural and effective way to mitigate the damage and flooding caused by hurricanes. However, half of the world's coastal mangroves have been destroyed, with only 32 million hectares remaining. Mangrove forests have six major threats: clearing, overharvesting, changes in freshwater flows, pollution, destruction of coral reefs, and climate change. Mangroves are often cleared for farmland and are harvested for fuel, construction wood, charcoal production, pulp, and wood chips. Mangrove trees are used to build homes and fences. The rate that mangroves were used for charcoal and firewood was sustainable for centuries in many places, but growing populations are overharvesting at a rate from which mangroves cannot recover. Agriculture and dams limit freshwater flow, which can starve mangrove forests. Since mangroves tend to do well in brackish waters where river deltas connect to the sea, dams can change the mix of fresh and salt waters, creating salinity levels that are too high for mangroves to survive. Pollution carried down rivers and dumped in oceans can destroy mangrove habitat as well. Coral reefs that are destroyed by aggressive fishing techniques or cleared for shipping lanes no longer protect mangrove forests by preventing soil from being washed away by storms. These continual stresses have caused much loss, but many groups are working to reverse the damage and recover mangrove forests where it is possible.

Cities and hotels have cleared mangroves that are unlikely to be replaced in the near future; however, there are many areas that are good candidates for restoration. If a global carbon market can be created, there will be even more areas that are financially incentivized to restore historic mangrove forests. Non-governmental organizations (NGOs) such as the World Wildlife Fund (WWF) and Nature Conservancy are working with governments across the world to recover and restore mangrove forests. They have identified 2 million hectares of land that is ideal for mangrove restoration and reforestation. Nations across the planet have realized the advantages of having these mangrove forests as a buffer against the double threat of rising seas and high-intensity storms. More developed nations have invested money into these efforts. It will be particularly important for the nations of the world along with international NGOs to invest in the mangrove restorations of developing nations. These forests have the potential to capture and store carbon as they also protect coastal communities. They will be particularly important in the future, as they are a foundational species needed to create sustainable coastal communities.

Shellfish take carbon out of the ocean. As shellfish grow, they secrete calcium carbonate ($CaCO_3$), storing carbon in their shells, where it remains for thousands of years. Oysters, which are popular with seafood lovers, obtain most of their diet by filtering out phytoplankton. For this reason, oyster farms need to be located in areas with adequate phytoplankton. An oyster creates no CO_2 during its lifetime.

Scientists estimate that the world's oceans have absorbed much of the CO_2 that human activities have put into the atmosphere. The downside is that higher CO_2 levels can cause the ocean to acidify. The ocean is currently absorbing four times as much carbon as it did on average from 1800 to 1994 (Gruber et al. 2019). Acidification can create conditions in which it is difficult for shellfish to create shells. According to NOAA (National Oceanic and Atmospheric Administration) researchers, this is already happening in the world's oceans. Fortunately, sea grasses can reduce CO_2 concentration and create a healthier environment for shellfish. Both shellfish and certain sea grasses can be grown for human consumption. The process of growing food in the ocean is called aquaculture. It has the potential to help create a more sustainable food system and improve conditions for ocean plants and shellfish as well.

Shellfish filter water, absorb nutrients, and permanently fix carbon into their shells. It is important that shellfish aquaculture is placed in an area in which it is not supplanting a native shellfish population. According to a study by the University of Florida, the hard clam industry in Florida sequesters 1.2 million pounds of carbon annually (Baker and Baker 2010). Overall, shellfish reefs have been shown to not only sequester carbon on their own but also create habitat for other forms of blue carbon (Fodrie et al. 2017). A shift in eating habits toward shellfish and away from land-based animals has excellent potential for decreasing CO_2 output caused by human eating habits. A growing shellfish industry has the ability to be CO_2 neutral at least and CO_2 negative at best (Avila, Grogan, and Larkin 2015).

Seaweed, seagrass, and kelp are terms used in relation to blue carbon. Seaweeds are macroalgae consisting of nearly endless varieties that fall into three main categories: green (Chlorophyta), red (Rhodophyta), and brown (Phaeophyta). Kelp is a form of brown seaweed. Macroalgae, along with other types of ocean algae, produce between 50 and 85 percent of the earth's oxygen. Kelp grows tall and thick and is a nursery for many fish species. Seaweed is not a vascular plant but resembles vascular plants and fills many of the same ecological roles. It converts CO_2 into oxygen, and it can also absorb minerals that can otherwise become toxic in the ocean. Seagrasses are exactly what they sound like: grasses that grow in the ocean. They convert CO_2 into oxygen and stabilize bays, inlets, and other shallow sea waterways. They also have the ability to help heal the damage caused by human runoff and waste. They have also been shown to create tiny granules of a carbon-based mineral (calcium carbonate) that acts to counter the acidification of the ocean. The positive ramifications of encouraging seaweed and seagrass have motivated nations and communities to plant and/or protect the places where they grow.

Algae are able to absorb nutrients from agricultural runoff and toxic spills. Seaweed can be grown in places that have nutrient runoff from agriculture. These nutrients can cause toxic blooms of dinoflagellates in saltwater and cyanobacteria in freshwater that result in a depletion of oxygen in the water. This can lead to die-offs of all aquatic plants and animals in an area. However, some of the worst effects can be ameliorated if there is enough seaweed. Seashores that have higher amounts of seaweed are more resilient to spills and runoff.

Kelp, specifically, of all the forms of seaweed, is an excellent ally in the fight against climate change and global warming. Kelp can grow especially fast (up to 2 feet per day). Underwater kelp forests can sequester up to twenty times as much carbon per acre as terrestrial forests (Hurliman and Zucker 2019). Kelp grows on rocky shores, much like other seaweeds and seagrasses; however, it is not able to decompose there when it dies. Currents take it out to the ocean depths, where the carbon it is made up of is unlikely to reach the atmosphere. This process can successfully sequester carbon naturally without much human intervention. Kelp, along with other macroalgae, currently sequesters an estimated 200 million tons of CO_2 annually (Krause-Jensen and Duarte 2016). Kelp can be grown using marine permaculture (permanent agriculture). This process can allow kelp to grow in areas where climate change has made conditions unsuitable for kelp to grow naturally. Where nutrient/water circulation has been disrupted due to rising global temperatures, nutrient-rich waters can be pumped into kelp beds using solar pumps. In this way kelp forests can be put in places that are not being used for anything else, and they can help sequester carbon and reduce the worst effects of climate change.

Seagrass beds are being replanted across the planet, but they are only successful in 37 percent of cases. Success tends to be correlated with the scale of replanting (van Katwijk et al. 2016). The difficulty here is that although larger plantings are more successful, it is more expensive to plant larger areas. The positive outcomes of restoring seagrasses are well worth the costs, however. In places where seagrasses are successfully reintroduced, the ocean becomes less acidic, which benefits all ocean life that needs to create a shell, from crabs to clams. In an area that has been successfully replanted with seagrass and kelp, the overall health of a coastal ecosystem is much improved. This in turn creates a more productive ocean. A more productive ocean combats climate change and helps feed human populations. Most of the restoration projects done to this point are in the United States, the EU, Japan, and Canada. South America, Africa, and South Asia have not done much work to rebuild seagrass beds. It is clear that wealthier countries have been spearheading these restoration efforts. The role of seagrass as a natural "antacid" means that even more seagrass restoration projects will be needed if there is any hope of stopping the CO_2 released by humans from acidifying the ocean and destroying entire ocean ecosystem (Borunda 2020).

Natural processes that are being restored or replicated through human intervention have the most potential to help humans address climate change with the fewest possible side effects. They encourage nature to do the heavy lifting of carbon storage. By quickly implementing a wide variety of these techniques, the world may be able to avoid the risks involved with climate engineering (geoengineering) strategies. The techniques discussed in the next section are more invasive, and many of them may have unknown consequences, as they have never been implemented on a large scale. If action is not taken soon to mitigate the effects of climate change and global warming, it may be necessary to implement these riskier techniques in the future to keep the planet habitable for humans and other life forms.

Geoengineering Carbon Storage: Ocean Iron Fertilization

Fertilizing the ocean with iron has been an idea that has excited many people over the years. The ocean appears to have low amounts of iron in many places. Iron brought into the ocean from dust storms is essential for entire local ecosystems. Theoretically, added iron can lead to the growth of phytoplankton, which is one of the keys to life in the ocean. Phytoplankton also takes CO_2 out of the water and sequesters it after the phytoplankton dies and sinks to the bottom of the ocean. The idea of fertilizing the ocean with iron is to stimulate the growth of phytoplankton at such a large level that it could partially mitigate some of the effects of anthropogenic climate change (Emerson 2019). This is one of a number of proposed geoengineering projects. Geoengineering is controversial because human interventions have had a history of creating unintended negative consequences.

Experiments both legal and illegal have not demonstrated that the process can be successful (Zhongming 2020). Recent studies have indicated that seeding the oceans with iron may actually cause toxic algal blooms that can kill off sea life. The ocean is an interconnected ecosystem. Making major changes to one part of the ocean is likely to have unintended consequences in other parts of the ocean, sometimes thousands of miles away. For example, research by scientists looking into seeding the Southern Ocean with iron has demonstrated that doing this would cause an uptake of macronutrients in places like the North Atlantic (Lauderdale et al. 2020). The effects could be catastrophic for the fisheries there on which many humans depend. Any geoengineering project that involves the ocean must be carefully studied and must honor the precautionary principle, which states that until a technology has been proven safe beyond all doubt, it should not be implemented. The difficulty that scientists and world leaders are facing is that climate change is happening now, and it is estimated that humanity may have only until 2030 to change course enough to avoid the worst effects of climate change.

INTENSIVE AND EXPENSIVE GEOENGINEERING

De-acidifying (Alkalizing) the Ocean by Adding Calcium Carbonate

The CO_2 that is added to the atmosphere can dissolve into the ocean near the surface where atmosphere and ocean interact. Some CO_2 remains in a state of dissolved gas, and some of the carbon in CO_2 is used for photosynthesis by tiny marine life like phytoplankton and large marine plants like seagrasses. These organisms, both big and small, release oxygen and eventually die with the carbon intact. The carbon in CO_2 becomes part of the organic matter of the oceans. Some organic matter near the surface will break down and the stored carbon will be released, but the majority of this organic matter settles in the depths of the ocean where it will stay for millennia. In this way the ocean is a carbon sink; it takes in more carbon than it releases. The dangerous part of this cycle is that too much dissolved CO_2 combines with ocean water to create a weak carbonic acid (H_2CO_3). However, there are three reactions that can break down this acid and balance the pH of the ocean. Unfortunately, a fourth (and final) reaction for balancing ocean pH relies on

CO_3^{2-} (carbonate), which has limited availability in the ocean (Harvey 2008). Carbonate has the ability to help the ocean absorb mass amounts of CO_2. When there is not enough carbonate to interact with CO_2, the water becomes acidic due to the increase in hydrogen ions. Carbonate has another very important role in the ocean; it is a building block of calcium carbonate. Calcium carbonate is found in two important mineral forms: aragonite, which is used as a building block for coral and some zooplankton (Harvey 2008), and calcite, which is a building block for some phytoplankton as well as foraminifera (tiny marine animals from 1 mm to 3 cm). These tiny animals are the foundation of the ocean ecosystem. Ocean organisms that create shells also use calcite. If too much ocean calcite is used in interactions with CO_2, these organisms will not be able to survive. This could eventually destroy the entire ocean ecosystem.

Over thousands of years, very low calcite concentrations will lead to dissolution of calcite from the depths of the ocean. Much ocean life would already be extinct at this point. In order to avoid the negative cascading effects of a lack of calcite, some scientists have suggested dumping ground limestone, which is mostly calcium carbonate, into the ocean so that it dissolves, releasing calcite and keeping the ocean from becoming too acidic for ocean life. Mass dumping of ground limestone could start to demonstrate effectiveness in as little as two to three years. If this technique were successful, the ocean ecosystem could continue to thrive as it continued to sequester tons of CO_2.

The downside of this theory are threefold: (1) it would require an enormous amount of energy to finely grind limestone and that energy would most likely come from fossil fuels, which add CO_2 to the atmosphere; (2) the grinding and application processes would take a long time and require international consensus; and (3) there could be many unintended consequences of large-scale geoengineering of the oceans. There is another, similar proposal that looks to dump quicklime, which is derived from limestone, into the ocean in order to sequester CO_2. However, it runs into the same three problems and is even more carbon-intensive. There may be a place for this if human activity continues to degrade ocean ecosystems at an increasing rate. Desperate times may require desperate measures. As of 2023, there needs to be more testing before any larger projects will be going forward.

Stratospheric Sulfate Aerosols

The idea here is simple, but the long-term effects are extremely complicated. By spraying reflective aerosols that contain metal compounds into the earth's stratosphere, scientists believe they can reflect light from the sun and cool the earth. In effect, the process involves human-made clouds containing compounds that will eventually fall to the earth, where their long-term effects are unknown. It is possible that the widespread use of these chemical compounds could create more problems than they solve. The precautionary principle dictates that until something is proven safe it should be treated as dangerous. Adding chemical compounds en masse to the earth needs long-term scientific study before going forward. Because of the need for long-term study, it is unlikely that the technology would be ready in time to combat global warming.

Space Sunshade

Creating a space sunshade is one of the more fantastic geoengineering ideas to combat climate change. As the name implies, the concept is to build a compactable and expandable shade that could be launched into space and then be spread out and held in place by satellites. Another idea is to have many small sun-blocking spacecraft that move in unison. Small spacecraft would be easier to get into space than a one-piece sunshade, which would have to be extremely large to decrease the amount of sunlight the earth receives. A space sunshade fitted with solar panels could potentially send energy down to earth for use. Science fiction authors as well as some futurists have proposed the idea that Fresnel lenses could be created in space, perhaps as small as 1,000 km in diameter. The concave shape of a Fresnel lens would naturally decrease the amount of light coming to the earth by about 1 percent. Any shade/lens would to be located at the L1 point between earth and the sun (1.5 million miles above the earth, directly between the earth and the sun). These ideas are all very hypothetical because they would affect all the countries on the planet and require immense resources. There would need to be an international consensus and international cooperation in order for this type of project to become reality. At this point, these proposed solutions are extremely unlikely to come to fruition. Rising temperatures would likely need to be spiraling out of control, threatening the habitability of the planet, for these ideas to be considered. If nations cannot agree on more natural and inexpensive methods to control climate change and global warming, eventually ideas like this may have to be considered.

ECONOMIC CHANGES

Decreased Use of Coal

The use of coal began a rapid decline in 2011 in the United States. The main reason for this decline has to do with the shale fracking boom in the country. The fracking boom created a glut of natural gas, driving the price of natural gas to all-time lows. (This changed during the Russian invasion of Ukraine and the subsequent shortage of natural gas in Europe.) This generally inexpensive energy source outcompetes coal and is responsible for the majority of coal's decline. Other factors are a decline in energy demand and the increasing competitiveness of renewable energy. The demand for electricity in the United States has been relatively flat since 2007. The economy has been decoupled from energy demand as it has become increasingly driven by information more than industrial output. Continual improvements in energy efficiency of everything from light bulbs to laptop computers have also played a role in flattening demand for electricity. Energy needs tend to only increase during hot summers and cold winters. Renewables, as discussed elsewhere in this book, are cheaper every year and have been less expensive than coal in most markets since 2015. Both wind (onshore and offshore) and solar (photovoltaic and concentrated) are able to outcompete coal. Another factor affecting coal, albeit less than the others, is that coal emissions have an uncertain future. A more liberal or progressive government will most likely require coal plants to limit emissions of both pollutants and CO_2. Most utilities and large businesses are investing in either natural gas plants or renewable energy. There are no new coal plants planned in the United States, but that is not the case across the planet. China continues to build more coal plants than the entire world combined but, conversely, is planning on decreasing coal use. Up and coming countries in Asia and Africa are turning to coal due to funding from coal-exporting countries like China. These countries fund the building of coal-fired power plants, thus ensuring a future export market. Still, the trend across the planet is toward fewer new coal plants and more renewable energy. If current trends continue, no countries will be building coal power plants in 2030.

Increased Use of Natural Gas

The fracking boom (described in detail in Chapter 2) has made natural gas abundant and affordable. Fracking still remains controversial because of its environmental effects, especially its effects on groundwater. Natural gas burns cleaner than coal, but it still releases CO_2 into the atmosphere, heating the planet. The fracking boom has quickly increased the production of natural gas in the United States, but the speed of the boom has led to many sloppy sites that are leaking methane (the main component of natural gas). Methane is a dangerous greenhouse gas because it is thirty times more effective than CO_2 at trapping heat in the earth's atmosphere. Despite these dangers, natural gas–powered plants continue to replace coal-fired plants across the United States and the globe.

Coal was once responsible for the vast majority of electricity production in the United States. As recently as 2010, it was responsible for over 45 percent of energy production in the country but now produces 22 percent. Coal's decline is mirrored by the rise of gas and renewables. Gas and renewables produced less than 30 percent of total energy in 2010, but now they produce over 50 percent of all energy. If Energy Information Administration (EIA) predictions are correct, natural gas will begin to lose ground to renewables. Many people in the energy field see natural gas as necessary for the transition toward renewables. The ability of certain types of gas power plants, also referred to as peaker plants, to quickly start up and quickly power down makes them ideal when paired with more intermittent renewable energy sources. Batteries along with pumped storage will likely eventually replace gas peaker plants as well. Battery technology is still too expensive to compete with gas peaker plant power, but if battery price declines continue their trends, batteries will soon be more affordable than gas.

Increased Use of Trucks and SUVs (for Personal Transportation)

Passenger vehicles have been required to continually increase their fuel efficiency in the United States. The Energy Policy and Conservation Act (EPCA) was passed in 1975 as a reaction to the oil crises faced by the United States in the previous years due to the country's support of Israel against its oil-rich neighbors. The subsequent oil embargoes against the United States caused high oil prices, shortages, and long lines to receive rationed gasoline. Trucks, minivans, and larger SUVs were not originally covered by government fuel efficiency regulations passed in 1975 because these types of vehicles were not regularly used for personal use. Vehicles that were classified as light trucks (i.e., working vehicles) were a very small part of the market in 1975. This loophole allowed light trucks to avoid stringent fuel-efficiency requirements. Sales of light trucks (including trucks, vans, and SUVs under 8,500 lbs.) have increased by over 600 percent since 1980. The average fuel economy for all personal vehicles increased during 1975–87, as light trucks remained a small percentage of total vehicles sold in the United States. That changed with the increased demand for trucks, minivans, and SUVs that accelerated in the 1990s and continues in the 2020s. Despite government regulation intended to increase fuel-efficiency standards, average fuel economy decreased from 1987 to 2004 (EPA n.d.). Gas prices spiked in 2004, which reversed the trend toward larger vehicles. Continually rising gas prices encouraged the administration of George W. Bush to pass legislation that increased the required fuel efficiency of vehicles, also called the Corporate Average Fuel Economy (CAFE) standards. CAFE standards were increased under the Barack Obama administration, including standards for light trucks. CAFE standards continued to increase until 2020 when the Trump administration worked to lower CAFE standards for the first time since the EPCA was passed in 1975. The Trump administration's attempts to lower CAFE standards to an average of 40 mpg were countered by the Biden administration, which moved the standard to 49 mpg, but they remain lower than the Obama administration's standard of 55 mpg. An increased use of electric vehicles may offset the effects of the overall lower CAFE standards (compared to Obama's standards). If that is the case, then average miles per gallon (mpg) will continue to increase, and the United States will continue to use less gas and create less pollution and CO_2.

INDUSTRY RESPONSES TO CLIMATE CHANGE

Some of the earliest research on anthropogenic (human-caused) climate change was funded and carried out by Exxon (now ExxonMobil). Their scientists discovered that the release of CO_2 from the burning of fossil fuels was going to heat up the planet by 2–3 degrees. The senior scientist for Exxon in 1977, James Black, announced that humans had 5–10 years before they would need to make some hard decisions regarding energy strategies (Hall 2015). The leadership of Exxon could see that a drastic shift away from fossil fuels would be disastrous to their current and future profits. In order to counter this information, Exxon began to fund pseudoscience to counter the narrative that global warming/climate change was the scientific consensus. In 1989, Exxon funded a group called the Global Climate Coalition. Despite the environmental sound to the name, the group was formed to question the scientific evidence for climate change. After a decade of working to muddy the waters, Exxon successfully lobbied the United States, China, and India to not sign the Kyoto Protocol. The 1991 Kyoto Protocol sought to globally limit greenhouse gas (GHG) emissions. Exxon spent at least $30 million funding climate change denial research. ExxonMobil spends millions more funding politicians that deny climate science (Parafiniuk and Smith 2019).

Seeding Doubt

Worldwide consensus on global warming came about by the early 1990s. George H. W. Bush, the Republican US president at the time, addressed the United Nations and said: "The

United States is strongly committed to the IPCC process of international cooperation on global climate change" (Bush 1990). He spoke of the need to address this real threat with "aggressive and thoughtful action." The research from Exxon in the 1970s had come to the conclusion that the burning of fossil fuels was heating the planet. Global warming/climate change looked like something that could be corrected using the same successful approach used to fix the hole in the ozone. Nations would gather and come to a common solution that would be funded by the wealthier countries and implemented by all nations. However, the big oil majors (ExxonMobil, Chevron, British Petroleum, and Royal Dutch Shell) realized that any solution scientists could imagine would mean these companies would lose billions of dollars. Oil majors began a well-funded campaign to question the science, seed doubt in the public imagination, and to fund politicians that would parrot the talking points that insisted that climate change was unsettled science and that global warming may be a naturally occurring phenomena. For example, 83 percent of ExxonMobil-supported, peer-reviewed papers along with 80 percent of internal company documents reaffirm that climate change is human-caused, but only 12 percent of their advertisements (that mention climate change) admit this. Of the ExxonMobil ads addressing climate change, 81 percent express doubt about the human origins of climate change (Supan and Oreskes 2017). This countermovement has put the nations of the world thirty years behind where they should be in facing what may be the greatest threat to the long-term existence of humanity as well as our fellow species. While British Petroleum and especially Royal Dutch Shell have pivoted away from climate denial and have made some forays into renewable energy, ExxonMobil and Chevron along with Koch Industries and Marathon Petroleum continue to work diligently to stop any action on climate change.

Buried Research

Private research does not legally have to become public. The researchers for ExxonMobil did not have the right to publish their work unless previously approved by ExxonMobil. Researchers who had discovered earth-shattering new information were muzzled by NDAs (non-disclosure agreements) that barred them from discussing their work. Research continued for decades, but it was buried or unreleased. This research eventually came to light decades later (Hall 2015). This duplicitous process of continuing to find results that reaffirm the growing consensus on climate change while funding those that dispute the veracity of climate change (Parafiniuk and Smith 2019) was effective in stopping any meaningful change in the world economy that would have come about with a move away from gasoline-powered automobiles.

Political Funding

In the late 1970s, corporations began funding political campaigns on both sides of the aisle. It became evident that if corporations supported candidates, the newly elected politicians were more likely to support legislation that in turn supported the corporation's agenda. Politicians would also kill legislation that would be bad for the bottom line of their corporate sponsors. There were two kinds of politicians that oil companies tended to support the most: those that expressed doubts about climate change (if it is happening or if it is human-caused) and/or those that denied it was happening at all. Many of these politicians would use a cold winter as proof that climate change was not happening, for example. To a lesser extent, the oil majors supported candidates who gave climate change lip service but would not pass legislation that addressed it. The standard reasoning was that anything that could negatively affect the economy was an impossible task. Climate change also had the disadvantage of not being very dramatic or obvious, so it did not appear as an immediate threat. In this way, it is easy for politicians and voters to deprioritize climate change in favor of immediate issues, such as the economy, or wedge issues such as guns, marriage equality, transgender rights, and abortion. In order to provide justification for politicians that continue to espouse the unscientific opinion that climate change is not a threat, oil majors began funding organizations that funded pseudoscience.

Funded Pseudoscience

There are scientists who buck the consensus on climate change. Most of them are not climate scientists. They have realized that there is ample funding for contrarian climate research or at least research that indicates that climate change might be a net positive for humanity. The list of researchers and their connections to big oil companies is long, but the system for funding them is similar in most cases. Originally, oil companies were willing to directly fund some pseudoscience ventures. (By pseudoscience, we mean work that claims scientific authority but is not accepted as legitimate scientific work by peer-reviewed journals in the climate field.) However, public pressure has led many companies to stop funding some of these climate change contrarians. Most companies now donate to other organizations that fund these same people (Dunlap and Jacques 2013). In this way, they can deny culpability, but they still get more contrarian views out there to provide a thin veil of legitimacy to those who promote doubting climate change.

Green Gilding and B Corporations

Green gilding is the process of making a company appear sustainable while actively pursuing policies that undermine the government's ability to enforce and/or legislate sustainable policy. This is seen most egregiously in the oil industry, where billions of dollars are spent lobbying against policies that many companies claim to support publicly (Parafiniuk and Smith 2019). It is also seen in less obvious sectors such as soda manufacturers that spend millions fighting against deposit schemes to enforce recycling but advertise campaigns to encourage recycling. Countries with expensive deposit programs have high recycling rates. Norway recycles 97 percent of its plastic bottles compared to just 29 percent in the United States (Tiseo 2020). Despite clear data on the effectiveness of deposit programs, companies know that the extra money paid for a deposit can potentially lower sales. This is why they fight against deposit programs. Green gilding can be seen in the auto industry as well. Companies tout their environmental records while fighting against fuel efficiency standards and EV mandates.

Green gilding might appear as just a cost of business, but it is much more dangerous than simply greenwashing, which is the process of doing a few small projects that are good for the environment while continuing business as usual. Green gilding processes deceive the public into believing that progress is being made while they are effectively moving sustainability backward. Green gilding has become more common since the turn of the century.

As some companies find they are better served by the duplicity of green gilding, there are other corporations that have become Certified B Corporations. These corporations embrace the triple bottom line of community, the environment, and profit. Community here is defined as both the people who work in the corporation and the people in the community that are affected by both the business and the business's product. There are 3,500 Certified B Corporations, and that number is continually growing. As the double threat of climate change and economic inequality has the potential to destroy communities, B corporations work to build stronger communities through practices that benefit the entire community and the ecosystems that are their support.

Certified B Corporations have been able to change the legal requirements that force CEOs and corporate boards to prioritize profit over all else. Certified B Corporations are structured to legally bind the leadership to make decisions that will balance the needs of the community, the workers, the environment, and profits. This principle is sometimes called the triple bottom line: people, planet, and profit. (For more information on B corporations visit https://bcorporation.net/about-b-corps.) This is another avenue in which forward-looking leaders have the opportunity to embrace the changes in energy that are needed to prepare their businesses for the future.

Sustainability

Sustainability is a word used often and has many different meanings. It has been used and abused by governments and business to green wash and green guild. For those that work within the field of sustainability the definition is clear: meeting the needs of the present without sacrificing the ability of future generations to meet their own needs (Brundtland 1987). Heather Farley and Zachary Smith have a more precise definition of what they call neo-sustainability: "The ability of an activity to sustain a system by improving its quality and operating within its limits" (2020). Many companies claim to be sustainable, but not all of them are actually engaged in sustainability. B corporations are one way that a company can actually be working toward a more sustainable society; another is signing the RE 100 pledge. This pledge commits a company to 100 percent renewable energy. These are voluntary commitments, but they are part of a bigger movement.

Offshoring Carbon

The other side of true sustainability or neo-sustainability is greenwashing or, worse, green gilding. One way a company or a country can do this is by having their dirty jobs done in other countries. Refineries, factories, and waste disposal sites can all be offshored to countries with loose environmental regulations and little to no oversight. Companies such as Walmart have been accused of offshoring their carbon footprint. While Walmart has been able to decrease waste and add solar panels to their roofs, the products that they sell can have a large carbon footprint. The products are then shipped across the planet, increasing the amount of GHGs released in the process. No one would argue against the local improvements that Walmart has made to its buildings. The argument is that the biggest need for change is in the production and transportation of the goods that are sold in the stores. Many companies are involved in similar processes. The need for a global agreement on the pricing of CO_2 with an enforcement mechanism would be a simple way to close this worldwide loophole.

NATIONAL RESPONSES

Despite the fact that capitalism has become the dominant economic system across the planet, the winners and losers in the world of energy are often chosen by governments. This may sound like a return to the mercantile system that the legendary pro-capitalist Adam Smith raged against, but in the field of energy, government control and/or interference is the norm. There are a number of reasons for this, but the most important ones are the following: energy is a national security issue; there is a lot of money and jobs tied up in energy systems; and national subsidizing of energy systems has been seen as an absolute imperative in order to protect local energy systems from foreign ones. Protectionist policies have been implemented to ensure local production. This is most evident in the coal, gas, and oil industries. As mentioned before, government subsidies of fossil fuel industries are over thirty times the subsidies given to renewable energy industries. This creates a system where older, less efficient technologies are propped up over newer, more efficient technologies. A few large companies are then able to succeed at the expense of an entire population that is denied access to more affordable energy. These policies are most important in China, India, the United States, and the European Union, which release the vast majority of CO_2 into the atmosphere. The fate of the entire world will depend on the decisions made by the leaders of these four entities.

China

China now produces more CO_2 than any other country on earth. Most of their energy comes from the burning of coal, but they also have more renewable energy than any other country on earth. This dichotomy is the fundamental crux of the move to renewable energy. The questions that are most important are: Can China continue its explosive economic growth while transitioning to renewable energy? How long will it take to make that transition? Will its belt and road initiative promote fossil fuel energy or renewable energy?

China appears to have hedged its bets with regard to its energy production. Coal continues to produce the majority of energy in China, but coal's share of electrical production shrank from 70 percent in 2011 to 58 percent in 2019. Continued tensions with the United States have pushed China away from cleaner natural gas. China has loosened restrictions on coal mining and the building of coal-fired power plants and has also rolled back subsidies for renewable energy. Despite these setbacks, renewable energy continues to grow, albeit at a slower pace than in the past. China has also green-lighted shale fracking projects as well as methane from coal operations (Standaert 2018). One of the biggest problems China faces is the lack of high-capacity electrical grid infrastructure. Many of the sunniest and windiest places in China are located in the sparsely populated interior. The majority of Chinese people live in the east and southeast areas of the country. The other problem is that many local economies are dependent on coal-mining jobs and coal-power-plant jobs. As China is now a hybrid economy embracing parts of communism and parts of capitalism, there are people making money in the status quo. In this way, China is running into some of the same local political pressures that capitalist democracies run into. Entrenched interests are making profits and pressuring the Chinese government to roll back the transition to renewable energy. The final difficulty China faces in the transition toward renewable energy is that with the United States vacillating on climate change and clean energy policies between administrations, China no longer feels the same pressure to continue to lead the world in renewable energy. They will be able to stay the leader without much effort, unless the United States is able to continue the momentum that began under the Biden administration with the Inflation Reduction Act.

India

India is a large and complicated country that is both wealthy and impoverished, steeped in ancient ways and looking toward a modern future. India has the ambition to become the world leader in renewable energy, but it also must keep up with a growing population moving to urban areas. There is an intense desire for modern conveniences such as cars, electrified kitchens, and appliances. It is sometimes easier to build a coal plant than to create a system that can harness renewable energy. Nevertheless, India has embarked on a mission to install 450 GW worth of solar and wind capacity by 2030. India's total energy capacity from all sources was 369 GW in 2020. While many energy consultants criticize India's goal as unachievable within ten years, India went from 1 GW of renewable energy in 2010 to 60 GW of solar and wind by the end of 2020 (Frangoul 2020) and is producing 120 GW of renewable energy in 2023. That change is nothing short of remarkable, a 12,000 percent increase. (Even as renewables make up a larger percentage of the energy produced in India, the overall use of coal still grows, albeit slower than it was in the past.)

India may become the most important player in energy, as it will soon be the most populous country on earth. India is still in the beginning of its economic and technological growth trajectory. In this way, it has an advantage over most mature economies that already have entrenched energy systems. India has, of course, had electrical systems and advanced manufacturing for decades. The difference is that India is improving and expanding its electrical grid at a time in which renewables are more affordable than most, if not all, fossil fuels. The only complication is devising a way to counter the intermittency of solar and wind power. There are many ways to do this, including pumped hydro, batteries, and concentrated solar power. These methods can bring up the cost of renewables to that of fossil fuel power plants. If India does meet its ambitious targets, it will have the distinct advantage of having an energy system that is not tied to the ups and downs of supply and demand. Renewable energy is always becoming less expensive. The ability to promise manufacturers inexpensive green energy that will likely continue to decrease in price will give India an advantage that most other countries will not have.

India, like all countries, has competing interests. At the same time that the country is pushing forward with renewable energy, the Make in India initiative has put tariffs on less expensive solar panels made in China. However, India has not wavered in its commitment to renewable energy. This puts Indians in a much better place than either China or the United States. Prime Minister Modi has swung the country to the ideological right but without the crony capitalism that many times accompanies such ideological swings. This may indicate that regardless of who leads India in the future, its path to renewables will continue. If the people of India are successful in becoming a renewable energy powerhouse, India may be the first major country since the Industrial Revolution to be disconnected from the conflicts over fossil fuel energy resources. This may put India into a position of power in the future.

European Union and Norway

The European Union (EU) is made up of twenty-seven countries, each with their own goals, agendas, and economic needs. The EU was an early adopter of renewable energy, but despite the early push it has struggled to achieve its goals. The EU has made progress in doubling the percentage of energy produced by renewable energy since 2004. Germany, in particular, has been pushing renewable energy while also attempting to decommission its entire nuclear power plant fleet due to safety concerns after the nuclear accident at Fukushima, Japan. This has proven difficult, and Germany has had to continue to rely on coal, especially brown coal or lignite, which has particularly dirty emissions. Germany currently has a plan to discontinue coal in 2038. The continued mining of lignite has led to the demolition of over sixty-six towns over time. Overall, Germany and the other twenty-six countries of the EU have made great strides in moving toward renewable energy.

Sweden has been the most impressive EU country. They produce over half of their electricity from renewable sources (54 percent). Norway, which is not a member of the EU, derives almost 100 percent (99+ percent) of its energy from renewable resources: mostly hydropower with some wind power as well. Finland, Latvia, and Denmark all produce nearly 40 percent of their electricity from renewable sources. Figure 3.2 illustrates the percentage of renewable energy achieved by EU states in 2018; the black dash on each bar represents that state's goal. The EU states have had variable success in meeting their goals, but underachieving member states are counteracted by overachieving states. The EU beat its target to reach 20 percent renewable electricity production by the end of 2020, achieving a rate of 22.1 percent.

The bigger goal of becoming carbon neutral by 2050 will depend on each country's willingness to support the burgeoning carbon-free industries such as electric vehicles (EVs), hydrogen fuel–cell vehicles, and renewable energy. Once again, the non-EU European country of Norway is leading the way. Norway incentivizes EV purchases by not taxing those vehicles and giving them free registration. Over 90 percent of Norwegian cars sold

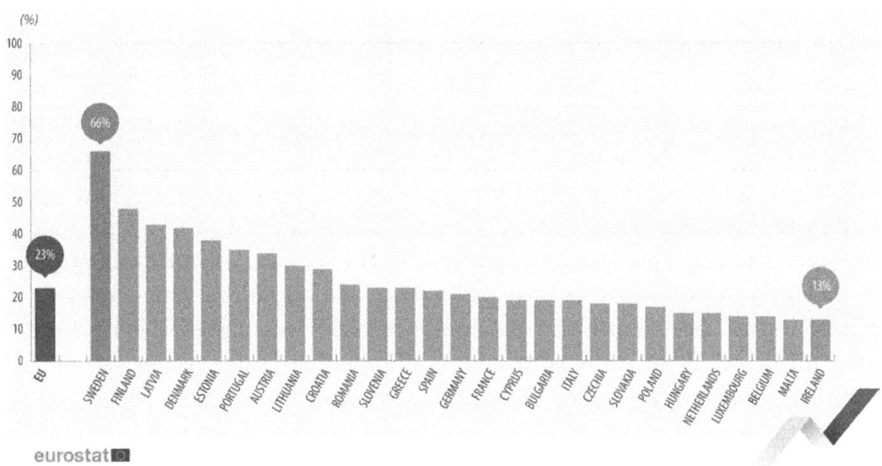

FIGURE 3.2 Overall share of energy from renewable sources in 2022. *Source*: Eurostat, https://ec.europa.eu/eurostat/statistics-explained/index.php?title=File:Renewable_energy_2022_infographic.jpg.

in 2022 had plugs. That means they were either an EV or a plug-in hybrid. Norway has taken the money that the country has made from its nationalized oil industry and parlayed that into becoming the European leader in de-carbonization. Not every country has Norway's resources, but there are many ways to successfully transition to renewables. Iceland, for example, runs on geothermal energy and hydropower. Chapter 4 will look at some of these examples of decarbonization and how they can be used as blueprints for countries across the globe.

United States

The United States has ebbed and flowed with its leadership of carbon-free energy. Nuclear power was used to create electricity starting in the 1950s. The United States leads the world in nuclear power produced, which is carbon-free. The problem with nuclear energy is that it is potentially extremely dangerous and currently extremely expensive. The low cost of natural gas and renewable energy sources makes it difficult for nuclear power to compete. A carbon pricing system would help, but it would still be unlikely to be enough for nuclear power to compete without significant government subsidies. Public fear of nuclear accidents also hampers public support for new nuclear plants. Insurance companies are afraid to insure nuclear plants, putting pressure on governments to intervene and provide guarantees. The United States has a couple of nuclear projects in the works, but there have been delays and cancellations due to expensive overruns and the bankruptcy of builder Westinghouse Electric Company. Unit 3 at Plant Vogtle in the state of Georgia is the only exception in recent years, and it came online in 2023.

Solar energy is another area in which the United States was a leader but has fallen behind other countries. Solar water heaters were popular in the United States in the 1970s and early 1980s. President Jimmy Carter put solar water heaters up on the White House, but they were removed by the following president, Ronald Reagan. These symbolic moves were mirrored by actual policy moves. The Reagan administration did not support the subsidies that the Carter administration had enacted to encourage people to install solar water heaters on their homes. By 1986, the Reagan administration had eliminated all federal incentives for solar water heaters. This caused the nascent solar industry in the United States to collapse with over 90 percent of all solar manufacturing and installation companies going out of business.

In the time that the United States stopped leading the world in solar technology, Germany and China emerged as leaders in solar technology and adoption. The US solar industry surged back to life after the Energy Policy Act of 2005 was passed. This act provides a tax credit for some of the costs of installing photovoltaic solar panels, which produce electricity instead of heating water. While the percentage covered has ebbed and flowed, the United States still has a tax credit that decreases each year. The incentives would have ended in 2023, but they were replaced by even stronger subsidies in the Inflation Reduction Act (IRA) of 2022. It is hoped that solar panels will eventually be competitive without subsidies, but currently subsidies help solar compete with subsidized fossil fuels. The United States lost almost twenty years of solar leadership because US companies had to compete with subsidized companies in other countries. The current renaissance in solar is focused more on installation than production, but US production has increased recently. The United States was just too far behind to compete with companies in Asia that can produce panels at a fraction of the price without some subsidies. The IRA has created enough incentives to revitalize US manufacturing of solar panels.

The United States has become a leader in wind energy production. This came about due to the growth of companies exploiting the wind energy that exists in Texas and Oklahoma. An oilman named T. Boone Pickens realized that it would be profitable to build wind power farms in Texas, and he helped usher in a new era in US wind power. The growth in wind power was further helped by US subsidies starting in 2010. These subsidies were scheduled to end in 2020, but subsidies will continue due to the IRA of 2022. The United States has not caught up to the European Union or China in wind energy production, but it has the potential to become a leader in wind energy production if the offshore wind market is tapped. There are many planned offshore wind farms off the East Coast of the United States. It is yet to be seen if these wind farms will lead to an offshore wind boom like the one experienced in Europe in the 2010s. There are still considerable hurdles for offshore wind, including not-in-my-backyard syndrome. East Coast states such as Massachusetts and New York have seen powerful, wealthy interests stop proposed projects because they would be in view from their summer homes. The only operating offshore wind farm is the Rock Island Wind Farm off New Shoreham, Rhode Island. There are only five turbines, creating 30 MW of electricity. There are many projects that are nearing completion, with over 41 GW of offshore wind turbines pledged by East Coast states alone.

How a country addresses the pressing problem of climate change is intertwined with how a country transitions to renewable energy. Politicians who champion the fossil fuel companies that fund their campaigns can put the United States behind the rest of the world. The United States, due to its advanced technologies and big tech companies, is poised to make a big move toward renewable energy. How successful the country is at this will depend on the technological advances from Silicon Valley and the political will of US leaders.

Russia

Russia is a petrol state. Over 52 percent of all Russian exports are oil, gas, and coal. The country exports more oil and gas than any other country on earth and is the third-largest exporter of coal. The only other energy export that Russia is involved in is nuclear power. Russia has very little renewable energy except for hydropower.

Perhaps due to its location and weather, Russia does not embrace solar and wind power. Much of Russia is cold and dark half of the year. In reality, Russian policy is only the will of one man, de facto dictator Vladimir Putin. Putin has maintained power by skirting the constitution and switching between being Russian president and Russian prime minister. The Russian legislative body changes laws so that the most power is always in whichever office Putin holds. Putin has mocked green energy. He claims net-zero energy production will cause humanity to go back to living in caves and that wind turbines harm birds and worms (Soldatkin 2019). Russia appears unlikely to warm to renewable energy without a big change in political leadership.

Despite these positions, Russia produces a lot of carbon-free energy. One bright spot is hydropower. According to the International Hydropower Association, Russia has the sixth-highest hydropower capacity (IHA n.d.). It also produces electricity using nuclear power. Russia is the third-largest producer of nuclear power, behind only France and the United States. Russia also produces almost half of its energy from natural gas that burns cleaner than coal, which makes up about 15 percent of its energy production (IEA 2023). Overall, Russia produces less GHGs per capita than the United States as well as Australia and most of the Arab States of the Persian Gulf. Russia continues to build nuclear reactors within Russia and abroad. There are many experts who see nuclear energy as part of the path toward net-zero energy production, notwithstanding the dangers. Russia continues to lead the way for net-zero nuclear energy. Russia's invasion of Ukraine and the subsequent sanctions put on Russia by many countries have caused Russian fossil fuel exports to decrease; however, China and India continue to import Russian oil, and Russian exports are rebounding.

Russia is both an exporter of fossil fuels and adopter of low-carbon energy sources such as hydro and nuclear. The infrastructure in Russia coupled with leadership that does not shy away from all forms of renewable energy could transform the country into a leader in low-carbon and zero-carbon energy. However, Russia's current dependence on fossil fuel exports locks the country into supporting international policies that continue and support the use of fossil fuels. Russia also uses a benchmark of 1990 to "decrease" the amount of CO_2 that is produced. In 1990, Russia was still part of the Soviet Union (USSR). The USSR had many smokestack industries that were not able to continue and/or compete in an open Russia. Because of this, the Paris Agreement goals that Russia put forward were actually an increase in CO_2 production from when Russia signed the agreement. Russia has tightened their goals since signing the accord, but these are still seen as baby steps by most observers. Nuclear power remains the silver lining to Russian policies, if climate change is to be slowed. Russia currently hopes to produce 70–80 percent of its electricity demands with nuclear power by 2100. This coupled with hydropower and some renewables could easily make Russia a net-zero energy producer, even if that is not the stated goal.

CONCLUSION

The conflict over energy resources is quickly changing. No longer do civilizations fight wars over access to hunting grounds, wood, coal, or petroleum. The new conflict is both international and internal. There are people, countries, and corporations that need fossil fuel sales to continue so they can profit. The entire world and all its inhabitants will suffer if fossil fuels are continually burned without the equivalent CO_2 that is released being removed from the atmosphere. Petroleum-dependent exporter countries will work with international oil majors to influence politics in other countries in order to ensure that policies and subsidies do not shift from fossil fuel extraction toward renewable energy adaptation. This creates strange bedfellows. Russia, ExxonMobil, Saudi Arabia, and Chevron all want the United States to have the same energy policies, for example. Chinese solar panel manufacturers, German wind turbine producers, and the US-based Sierra Club all want the opposite policies.

The United States, China, EU, and India are the four most important players in this new conflict. The policies adopted by these countries will lead the world in one direction or the other. If all work toward creating policies that promote renewable energy, then the world can and will quickly transition to renewable energy and renewable transportation. If the fossil fuel exporting countries and corporations are able to bend these four countries' policies toward the fossil fuels status quo, then the transition will come much later if at all. For many species and millions of people, this conflict will have life-and-death consequences. In Chapter 4, we will discuss the possible outcomes of this conflict and ramifications for people, countries, species, and life on the planet as a whole.

FURTHER READING AND REFERENCES

Ambrose, Jillian. 2020a. "UK Firms Solar Power Breakthrough Could Make World's Most Efficient Panels by 2021." *The Guardian.* August 14. Accessed July 10, 2024. https://www.theguardian.com/business/2020/aug/15/uk-firms-solar-power-breakthrough-could-make-worlds-most-efficient-panels-by-2021.

Ambrose, Jillian. 2020b. "World's Wind Power Capacity Up by Fifth after Record Year." *The Guardian.* March 25. Accessed July 11, 2024. https://www.theguardian.com/environment/2020/mar/25/worlds-wind-power-capacity-up-by-fifth-after-record-year.

Avila, Jorge, Kelly Grogan, and Sherry Larkin. 2015. "Estimating the Value of Ecosystem Services Generated by the Hard Clam Industry in Florida." August 31. https://shellfish.ifas.ufl.edu/wp-content/uploads/valuation-summary-083115.pdf.

Baker, Patrick, and Shirley Baker. 2010. *Carbon Fixation by Florida Cultured Clams.* November 4. https://shellfish.ifas.ufl.edu/projects/shellfish-farm-environment/carbon-fixation/.

Barron-Gafford, G. A., M. A. Pavao-Zuckerman, R. L. Minor, L. F. Sutter, I. Barnett-Moreno, D. T. Blackett, M. Thompson, K. Dimond, A. K. Gerlak, G. P. Nabhan, and J. E. Macknick. 2019.

"Agrivoltaics Provide Mutual Benefits across the Food–Energy–Water Nexus in Drylands." *Nature Sustainability* 2, no. 9: 848–55.

Borunda, Alejandra. 2020. "In the Chesapeake Bay, Saving Seagrasses Can Fight Ocean Acidification." *National Geographic,* June 2. https://www.nationalgeographic.com/science/2020/06/chesapeake-seagrasses-fight-ocean-acidification/.

Brundtland, G. H. 1987. *Our Common Future: Report of the World Commission on Environment and Development*. Geneva, UN-Document A/42/427.

Bush, George. 1990. "Remarks at the Intergovernmental Panel on Climate Change." Washington, DC: Office of the Press Secretary.

Carlson, Kimberly M., and Lisa M. Curran. 2013. "Refined Carbon Accounting for Oil Palm Agriculture: Disentangling Potential Contributions of Indirect Emissions and Smallholder Farmers." *Carbon Management* 4, no. 4: 347–9.

Chatti, M., J. L. Gardiner, M. Fournier, B. Johannessen, T. Williams, T. R. Gengenbach, N. Pai, C. Nguyen, D. R. MacFarlane, R. K. Hocking, and A. N. Simonov. 2019. "Intrinsically Stable In Situ Generated Electrocatalyst for Long-Term Oxidation of Acidic Water at up to 80°C." *Nature Catalysis* 2, no. 5: 457–65.

Coady, David, Ian Parry, Nghia-Piotr Le, and Baoping Shang. 2019. "Global Fossil Fuel Subsidies Remain Large: An Update Based on Country-Level Estimates." International Monetary Fund.

Dincer, Ibrahim. 2012. "Green Methods for Hydrogen Production." *International Journal of Hydrogen Energy* 37, no. 2: 1954–71.

Dunlap, Riley E., and Peter J. Jacques. 2013. "Climate Change Denial Books and Conservative Think Tanks: Exploring the Connection." *American Behavioral Scientist* 57, no. 6: 699–731.

EIA. 2021. "After 2020 Decline, EIA Expects Energy-Related CO2 Emissions to Increase in 2021 and 2022." January 26. https://www.eia.gov/todayinenergy/detail.php?id=46537.

EIA. 2022. "Hydropower Explained, Tidal Power." Updated August 10.https://www.eia.gov/energyexplained/hydropower/tidal-power.php.

Eldardiry, Hisham, and Emad Habib. 2018. "Carbon Capture and Sequestration in Power Generation: Review of Impacts and Opportunities for Water Sustainability." *Energy, Sustainability and Society* 8, no. 1: 1–15.

Emerson, D., 2019. " Biogenic Iron Dust: A Novel Approach to Ocean Iron Fertilization as a Means of Large Scale Removal of Carbon Dioxide from the Atmosphere." *Frontiers in Marine Science* 6: 22.

EPA. 2024. "Green Power Pricing." https://www.epa.gov/green-power-markets/green-power-pricing.

EPA. n.d.a. "Hydrogen Storage: Hydrogen and Fuel Cell Technologies Office." Accessed December 31, 2019 and updated August 30, 2021. https://www.epa.gov/ozone-layer-protection/international-treaties-and-cooperation-about-protection-stratospheric-ozone.

EPA. n.d.b. *Highlights of the Automotive Trends Report*. Accessed June 16, 2020 and last updated November 19, 2021. https://www.epa.gov/automotive-trends/highlights-automotive-trends-report.

EPA. n.d.c. "Inventory of U.S. Greenhouse Gas Emissions and Sinks." Accessed August 11, 2020 and last updated April 14, 2022. https://www.epa.gov/ghgemissions/inventory-us-greenhouse-gas-emissions-and-sinks.

EPA. n.d.d. "Electricity Storage." Accessed November 11, 2020 and last updated November 22, 2021. https://www.epa.gov/energy/electricity-storage.

Farrell, Mark, Thomas K. Kuhn, Lynne M. Macdonald, Todd M. Maddern, Daniel V. Murphy, Phillip A. Hall, Bhupinder Pal Singh, Karen Baumann, Evelyn S. Krull, and Jeff A. Baldock. 2013. "Microbial Utilization of Biochar-derived Carbon." *Science of the Total Environment* 465: 288–97.

Fodrie, F. Joel, Antonio B. Rodriguez, Rachel K. Gittman, Jonathan H. Grabowski, Niels L. Lindquist, Charles H. Peterson, Michael F. Piehler, and Justin T. Ridge. 2017. "Oyster Reefs as Carbon Sources and Sinks." *Proceedings of the Royal Society B: Biological Sciences* 284, no. 1859: 20170891.

Frangoul, Anmar. 2020. "India Has Some Huge Renewable Energy Goals. But Can They Be Achieved?" *CNBC: Sustainable Energy.* March 3. https://www.cnbc.com/2020/03/03/india-has-some-huge-renewable-energy-goals-but-can-they-be-achieved.html.

Gautam, Aditya. n.d. "Solar Paint: The Next Big Thing in Renewable Energy?" *Solar Reviews.* Accessed December 3, 2020 and last updated May 9, 2021. https://www.solarreviews.com/blog/solar-paint-hydrogen-quantum-dot-perovskite-solar-cells.

Glaser, Bruno, Mike Parr, Christelle Braun, and Goodspeed Kopolo. 2009. "Biochar Is Carbon Negative." *Nature Geoscience* 2, no. 1. https://doi.org/10.1038/ngeo395.

Gruber, Nicolas, Dominic Clement, Brendan R. Carter, Richard A. Feely, Steven Van Heuven, Mario Hoppema, Masao Ishii et al. 2019. "The Oceanic Sink for Anthropogenic CO_2 from 1994 to 2007." *Science* 363, no. 6432: 1193–9.

GWEC. 2019. *Global Wind Report 2019: The Flagship Annual Report for the Global Wind Industry from GWEC Market Intelligence!* Accessed December 28, 2020, https://gwec.net/global-wind-report-2019/.

Hall, Shannon. 2015. "Exxon Knew about Climate Change Almost 40 Years Ago." *Scientific American* 26. https://www.scientificamerican.com/article/exxon-knew-about-climate-change-almost-40-years-ago/.

Harris, M. 2018. "Seabased Signs Deal to Install 100 MW Wave Energy Park in Ghana." *Hydro Review.* March 21. https://www. hydroreview. com/2018/03/21/seabasedsigns-deal-to-install-100-mw-wave-energy-park-inghana.

Harvey, L. D. D. 2008. "Mitigating the Atmospheric CO_2 Increase and Ocean Acidification by Adding Limestone Powder to Upwelling Regions." *Journal of Geophysical Research: Oceans* 113, no. C4. https://agupubs.onlinelibrary.wiley.com/doi/full/10.1029/2007JC004373.

Hawken, P., ed. 2017. *Drawdown: The Most Comprehensive Plan Ever Proposed to Reverse Global Warming.* London: Penguin.

Ho, Clifford K., and Brian D. Iverson. 2014. "Review of High-Temperature Central Receiver Designs for Concentrating Solar Power." *Renewable and Sustainable Energy Reviews* 29: 835–46.

Hurliman, Sylvia, and Hannah Zucker. 2019. "How Kelp Naturally Combats Global Climate Change." *Harvard University, The Graduate College of Arts and Sciences: Science in the News.* July 4. http://sitn.hms.harvard.edu/flash/2019/how-kelp-naturally-combats-global-climate-change/.

IEA. 2023. *World Energy Investment 2023*. Paris: IEA. https://www.iea.org/reports/world-energy-investment-2023.

IHA. n.d. "Country Profile Russia, Russia Ranks Second in the World in Undeveloped Hydropower Resources." Accessed October 4, 2020. https://www.hydropower.org/country-profiles/russia.

Kim, K. H., and L. Zsuffa. 1994. "Reforestation of South Korea: The History and Analysis of a Unique Case in Forest Tree Improvement and Forestry." *Forestry Chronicle* 70, no. 1: 58–64.

Krause-Jensen, D. and Duarte, C.M., 2016. Substantial role of macroalgae in marine carbon sequestration. *Nature Geoscience*, 9(10), pp.737-742.

Landgraf, Matthias. 2019. "August 2018 Press Release: Carbon Neutral Fuels from Air and Green Power." *Climeworks.* Accessed September 15, 2019. https://www.climeworks.com/wp-content/uploads/2019/08/Press-Release_Carbon-neutral-Fuels-from-Air-and-Green-Power.pdf.

Lauderdale, Jonathan Maitland, Rogier Braakman, Gaël Forget, Stephanie Dutkiewicz, and Michael J. Follows. 2020. "Microbial Feedbacks Optimize Ocean Iron Availability." *Proceedings of the National Academy of Sciences* 117, no. 9: 4842–9.

Lewis, Sophie C., Sarah E. Perkins-Kirkpatrick, Glenn Althor, Andrew D. King, and Luke Kemp. 2019. "Assessing Contributions of Major Emitters' Paris-Era Decisions to Future Temperature Extremes." *Geophysical Research Letters* 46, no. 7: 3936–43.

Maxwell, James, and Forrest Briscoe. 1997. "There's Money in the Air: The CFC Ban and DuPont's Regulatory Strategy." *Business Strategy and the Environment* 6, no. 5: 276–86.

NRCS. n.d. "Air." Natural Resources Conservation Service. https://www.nrcs.usda.gov/wps/portal/nrcs/detailfull/national/air/?cid=stelprdb1044982.

NREL. 2019. "Benefits of Agrivoltaics across the Food-Energy-Water Nexus." *National Renewable Energy Laboratory.* September 11. https://www.nrel.gov/news/program/2019/benefits-of-agrivoltaics-across-the-food-energy-water-nexus.html.

O'connor, P., B. Saulsbury, B. Hadjerioua, B. T. Smith, M. Bevelhimer, B. M. Pracheil, S. C. Kao, R. A. Mcmanamay, N. M. Samu, R. Uria Martinez, and M. Bonnet Acosta. 2016. " Hydropower Vision A New Chapter for America's 1st Renewable Electricity Source (No. ORNL/TM-2016/688)." Oak Ridge National Lab, Oak Ridge, TN (United States). https://www.energy.gov/eere/water/articles/hydropower-vision-new-chapter-americas-1st-renewable-electricity-source.

Parafiniuk, Adrah, and Zachary A. Smith. 2019. "Green Gilded Oil: How Faux Sustainability by US Oil Companies Is Undermining Neo-sustainability." *Sustainability* 11, no. 14: 3760.

Popovich, N., Albeck-Ripka, L., and Pierre-Louis, K. 2020. "The Trump Administration Is Reversing 100 Environmental Rules. Here's the Full List." *New York Times*, October 15. Accessed July 11, 2024. https://www.nytimes.com/interactive/2020/climate/trump-environment-rollbacks-list.html.

Richter, Alexander. 2020. "The Top 10 Geothermal Countries 2019—Based on Installed Generation Capacity (MWe)." *Think Geoenergy.* January 27. https://www.thinkgeoenergy.com/the-top-10-geothermal-countries-2019-based-on-installed-generation-capacity-mwe/.

SEIA. n.d. "Solar Industry Research Data: Solar Industry Growing at a Record Pace." *Solar Energy Industry Association.* Accessed July 4, 2020. https://www.seia.org/solar-industry-research-data.

Schwalm, Christopher R., Spencer Glendon, and Philip B. Duffy. 2020. "RCP8. 5 Tracks Cumulative CO2 Emissions." *Proceedings of the National Academy of Sciences* 117, no. 33: 19656–7.

Snæbjörnsdóttir, Sandra Ó., Bergur Sigfússon, Chiara Marieni, David Goldberg, Sigurður R. Gislason, and Eric H. Oelkers. 2020. "Carbon Dioxide Storage through Mineral Carbonation." *Nature Reviews Earth & Environment* 1, no. 2: 90–102.

Soldatkin, V. 2019. "Russia's Putin Says Wind Power Harmful to Birds and Worms." *Reuters*. July, 9. https://www.reuters.com/article/us-russia-putin-renewables/russias-putin-says-wind-power-harmful-to-birds-and-worms-idUSKCN1U422X.

Standaert, Michael. 2018. "Why China's Renewable Energy Transition Is Losing Momentum." *Yale Environment* 360. https://e360.yale.edu/features/why-chinas-renewable-energy-transition-is-losing-momentum.

Tiseo, I. 2020. "PET Plastic Bottle Recycling Rates in Select Countries 2018." Statista, November 26. https://www.statista.com/statistics/1166550/plastic-bottle-recycling-rates-in-select-countries/.

Unies, N. 2014. "Collection des traités." *En ligne*, 3.

van Katwijk, M. M., A. Thorhaug, N. Marbà, R. J. Orth, C. M. Duarte, G. A. Kendrick, I. H. Althuizen, E. Balestri, G. Bernard, M. L. Cambridge, and A. Cunha. 2016. "Global Analysis of Seagrass Restoration: The Importance of Large-Scale Planting." *Journal of Applied Ecology* 53, no. 2: 567–78.

Weinstein, L. A., J. Loomis, B. Bhatia, D. M. Bierman, E. N. Wang, and G. Chen. 2015. "Concentrating Solar Power." *Chemical Reviews* 115, no. 23: 12797–838.

World Nuclear Association. n.d. "Nuclear Power Reactors." *World Nuclear Association*. Accessed December 17, 2020 and updated October 2022. https://www.world-nuclear.org/information-library/nuclear-fuel-cycle/nuclear-power-reactors/nuclear-power-reactors.aspx.

World Nuclear News. 2020. "First Serial Batch of MOX Fuel Loaded into BN-800." *World Nuclear News*. January 28. https://world-nuclear-news.org/Articles/First-serial-batch-of-MOX-fuel-loaded-into-BN-800.

Zhongming, Zhu, Lu Linong, Yao Xiaona, Zhang Wangqiang, and Liu Wei. 2020. "Seeding Oceans with Iron May Not Impact Climate Change." Science Daily. www.sciencedaily.com/releases/2020/02/200217162348.htm.

CHAPTER 4

Future Implications

INTRODUCTION

Earlier chapters discussed how humans have struggled over resources from the days of Paleolithic hunters fighting over hunting grounds and how human hunters outside of Africa were unable to balance the immediate need to hunt and eat megafauna (such as mammoths and mastodons) with the need to hunt and eat them in the future. This led to multiple extinctions across six continents (including parts of Northern Africa). Struggles over access to forests for hunting and lumber affected many ancient empires. Empires that were not able to manage their resources had to continually conquer in order to feed the needs of their people, their cities, and their armies. The industrial era saw people struggling over access to coal and iron. Those able to locate and harness the power of these resources conquered the planet and continue to take energy resources from the places they conquered. These resources included grains, lumber, coal, metals, and eventually oil. Oil came to dominate the last century and a half. The countries that either had oil or controlled access to oil were able to see success and wealth beyond the imaginations of ancient man. Millions died in the battle for control of the "black gold." Empires were laid waste, dictators were propped up, and the countries that controlled the flow of oil spread their cultures and languages across the planet.

Two intertwined but not necessarily related systems have sprung up and grown to prominence in this era: capitalism and democracy. While sometimes coexisting, they are at many times at odds with each other. In an ideal, functioning democracy, citizens are well informed, have access to all information equally, and are not being swayed by nefarious forces that are warping facts and information in order to serve their own purposes. An ideal democracy would put limits on business in order to protect citizens from monopolies and the price gouging that comes with them. It would protect the water, air, and land that its citizens need for their health and well-being. It would protect the other species that coexist with humans because humans are reliant on other species for their very survival. There are times that these basic protections run counter to the needs of a large business or corporation or of an entire industry (made of many small and large businesses and corporations). In these times, the needs of democracy and the needs of some of the large entities that exist in a capitalist system come into conflict.

It is in these times that a democracy can become subservient to the capitalist system. Large, wealthy, and powerful corporations and industries can influence both voters and politicians and convince them to create policies that do not protect the needs of the people. Clean air, water, and land can be and has been defiled and befouled. If nefarious capitalist powers are able to subordinate the needs of the many for the profits of a few, a democracy is not serving its intended goal of rule by the people and for the people. The power of capitalism within democracies has ebbed and flowed over the centuries since modern democracy began in the late eighteenth century. Reactions against what may become a corporate state can lead to socialist and/or communist governments that remove or limit democracy in order to control rampant capitalism. The opposite can happen as well: democracies can cede so much power to corporations that the people no longer have any power and the kleptocrats who control a country's industry control its government as well.

This balance between democratic power and corporate power is more important than ever before. Climate change has created a situation in which it is imperative and necessary to create guardrails and limits on capitalist growth if and when it comes at the expense of the environment and the future well-being of humanity. However, many corporations have grown so large that it is difficult for one country to control them. In many cases these corporations have so much control over what are supposed to be democratic governments that they are actually more in charge than the citizenry. In other situations, such as in China, there is no democracy, so the control of state corporations depends on the honesty and ethics of people who are not subject to votes or the will of the people. There are solutions to this wicked problem, but it means changing some of the rules of the world economy. New rules mean new winners and new losers. Many corporations that are currently successful may fight these changes because there are no guarantees on who will win and who will lose in a new system.

The world must create a system in which sustainability is valued and in which carbon emissions will become very expensive. Large-scale carbon extraction from the air and sea may become necessary in order to limit the damages caused by carbon emissions. Capitalism can exist within this system, but it can no longer count destruction of natural resources as an externality. The true value of the planet's air, water, soil, and natural resources must be paid by anyone who uses resources or pollutes the environment. These monies must be used to repair and remediate any damage done. Another option is that businesses can be sure to cause no harm to these essential resources in the first place and avoid having to pay for remediation. A sustainable system would be one that ensures that no more carbon is added to the atmosphere and that the environment is equal or better than it was before a process began. This is possible, but the conflict between those that already benefit from the current destructive and extractive systems and those that aim to create a sustainable system will determine when and if such a system is implemented.

There are many unknowns as humanity lurches toward this transition. The ability of humans and nature to be resilient to the changes that are already happening and to the changes that will soon come will determine how well humanity survives. It is not hyperbole to state that we are entering an epic that is unlike any other. In the past, human problems were, to some extent, localized. This is one of the first times that the problem is truly global (except for the threat of nuclear war). The actions of each country will help determine the fate of every other country. It will be a time of incredible change and it is yet to be seen if these changes will be profoundly positive or incomprehensibly negative. This chapter will discuss the conflict over the changes to come and their possible effect on both people and the environment of which we are all a part.

CONSEQUENCES OF LITTLE TO NO CHANGE IN ENERGY RESOURCE USE

This section could well be called, "How We Destroy the Planet." If there is no change in human behaviors, there will be drastic and relatively quick (within one human lifetime) environmental changes/deterioration. For example, rising sea levels is one of the most talked about effects of current human behavior. A business-as-usual scenario (one in which there is no GHG-related behavior change) will put 630 million people below annual flood levels (Kulp and Strauss 2019). To put that into perspective, 250 million people were estimated to have lived at that level in 2020. Flood events are predicted to be more intense and more frequent; this will likely lead to the greatest migrations of humans in our history. Human migration, on a crowded planet, will likely lead to conflict. The situation will be even worse for many of the other species on the planet. Extinctions that are already happening at alarming rates will accelerate due to the fact that the speed of planetary warming will exceed the speed at which species can adapt. Species that are moving toward the poles will run out of room to move.

The iconic polar bear is already an example of a species that cannot find a colder region to live in. There is not an icier place (accessible to polar bears) than the North Pole. The polar bear needs ice to float on while it hunts for seals. Without that ice, they will disappear. The polar bear is just the most visible species that is going through this process. There are thousands of species, big and small, plant and animal, that are not able to adapt or migrate their way out of this predicament. Human migration will also run into political and physical danger as cultures clash and arable lands decrease and change.

Rising Seas and Flooding

Rising seas are one of the most visible effects from global warming. Global warming causes the sea to rise in two ways: melting ice from glaciers and ice caps adds volume to the ocean that is not replaced by new ice or snow pack in the winter, and warmer water expands and takes up more room. Cities that cannot spend billions of dollars creating locks, levies, dams, and dikes will have to move or disappear. Most major cities are built on coasts or along rivers. Water was the easiest way to travel for most of human history. Many humans live in areas near the sea and on plains that are prone to flooding. According to NOAA, sea level rise has risen from 1.4 mm per year in the past century to 3.6 mm per year in the past decade (Lindsey 2022). While humans probably cannot stop sea level rise, they can limit it. NOAA projects that sea level rise will be between 1 ft (0.3 m) and 8.2 ft (2.5 m) by the year 2100. The difference here is enormous. One foot of water would be difficult to deal with, but adapting may still be within the reach of human power and imagination. Eight feet of water would mean up to a billion people would have to find new homes. Bays and estuaries that support unique ecosystems in their brackish water would be claimed by the sea, and the life that once thrived there would be wiped out. Low-lying coastal farmlands, especially rice farms across Asia, would be under the ocean. For example, Bangladesh and Vietnam face threats from saltwater invading their rice fields. In a worst case scenario, the hundreds of millions of people who live in these countries would not have enough food, and mass starvation would ensue.

Flooding can be an advantage to a farmer. The annual floods in ancient Egypt brought fertility to the Nile River Valley. The melt from the Himalayan glaciers brings needed water and fertile soil to India and Bangladesh. Monsoon rains bring needed moisture to these areas as well as to Pakistan. However, formerly once-in-a-lifetime flood events are becoming nearly annual events, as flooding increases in both severity and frequency. These floods take lives, destroy personal property, and wash away fields when flooding is extreme. The flooding in Bangladesh and India has been extremely severe, with major flood events in 1998, 1999, 2004, 2005, 2015, 2017, 2020, 2021, 2022, and 2023. There has been some flooding every year since 2015. Flooding has morphed from an unlikely event to a certainty (Ahmad 2020). On top of this, cyclones have become more severe in the area as well. Between global sea level rise, flood events, and storms, poor low-lying countries are forced to face the brunt of climate change consequences. Unless action is taken to curtail GHG emissions, countries like Bangladesh will eventually become almost uninhabitable.

Crop Losses

Climate change is already affecting crops across the globe. While overall the effect is negative, there are areas that will benefit. These areas will be able to grow more food in wetter and warmer weather. Research has demonstrated that climate change has led to an overall annual reduction of 1 percent of consumable food calories per year (Ray et al. 2019). For wealthy countries, this will not be a problem in the short term as most of these countries produce much more food than is needed. For poorer countries, this is very dangerous as this leads to more food insecurity. The number of undernourished people in the world had been decreasing annually for decades, but that changed in 2015; the number of undernourished people has increased each year since then (UN 2020).

Crop losses are also caused by major weather events such as floods and large storms (hurricanes/cyclones/typhoons). (These extreme weather events and their increased likelihood are covered in more detail in the "Extreme Weather" section of this chapter.) Long-term weather anomalies such as droughts are also affecting crop losses. Subsistence farmers are highly vulnerable, as they not able to survive a year in which they lose their entire crop or a large part of it. In poorer countries, the government does not have the resources to rescue these farmers. In these situations, farmers and those dependent on them may lose their lives.

There is no silver bullet to solve this problem. The most obvious solution is to slow or stop human-caused climate change. However, many climate change effects are caused by

GHGs that were released years and/or decades ago. This means that farming will need to adapt, as best it can, to the conditions that exist. Adaptation to storms and flooding is extremely difficult, if not impossible. However, it does work when drought, warming weather, or wetter weather is more achievable. Sorghum, for example, is one of the few crops that has done better with the effects of climate change in drier, warmer climates in which it is grown widely. Creating more drought-tolerant varieties of grain crops is one way in which humans can be ready for these situations. Another possibility is that, through international agreement, countries that produce more grains than they need can create a ready supply to be used to bring relief to people who are experiencing hunger due to storms and flooding. The world produces more than enough food, as rising obesity rates can attest. In preparing for the acute and long-term effects of climate change, using international resources can help the world cope with current and future climate-caused crop losses.

Species Extinctions and Spots of Hope

The Anthropocene, or the human epoch, is the time in which humans have begun to recreate the earth by changing the air, water, and land of the planet. In this period, humans have directly and inadvertently caused the extinction of thousands of species. This is sometimes referred to as the sixth mass extinction or the Holocene extinction. (Holocene is the current geological epoch.) The term we will use is the Anthropocene extinction. The Anthropocene extinction event began with the spread of humans out of Africa. Early human hunters were able to hunt megafauna to extinction across five continents: Europe, Asia, North America, South America, and Australia. The continual coexistence of megafauna and humans in Africa is theorized to be due to humans evolving in Africa with the megafauna. African megafauna also evolved with a fear of humans and were able to avoid extinction-level hunting. Overall, 83 percent of mammals have gone extinct since the rise of human civilization as well as half of plants (Bar-on, Phillips, and Milo 2018).

While humans have been hunting megafauna to extinction for millennia, it was a slow-paced affair at first. Not until the end of the last ice age, at about 10,000 BCE, did human hunters begin to wipe out megafauna, such as wooly mammoths, mastodons, and giant sloths, at an especially fast rate. The speed of extinction increased locally as empires expanded across the planet over the past 5,000 years with advanced weaponry. Species such as the Nile hippopotamus, straight-tusked elephant, and Chinese rhinoceros fell to the advance of local empires. That speed of extinction changed again with the invention of firearms. The once impenetrable hides and fleet feet of the surviving megafauna were no longer enough to protect them, and species like the Steller's sea cow, blue antelope, Honshū wolf, Atlas bear, and giant elephant bird were killed off. Hunters picked off bird species that were once safe in flight, such as the passenger pigeon. As humans began to alter ecosystems through agriculture, logging, and grazing, many species became extinct due to lack of native habitat. The Industrial Revolution started to change not only land but water and air as well. Currently, all of these forces are putting pressure on the species of this planet. Climate change combined with habitat loss puts all species in a precarious situation. Animals that are large enough to be made into a good meal also face a third threat from hunting and fishing.

Humans, which despite our numbers make up only .01 percent of the total biomass of the earth, have had an outsized effect on the planet. This ability can also be seen as an advantage. Humans, if they dedicate themselves to it, are capable of slowing the effects of climate change and saving species from extinction. There are many examples of areas that have seen an increase in biodiversity after an abrupt or slow change in human use. An example of abrupt change is the area around Chernobyl; deemed unlivable for humans after the nuclear power plant accident, it now abounds with wildlife, complete with large mammals and predators (Deryabina et al. 2015).

An example of a more gradual change in human land usage can be found in New England in the northeastern United States. New England was once the industrial center of the country but has become increasingly less industrial and more residential. The people of New England (Maine, New Hampshire, Vermont, Massachusetts, Connecticut, Rhode Island, and upstate New York) have passed a legislation that protects animals and natural areas. One of the most important indicators of ecosystem health is the health, as well as the number, of apex predators in an area. New England has seen bobcat populations rebound; wolves have returned; bear populations have increased; eastern coyote packs have flourished; and great white sharks have returned to New England's ocean waters (Liebowitz 2023; Monzón, Kays, and Dykhuizen 2014). The occasional mountain lion has been seen and documented in the area as well. All these predators demonstrate that there is a healthy and balanced ecosystem in an area that is heavily populated with humans. Deer, beavers, moose, turkeys, and seals have also returned in large numbers. This explosion of wildlife happened during the same period of time in which the population of New England doubled (1920–2020). The human population density is 236 inhabitants per square mile or 92 per square kilometers. How is this possible?

New England let its woods grow back. Farming has always been difficult in the area, and the internationalization of the food system has made it even less profitable. New Englanders are generally well educated and politically bent toward conservation. While human population density increased, most of that increase is in the city and towns. Rural areas have actually lost population. The regrown woods provide more habitat for animals, allowing their populations to rebound, and local/state regulations limit hunting and trapping of these animals. Limits on fishing have allowed seals to return, and the seals in turn brought the sharks. Limits on pesticides have brought a return of raptors; hawks, eagles, and ospreys now take their place in the skies. Decommissioned dams have allowed herring runs up New England's rivers to multiply exponentially.

The lesson that New England teaches the world is that if there are viable "seed" populations nearby, an area can be rehabilitated and humans can thrive alongside animal species. We are compatible because we are in the end a product of nature ourselves. Thoughtful policies that provide enough clean land, air, and water can help species that were driven to the brink rebound incredibly. However, if species are driven toward extinction and their populations dwindle too low, repopulation may be impossible. Given a chance, nature will find a way, as the example of New England's "coywolves" demonstrate: they are a mix of four distinct canine groups (coyotes, eastern and western gray wolves, and domestic dogs). New England is a stellar example of how nature can reorganize and succeed when given enough space and time.

Wildfires and Ecosystem Disruption

The wildfires of 2019 and 2020 were the most destructive in human recorded history. (The years 2021 and 2022 had similar raw numbers of fires, but acres burned in the United States were about 7 million compared to 10 million acres in 2020.) The western half of the United States saw fires unlike any other fire season. Fires started early, burned more intensely, and burned for longer than ever before. The fires in Australia could be seen from space, and the smoke from these fires changed the weather in New Zealand. Not only were forests lost, but billions of invertebrates, mammals, plants, and birds perished in these fires as well. In 2023, wildfires were below average in the United States, however; Canada had the worst wildfire season ever. Canada lost an area larger than the entire state of North Dakota or over 18.4 million hectares.

Climate change causes global warming, but it also changes weather patterns. Extremely dry seasons that were once rare occurrences become the norm and add up to become intense droughts. These droughts cause forests to become tinderboxes just waiting for a spark. Natural fire cycles that would normally clear out the underbrush of the forest become deadly crown fires (fires that burn to the top or the "crown" of the tallest trees). These fires destroy ecosystems in ways that cannot be repaired, at least not on a human timescale. Human forestry practices can exacerbate these fires by leaving behind forests that had at one time been clear-cut. The forests that return to clear-cut areas are much denser, less healthy, and younger than the older forests they replaced. These forests are not fire-resilient and can burn completely. A century of fire suppression also adds to the fuels in the forest. Yet, the biggest driver of these fires is climate change, and the biggest driver of climate change is the release of CO_2 into the atmosphere through burning of fossil fuels.

Some wildfires burn so hot that they destroy not only the life aboveground but also the life underground. In some of these cases, where the mycorrhizal fungi in the soil is destroyed, the trees and the plants do not come back. Mycorrhizal fungi support life above ground by transferring needed nutrients to trees in exchange for sugars that the trees make through photosynthesis. These mutualistic relationships are needed for the survival of both (Johnson et al. 1991). Common dominant forest species such as pine, eucalyptus, beech, spruce, and oak trees, as well as plants in the rose family and orchids, need these fungi to survive. These hot, intense fires can create landscapes that no longer support the species that once thrived in these areas (Peréz-Izquierdo et al. 2020), forever altering the ecosystems. New ecosystems can develop over long periods of time, but the original forest ecosystem will not return.

Human Health Impacts

The World Health Organization estimates that there are 4.6 million deaths every year that are directly attributable to air pollution. So, if "business as usual" is the model used, it can be expected that the number will continue to grow. If fossil fuels are replaced with renewable energy, that number can begin to shrink. It is simple math with life-and-death consequences. These numbers are verifiable and knowable, but most importantly they are changeable.

The future human health impacts due to climate change would build on the already staggering number of deaths due to pollution. We have discussed how many people (over one billion) could be forced to move due to sea level rise. Warmer and drier weather that leads to drought could cause hundreds of millions of people to search for new places to farm and to live. The world has had a preview of mass migration with the civil war in Syria.

Beginning in 2011 in Syria, drought and crop losses sent many people from the country into the cities in search of food and work. Sectarian issues were exacerbated by the fact that the Syrian dictator, Bashar al-Assad, belongs to a minority Alawite sect of the population. Al-Assad was only able to hold on to power with Russian, Iranian, and other foreign military aid. The Russian air force's ability to destroy rebel strongholds is the main reason that al-Assad still holds power. This war became international with belligerents from dozens of countries. Total deaths are estimated to be over half a million men, women, and children. Five to seven million displaced persons spilled out across Eurasia. The influx of refugees caused internal strife throughout the Middle East and Europe.

Countries have dealt with the influx of refugees with varying degrees of kindness and coldness. Lebanon took in over 600,000 people. With a population of only 4.8 million, they were the most generous on a per capita basis. Turkey has the most refugees of any individual country with 1.7 million. Germany took in over 1 million refugees, but Hungary, Sweden, Norway, and Finland had more refugees on a per capita basis. The number of people applying for refugee status added up to almost 2 percent of the Hungarian population (UNHCR 2020). These numbers caused Hungary to close its border with Croatia. (They later accepted more refugees that were transferred from Italy and Greece.) Many refugees hoped to get into any EU country and achieve refugee status. Some would then head to Germany, the UK, and other Western European countries that had strong economies in which to look for work.

All of this new immigration led to the growth of far-right political organizations across Europe. These organizations have the basic stance of "keep Germany for Germans," for example. Unlike colonized countries that have had centuries-long influxes of immigrants after the genocide of local peoples, most European countries had a population that is genetically tied to their country's namesake. The majority of the population of Sweden is Swedish, for example. (Although, Sweden does have a sizable immigrant community.) In order to please those in the population that fear being overrun by refugees, Northern Macedonia, Serbia, Hungary, Monte Negro, and Bosnia and Herzegovina have built fences and enacted policies to stem the flow of migrants out of Syria and Iraq. Austria, Italy, and Greece have built fences and stepped up patrols as well. Nonetheless, refugees are desperate people who have lost everything, and when countries block migration paths, refugees will risk more dangerous routes to get their families to safety. This leads to refugee deaths in overcrowded boats and fast-flowing rivers.

While the crisis in Syria started due to drought caused by climate change and the subsequent migration of people into urban areas, politics and civil war are the reasons that millions have left the country. In Iraq, the refugee crisis started due to a war over the control of oil, a classic energy-resource conflict. Current conditions forcing people out of the country are related to warfare originating from religious and political differences. These countries demonstrate how refugee crises can be instigated by conflict over fossil fuel resources and by the effects of burning those fossil fuels for electricity, heat, and transportation. However, these examples involve, at most, 7 million refugees. It is difficult to comprehend the human health costs of 700 million refugees. The wars these issues might cause and the possible loss of human life is staggering. The costs of not addressing climate change are almost unimaginable. Through substantial investment in solutions such as renewable energy and electric vehicles, as well as energy efficiency, many of these outcomes may be avoided.

Extreme Weather

Climate change and global warming have created a situation in which extreme weather thrives. One form of extreme weather can affect others: droughts cause forest fires and paradoxically can also lead to flooding when dried and burnt lands are unable to absorb precipitation. In areas that still support subsistence farming, these extreme weather events can have life-and-death consequences.

Drought can be caused or exacerbated by global warming (one of the main effects of climate change). Warmer weather can lead to drier soils and less vegetation cover, which in turn leads to less moisture being evaporated and less moisture available for the creation of clouds. This "positive feedback loop" can lead to prolonged droughts and eventually the unwelcome transition of a once verdant area into an arid or semi-arid land. These droughts destroy forests and farmlands alike. While some areas dry out, other areas may see increases in rainfall, and increased rainfall can have both positive and negative effects (NOAA 2020). Climate change has led to more forest fires and crop failures.

Wildfires can become their own form of extreme weather. Clouds of ash spread out over thousands of miles. Unhealthy air full of particulate matter descends on cities near and wide. This "weather" causes people to lock themselves indoors and leads to increased deaths for those with asthma and other lung issues. It darkens the skies and blocks out the sun, changing the weather throughout the duration of the fires. These fires can last for months, changing the local weather for entire seasons.

Floods have many causes, but two of the most common are too much rain and rain after a forest fire. In the United States, the Western half of the country is drying out, but the Midwest has seen increased flooding due to rainfall that saturates the soil and fills reservoirs. These waters eventually flow over dams and break levies. These floods cause destruction of farmland, property, and human life. Forest fires leave behind a parched landscape that cannot absorb moisture, and small rain events can become destructive floods. Many areas that are still recovering from forest fires will see flooding within a year after the fire.

Hurricanes, cyclones, and typhoons are all names for the same type of storm. These storms become more powerful in warmer weather, so global warming can increase their intensity. Stronger storms are more destructive and bring in more rain that leads to more flooding. They have higher wind speeds that destroy more structures, and they result in higher storm surges that flood coastal areas. In 2020, the Caribbean Sea and Southeast United States saw a record number of named storms (storms that reached sustained winds of over 38 miles per hour).

Extreme events that are not climate change or global warming related, such as earthquakes and volcanoes, will continue to exist, but on a planet that is already stretched thin by increasingly frequent extreme weather events, there may be less help coming to the victims of these other disasters as well. Warmer weather is already bringing more intense and destructive storms. If these extreme weather events create a cyclical pattern, it is likely that some areas that have been populated by humans for millennia will have to be abandoned. The areas most vulnerable are coastal communities, island communities, and communities that are built in flood plains. It is likely that governments will grow weary of spending money to repeatedly rebuild these vulnerable communities, so people will have to move. This process is already happening on islands off the coast of the US state of Louisiana. It is a process that is likely to accelerate if climate change continues unabated.

Feedback Loops and Abrupt Climate Change

When looking back across the geological climate record, there is much that we understand. The growing field of historical ecology (or paleoecology) has done an excellent job of uncovering the climate history of the earth (Swetnam, Allen, and Betancourt 1999). We understand the natural cycles in the relationship between the earth and the sun that affect climate. The brilliant Milutin Milankovíc discovered these cycles (now called the Milankovíc cycles). These planetary cycles explain climate cycles that have existed over millions of years. They are based on three things: the shape of the earth's orbit, which is called eccentricity; the angle at which the earth's axis is tilted compared to the earth's orbital plane, which is called obliquity; and the direction in which the earth's axis is pointed, which is called precession (Buis 2020). At its present location within these cyclical events, the earth should be entering a cooling period. These planetary cycles result in slow changes over long periods of time. The earth should be starting to move toward the next ice age, which should take thousands of years. Places such as the Southwestern United States should be getting wetter and cooler, but human effects on the climate have changed that (Fawcett et al. 2011). There are some climate changes in the geological climate record that were remarkably fast. The earth may have hit certain tipping points (discussed in the next section) that caused rapid and irreversible climate change due to events such as super volcanoes and asteroid impacts. There are also feedback loops and climate forces, both internal and external, that accelerate the speed of changes to the climate (Anderson 2015). Some of these feedback loops have been observed during the current human-caused climate change event that we are living through. A good example of this is in the ice in the Arctic Ocean. Warming waters underneath the Arctic ice increase the melting on top of the ice, which has already accelerated due, in part, to melt water pools on top of the ice. Melt water is darker than ice and so retains more heat and reflects less light than solid ice. These feedback loops combined with an external forcing (in this case, a human-caused increase in atmospheric CO_2 concentrations) can cause abrupt changes in climate.

Another smaller example of a similar process has been seen in Central and South America. There are areas where the rainforest creates rain through evaporation. This is especially true in places where weather moves in from the coast and travels into the hills above. In some of these coastal areas, the rainforest has been cleared for tourism, settlement, farming, and/or cattle. Once these areas are denuded of forest, clouds no longer form on the coast and the rainforests in the interior do not receive the rain they need. These forests can experience drought and then major forest fires. Soon an entire ecosystem is transformed from a rainforest to a savannah.

Other feedback loops have both positive and negative effects. A warming earth will have more moisture in the air and create more clouds. Clouds can reflect sunlight. More reflected sunlight could slow the warming of the earth; however, because warmer air holds more water vapor, which in turn traps heat, the earth will continue to warm. As the climate changes and warms, rain will increase in some places and decrease in others. This could increase plant life in some areas, which can decrease CO_2 in the atmosphere, but it can also lead to more floods, which can remove large trees that store carbon. In formerly wet areas that see a decrease in rainfall, such as the West Coast of the United States and Australia, multiple dry years result in major droughts, droughts become forest fires, and eventual rains turn into floods as the burnt lands cannot absorb water. Many of these forests are forever changed. Forests are replaced by scrub and brush; scrub and brush are replaced by grasslands.

The most important and impactful of these feedback loops is the one involving ocean ice and glaciers. As ice becomes water, water absorbs heat. The warmer planet melts more ice and increases the heat of the planet, which in turn melts more ice. In this scenario, triggered by increased GHGs, the earth gets warmer and the oceans rise, leading to coastal flooding and what may be the greatest human migration away from the seashore that has ever happened.

Tipping Points

A tipping point is a term that was popularized by author Malcolm Gladwell. It is the point at which a trend cannot be stopped or reversed. For example, it is the point when someone jumps off a cliff or starts down a slide. There are a number of ecological tipping points that will most likely occur if human production of GHGs is not significantly curtailed. The two most talked about tipping points are the disintegration of the Greenland ice sheet and the disintegration of the West Antarctic ice sheet. The thawing of the permafrost that surrounds the Arctic Circle and the ensuing release of massive amounts of methane is another major tipping point that causes international concern. Coral reef die-offs, ocean acidification, and an Amazon forest dieback are other distinct possible outcomes of climate change and direct human behavior such as overfishing and deforestation. A change in ocean circulation, such as the Atlantic meridional overturning circulation (AMOC) breakdown, could be caused by a melting Greenland ice sheet and could change weather patterns and temperatures across Western Europe. Shifting monsoons could also affect Africa and the Indian subcontinent, which would change life for all the humans and animals that inhabit those areas. Some of these tipping points appear to be statistical certainties, but others depend on future human behaviors.

In 2020, researchers announced that the Greenland ice sheet had crossed a tipping point in 2018–19: it will most likely never again receive enough snow to balance the amount of mass lost to melt and iceberg calving. Further research using historical satellite data has demonstrated that the Greenland ice sheet passed the tipping point in the year 2000 (King et al. 2020). The ice sheet has an average loss of 250 billion tons per year. Its rate of melt has increased 700 percent since the 1990s. If these trends continue, the melt will contribute to 1.5 meters of sea level rise by the year 2100. While it is unlikely that the melting of the Greenland ice sheet can be stopped, it can be slowed. Severe cuts to GHG emissions across the planet can limit the amount of annual melt to a level that could allow countries to adjust to the annual sea level rise.

Until recently, the Antarctic ice sheet, and especially the East Antarctic ice sheet, was seen as generally safe from major melt events, but that assessment has changed. The rate of annual melt has grown from 40 billion tons lost per year between 1979 and 1990 to 252 billion tons lost per year between 2009 and 2017 (Rignot et al. 2019). Antarctica is melting six times faster than it did thirty years ago. Antarctica and Greenland currently only account for about a third of sea level rise, estimated to reach over 71 centimeters (28 inches) by 2100. The other culprits are melting glaciers across the planet and thermal expansion due to the warming ocean waters. Sea level rise would be worse if it were not for all the dams that have been constructed for hydropower and agriculture. There are few major dam projects in the works that would hold back additional water. The real fear is that the ice sheets of Greenland and Antarctica will start melting at a speed that will surpass human's ability to counter rising sea levels. Current research has demonstrated that in the past, Antarctica melted at much faster speed than today with less CO_2 in the air (Blackburn et al. 2020). When that melt happened, sea levels rose over 30 feet. That kind of sea level rise would put many of the world's major cities underwater and force over a billion people from their homes.

The melting of permafrost near the Arctic Circle could be another tipping point that makes keeping global warming under 1.5°C unlikely. Locked inside of the permafrost (ground that is frozen for at least two years straight) is twice as much carbon as has already been released into the atmosphere. Some of that carbon is in wet soil in an anaerobic condition (without oxygen) and will be converted into methane gas when the permafrost thaws. Methane gas is up to thirty times more powerful than CO_2 as a GHG when released into the atmosphere. The Arctic is warming twice as fast as the rest of the planet, and for that reason scientists fear that a warming climate will push the permafrost past a tipping point and that a positive feedback loop will accelerate the warming. Scientists estimate that if the planet warms 1.5°C, the newly thawed permafrost will release between 68 and 508 billion tons of carbon. (Human activities in the newly warmed areas could also increase the amount of CO_2 released.) This carbon would lead to a warmer climate, which in turn would thaw more permafrost, and the feedback loop would continue. The severity of thawing permafrost will depend on how quickly humans stop burning fossil fuels for energy and transportation.

Coral reef diebacks have already happened many times, and the occurrences are increasing. Coral reefs are sensitive to many human activities, including trawling, fertilizer run-off (and resulting algae blooms), cyanide fishing, pollution (especially oil and chemical spills), use of sunscreen (oxybenzone and other chemicals), rising sea levels, and, most commonly and importantly, increased water temperatures and CO_2 levels. For the purposes of this book, we will focus on the last two issues.

Coral bleaching is the process in which coral polyps eject the photosynthesizing algae (zooxanthellae) that live inside of their tissues. These species normally have a mutualistic relationship: the zooxanthellae provide 90 percent of the coral's energy through photosynthesis, and the coral provides the CO_2 that is needed for photosynthesis. Extreme heat can cause this emergency ejection, but if the zooxanthellae are not able to recolonize the coral, the coral will die. Coral reefs suffered an extreme bleaching event from 2014 to 2017 that affected 70 percent of coral reef systems on the planet (NOAA 2020).

Coral reefs are home to over a quarter of all the fish in the ocean. Most of these fish cannot live anywhere else. Many other species use the reefs for reproducing. The loss of coral reefs would be devastating to the ocean ecosystem. Estimates indicate that 90 percent of coral reefs will be endangered in the 2030s and that all of them will be in jeopardy by the 2050s. This is due not only to rising sea levels and rising temperatures but also to the ocean acidification caused by dissolved CO_2 in the ocean.

Ocean acidification is the process of CO_2 dissolving into the ocean and changing the natural pH balance of the ocean. Ocean acidification "lowers the calcium carbonate saturation state of seawater" (Doo, Edmunds, and Carpenter 2019). Calcium carbonate is the building block that forms the hard structure of coral that is the base of a coral reef. It is also the building block of all shellfish. Ocean acidification could be detrimental for clams, oysters, mussels, scallops, starfish, squid, octopi, and many small organisms like plankton that form the foundation of food chains in the ocean. Ocean acidification can be stopped by putting less CO_2 into the atmosphere. One proposed geoengineering deacidification solution is to put 500 gigatons of ground alkaline rock (most commonly limestone) into the ocean. To put the scale of this undertaking into perspective, the total amount of mined rock (of all types) annually is 50 gigatons. This would be a project of immense scale with potential for currently unknown, unintended consequences. It is unlikely a project like this would be agreed upon and funded appropriately. However, the idea has been given serious thought by scientists and politicians alike (Renforth and Henderson 2017).

Another tipping point concern is related to the Atlantic meridional overturning circulation (AMOC). The AMOC brings warm salty water in a northward flow from the south and sends colder, deeper waters southward from the Northern Atlantic. The AMOC is a vital regulator of earth's climate. This circulation of waters provides nutrients to ocean life throughout the Atlantic. It also keeps Europe much warmer than it would be at the latitude it is located without it. If we look at a globe or a world map, we can see that Northwestern Europe is at the same latitude as Canada, but it is significantly warmer. That is because the flow of warm water also warms the atmosphere and brings with it milder weather than there is on the western side of the Atlantic.

The AMOC not only provides nutrients and warm weather, it also acts as a significant carbon sink (Gruber, Keeling, and Bates 2002). This is important, as it has helped sequester some of the CO_2 produced by human activity over the past two centuries. Frighteningly, the latest scientific observations have shown that the AMOC is slowing down faster, by a factor of 10, than climate models had predicted. A warmer climate was expected to slow the AMOC, but not at the current rate. Increased freshwater that is melting off of Greenland may be to blame for this change (Haskins et al. 2019). The Greenland Ice Sheet may have already reached a point of no return. (If this is true, the rate of melting can be slowed by human actions, but it cannot be stopped.) If the AMOC slows dramatically or stops, there will be a threefold fallout: northwestern Europe will become significantly cooler; the nutrient stream that supports much ocean life will stop, causing many species to decline and some to go extinct; and the CO_2 that is stored through the AMOC flow will remain in the atmosphere or in the ocean where its effects will continue to be destructive to ocean health. Scientists have been able to connect past slowing of the AMOC to a period in Europe called the Little Ice Age. There is still some uncertainty as to whether the accelerated rate of a slowing AMOC will continue, as direct observations only go back to 2004. The feedback between fresh melt water and the slowing of the AMOC could possibly slow some of the melting of the Greenland Ice Sheet. If the slowing of the AMOC takes centuries, it could be a process that humans have enough time to react to; however, if the current rate of slowdown continues, there could be serious repercussions that have to be addressed in the coming decades.

Some fear that climate change may directly cause forest diebacks in the Amazon. However, most climate models do not see the Amazon forest dying off under most scenarios. According to Dr. Peter Cox, the Amazon rainforest is directly threatened by human clearing of the forest and human-caused wildfires, but not from climate change. One and a half acres of Amazon rainforest are cleared every second. That is 90 acres a minute or 540 acres an hour. Forest is mostly cleared for grazing and farming although sometimes for timber. The good news, then, is that changes in human behavior can have an immediate effect on the health of these forests, whereas many climate-based catastrophes cannot be stopped. The Amazon rainforest stores 15 percent of the worldwide carbon that is stored in vegetation. However, if the entire Amazon forest were lost, atmospheric oxygen on earth would only decrease by 0.01 percent (Cox 2020). Recent research has demonstrated that earlier estimations of Amazon forest diebacks were incorrect in part due to the process of photosynthesis. More CO_2 in the atmosphere increases the efficiency of photosynthesis and decreases the water needs of many of the plants in the Amazon. The Amazon will have the ability to "save itself," as long as human behaviors are changed to stop active deforestation. This will require political leadership, not just in Brazil, but in all the countries that contain the Amazon. A worldwide carbon pricing system could pay countries that preserve and grow their forests. A system like this might motivate the Amazonian countries to preserve the natural beauty and functionality of the forest that is nicknamed the "lungs of the earth."

Tipping points are frightening, but many of these scenarios can be avoided and/or remedied. It will depend on the work of many in the political realm. People must elect those that will prioritize the environment that supports all humanity.

FUTURE ENERGY PROJECTIONS BASED ON CURRENT TRENDS

This section looks at the major sources of electricity generation and what their likely future will be. (For information on past and present use of these energy resources, see Chapters 1 and 3.) However, there is always the possibility (and hope) of a breakthrough technology that can alter predictions.

Coal

The future of coal is closely tied to the future of China. China consumes over half of the world's coal and more than quadrupled its use of coal from 2000 to 2023. China also exports coal and coal-plant technology. It has pledged funding for or funded over 70 percent of all currently planned coal-power plants. Without Chinese funding, many of these plants would not be built, as many major banks are moving away from coal. (China has pledged to stop building coal plants, but many plants are in the process of being built.) China also continues to build coal plants within its borders, despite having an already overbuilt coal-power-plant capacity and pledging to transition to renewable energy. The current building spree has more to do with local/provincial politics than national directives. Building these coal-power plants adds to GDP, employs a lot of people, and stimulates the economy. It may be that these new coal plants will not add much to Chinese coal usage in the long run, due to the Chinese build-out of renewable energy. Chinese coal consumption reached a peak in 2013 and began to decrease, but that trend has since reversed. China is reaching new highs in coal consumption, with 2023 the highest on record. It continues to build new coal-power plants in response to increased energy demands for cooling as temperatures increase and summer heat waves become the new normal.

While natural gas has been used by many countries to transition away from coal, China's natural gas reserves are not sufficient to meet its current needs. This means that China would have to depend on imports if it were to use natural gas instead of relying on its own plentiful supply of coal. China has created the world's largest solar power, wind power, and hydropower systems. The rate of growth of these renewable systems is likely to eventually cut into coal-power production. The IEA (International Energy Association) had estimated that China's coal use would plateau in 2022, as it had ticked up slightly in 2016, 2017, and 2018 after falling in 2014 and 2015. China saw lower coal consumption in 2020 due to the pandemic, but coal use was up in 2021, 2022, and 2023. Chinese coal consumption will likely plateau in 2025, barring any unforeseen economic changes.

The IEA has mapped out two possible trajectories for coal consumption worldwide. One scenario is business as usual, in which coal use will stay about the same across the planet. Advanced-economy countries will move away from coal, which will keep coal prices low. Developing countries, in Asia especially, will replace those losses by increasing their use of coal as they modernize their economies. The second scenario is one of sustainable development, in which nations with advanced economies will drastically reduce their use of coal. Developing nations will not build coal plants at all and instead move to a renewable energy model that will be a mixture of solar power, wind power, and hydropower, supplemented by pumped hydro and natural gas. In this scenario, coal use will drop by over 60 percent by 2040. It seems most likely that reality will fall between these two extremes. Most advanced countries will continue the decarbonization of their electric power grids. Affordable battery storage as well as pumped hydro will accelerate that trend. China will most likely hedge their bets and continue to lead the world in both coal and renewables. According to their fourteenth five-year plan, China will continue to increase CO_2 emissions and possibly coal use until 2025, after which it will begin to aggressively decarbonize its economy through the use of renewable energy and an economic transition from manufacturing to banking and tech. Most countries will choose renewables due to their low cost, but coal will remain the most affordable option for a few countries that have already built new plants. The bigger problem is that many of the coal-power plants that are already built are relatively young. These plants will need to provide a return to their investors. When the investor is the state, it is almost guaranteed that the plant will continue to operate for most of its potential lifespan. In the United States, coal power plants are mostly old and ready to be retired. In Asia, majority of coal plants have been built this century. It will take a number of local, national, and international incentives to accelerate the transition away from coal. The transition will depend on international political will and the economics of energy, including the affordability of new technology. Coal use is on the decline in many countries, but the speed of that decline is uncertain.

Oil

Oil is the "black gold" that the world has shed so much blood over (see Chapter 2). After a century and a half of war and political maneuvering, we have never hit "peak oil." Peak oil is the long-standing idea that demand for oil would continue to increase and new oil discoveries would continue to decrease until the world reached peak oil production. At that point, world oil production would not be able to meet demand. Each year thereafter, less oil would be produced and costs would skyrocket. While this scenario did indeed appear to be likely, according to the IEA, as conventional oil peaked in 2006, it was replaced by "tight oil." Tight oil is oil that is obtained by fracking (see Chapter 2). It is located in porous rock formations of shale or sandstone. The fracking boom has led to a glut in oil, led by American producers. Deepwater drilling in oceans has also added to the abundance of oil.

The real peak for oil, then, is likely to be peak oil demand. Peak oil demand has either already occurred in 2019, as there was a big drop off in demand in 2020 caused by the Covid-19 pandemic shutdown, or it will occur sometime over the next decade as EVs (electric vehicles) change the oil market. EVs had a worldwide market share of 2.6 percent in 2019 (Wheeler 2020) and 4.2 percent in 2020, 8.7 percent in 2021, and 14 percent in 2022. Estimates indicate that EV sales made up about 18 percent of market share internationally in 2023. Almost every major car company either produces an EV or is planning to produce one in the near future. Sales in the United States continue to climb. Internationally, sales are increasing even faster. In Norway, 90 percent of all vehicles sold are EVs (a figure that includes plug-in electric hybrid vehicles, PHEVs); this leads all countries in EV market share. In China, about 36 percent of all new car sales are EVs, and this percentage is trending higher every year. Because China is such a large market, this means that more EVs and PHEVs are sold there than anywhere else in the world.

The rate of decline in the demand for oil will depend on government subsidies of both EVs, PHEVs, and, more importantly, oil companies. Worldwide subsidies of oil companies are near $1,000,000,000,000 (1 trillion). Additionally, there are significant environmental costs, health-care costs, and military costs associated with international oil. Subsidies for EVs are comparably just a tiny fraction of those for the oil industry, yet they are continually under pressure by oil companies and manufacturers of ICE (internal combustion engine) vehicles. If countries even out subsidies between oil and EVs, the market share for EVs will grow exponentially. If countries drop subsidies for oil companies entirely, EVs will quickly become the economical choice for drivers as low oil prices are often derived from government subsidies. It is unlikely that governments will drastically decrease subsidies for oil companies in a peak oil demand environment. A more likely scenario is that governments will slowly phase out oil company subsidies as oil becomes less important as ICE vehicle sales decrease and EVs sales increase. Countries that do aggressively incentivize EVs will see their GHGs decrease significantly and see pollution-related health-care costs decrease proportionately. Once again, this is both an economic and a political issue. Economically speaking, EVs (with current government subsidies of oil companies) are still usually more expensive than ICE vehicles upfront but less expensive when the costs of fuel is included. If the political will is there to create the subsidies needed to bridge the extra costs of EVs until they are economically competitive with ICE vehicles, EV sales will easily and quickly overpower ICE vehicle sales. If governments agree to put a price on carbon or demand that companies that sell fossil fuels remove the equivalent amount of CO_2 out of the atmosphere, then EVs will already be more affordable than ICE vehicles. Finally, if governments do nothing at all, EVs will become competitive with ICE vehicles within a decade and are already less expensive when accounting for the lifetime cost of buying gasoline (petrol). The big question that remains is not "Will oil decline in worldwide use and importance?" but "How soon will oil decline in use and importance for the world?"

Natural Gas

Natural gas has been the fossil fuel winner in the renewable energy transition. Natural gas is a fossil fuel, but it burns cleaner than coal and oil, meaning less particulate matter is released. The fracking boom that catapulted the United States to the top of the list of oil-producing nations has also accelerated the move from coal to natural gas plants. Low natural gas prices have encouraged many companies and countries to make the switch. The switch from coal to gas is an improvement locally and possibly internationally depending on a few factors.

Natural gas is made up of a number of different flammable gases. Shale gas, which is the predominant type of gas drilled for in the United States due to the fracking boom, is made up of up to 90 percent methane; the rest is ethane, propane, and butane. Burning methane releases 50–60 percent less CO_2 than coal depending on the modernity of the gas-power plant. The real danger of methane is that it can leak out and into the atmosphere. Methane is really good at retaining heat; it is 90 times (that is 9000 percent) more powerful than CO_2 at trapping heat in the atmosphere. In the process of drilling, transporting, and burning methane, 1–2 percent of the gas is lost into the atmosphere, which can add up to a lot of heat-trapping gas. The upside to methane is that it only remains in the atmosphere for nine years. If natural gas-power plants can be used to bridge the gap between a coal-dominated energy system to a renewable one, they can be a viable energy option. Conversely, if countries and companies invest in new gas plants, they will want to see a return on their money, and these "bridge" plants could become bridges to nowhere but a continually warming planet.

Nuclear Energy

Nuclear energy has held the promise of a carbon-free and pollution-free energy future for seventy years, but the reality is much more complicated and nuanced. Nuclear energy does produce electricity through a nuclear reaction that creates no air pollution and releases almost no CO_2. Most nuclear power plants are extremely inefficient in their use of fuel in that there is always nuclear waste leftover in the process. Yet, in one sense, they are the most efficient form of nonrenewable energy. The amount of energy that can be released from a nuclear reaction of either uranium or plutonium is between 2 and 3 million times more powerful than what could be gained from burning any fossil fuel (ENS 2020). That said, many nuclear power plants only "burn" a small amount of the nuclear fuel they use in the reaction, leaving behind up to 98 percent of the radioactive fuel. Some of these spent fuels continue to be radioactive and dangerous to human and nonhuman life for tens of thousands of years. As no political entity in human history has ruled for that time span, it is unknown what will become of this dangerous material. Most communities do not want nuclear waste stored or transported in their areas. An accident is unlikely, but the effects of an accident can be and have been extremely dangerous to human life. The accidents that occurred at nuclear power plants, which have been discussed in previous chapters, have been intense and have killed many people. The Chernobyl nuclear accident destroyed a large area and made the villages within that area uninhabitable for the foreseeable future. That is a risk that most places and people would not take, given a choice.

For these reasons, nuclear energy has disparate futures depending on how much power the citizens of a country have and how deeply invested a country is in nuclear energy. In countries like Russia and China, which have either only a nominal democracy or no democracy at all, nuclear plants will continue to be built. The will of people is not part of the equation in these countries, so nuclear power can and will likely continue to be a larger part of their energy mix. In countries such as France and the United States that are deeply invested in their nuclear power plants, energy from nuclear power would be expected to continue to remain even. However, nuclear power plants retired due to the expiration of their safe life expectancy are unlikely to be replaced in the United States. This is due to local opposition and economics that favor renewable energy in most locations and natural gas power plants in other locations. The United States produces more nuclear energy than any other country as of 2023. France, which produces almost all of its power from nuclear, is likely to only phase out nuclear as the country increases renewables.

Germany and Japan both have constitutions that empower their citizens. These constitutions were heavily influenced by the United States after Germany and Japan were defeated in the Second World War. In the 1940s, the US government was dominated by a liberal democratic notion of democracy that protected the rights of citizens and expected government to be subservient to the will of the people. In these countries, citizen reactions to the 2011 Fukushima nuclear disaster in Japan have caused both countries to turn away from nuclear energy. From a CO_2/global warming standpoint, this step away from nuclear energy is a setback. However, the people of these countries feel safer without the threat of a nuclear accident.

Unless there is a new breakthrough in nuclear technology that offers guaranteed safety for the plants, it is unlikely that there will be a significant growth in nuclear power. Due to financial and safety concerns, it is unlikely that any capitalist democracies will be increasing their fleet of nuclear power reactors. Some countries will slowly decrease their number of nuclear plants, while others will do so relatively quickly. The increase in nuclear plants in autocratic and communist countries will most likely offset those reductions. The future for nuclear energy is likely to be a slow and small increase, at least with the political and economic conditions that exist in 2024.

Renewable Energy

Chapter 1 and Chapter 3 cover the past and the present of renewable energy, respectively. This section discusses the possible future of renewable energy. Hydropower is still the major player but has the least growth potential of all forms of renewable energy. Wind power produces the second most gigawatt hours of renewable energy and will continue to grow. Solar power produces the third-largest amount of renewable power, but it is growing at the fastest rate and will likely be the number one producer of renewable energy by 2040. Other forms of renewable energy, such as geothermal, tidal power, and wave power, have tremendous possibilities, but their future is uncertain due to the costs of geothermal and the nascent technologies of both tidal and wave power. Biomass power plants currently produce substantial energy, but there is little movement to tap into that potential and create new plants due to air pollution concerns. Overall, it is clear that all types of renewable energy combined will make up the majority of future electric energy production. The questions are: How much electricity will come from renewable energy? And, how soon will renewable energy surpass fossil fuels in electric energy production?

Hydropower currently makes up the majority of all renewable energy, but there is very little "low-hanging fruit" in the hydropower industry. The United States is predicted to remain at 100 GW a year into the foreseeable future. Dam decommissioning will be offset by small-scale hydropower and more efficient generators installed in older hydropower dams. There are a number of large projects being constructed in China, Pakistan, Ethiopia, Tajikistan, and Nigeria. Smaller but still significant projects are planned in Columbia, Angola, and India. These projects, as well as updates to older dams, will push hydropower from 14 percent of the total energy produced to 18 percent of energy produced by 2040. It is important to note that the IEA (International Energy Agency) has had a history of underestimating renewable energy growth. The World Energy Council (WEC) estimates that hydropower may double by 2050, due to the increasing scarcity of water (dams can both produce energy and control the flow of water) and also due to the continued push toward sustainable energy solutions for climate change. Hydropower will be used as a form of energy storage as well. Pumped hydro, using already existing hydropower dams, will greatly increase the reliability of both solar and wind. Dams can also slow the rise of seal levels. Both agencies agree that the vast majority of growth will come from the Asia Pacific region and the Central and South America region.

Wind power has been increasing across the globe at exponential rates. The International Renewable Energy Agency predicts that pace will continue and that by 2050 there will be 5044 GW of onshore wind power and 1,000 GW of offshore wind power, for a total of 6,044 GW. (The IEA agrees that this is a possible scenario but only under what they call a sustainable development scenario.) To put those numbers into perspective, today there are 651 GW of installed wind power capacity (counting both onshore and offshore). Wind power is less expensive every year, as companies streamline installation and production. New technology is always improving the efficiency and cost of wind turbines. Economies of scale also play a part in decreasing the costs of large projects. Like all renewable energy, wind power gives countries energy independence. They are no longer tied to the whims of energy markets. The US state of Texas has demonstrated that if energy markets are not hampered by government subsidies of fossil fuel energy production or political cronyism that entrenches outdated, dirty, and expensive fossil fuel power plants, then wind can and will win out on economics alone. As the world deals with the effects of climate change, wind's advantage will only continue to grow. Most models predict wind power to account for about 20 percent of the energy market in the near future.

Solar power is growing faster than any other form of energy. It can be installed anywhere that has sunshine. As solar panels become more efficient, solar energy is becoming a viable option in areas that do not have a lot of sun, as they can still produce electricity in cloudy weather. The International Renewable Energy Agency (IRENA) estimates that solar power will grow by 9 percent on average until 2050, when a full 25 percent of the world's power (8,519 GW) will come directly from the sun. One of the biggest reasons for the growth in solar PV (photovoltaics) is the dramatic drop in costs. Internationally, the price to install solar had dropped by over 70 percent on average between 2010 and 2018. These drops in costs are expected to continue. Solar also has the advantage of being able to produce efficiently and affordably at any scale. From rooftop solar to solar farms the size of small cities, solar power makes economic sense. While solar has the disadvantage of not producing in the dark, advances in energy storage are continually filling the gap at prices that decrease annually. New technology, such as solar windows that can be installed in homes and skyscrapers alike, will likely usher in a new era of solar energy that is not even currently accounted for in predictions.

It may be that in the future solar energy is so affordable that the thought of building a huge power plant to create electricity will seem as practical as burning wood or coal in a steam engine for transportation. Solar power is not limited to solar PV; concentrated solar power (CSP) is another form of solar that has been shown to create power that lasts long into the night. It will grow as well, but due to the size and expense of CSP, it is unlikely to grow at the projected rates of solar PV. Because solar is so inexpensive, cities, provinces/states, and countries are considering making solar PV mandatory on all new houses. The US state of California has already passed legislation that does exactly that. More governments will likely follow suit. The future of solar power remains bright.

Other renewables, such as geothermal, wave power, and tidal power, still have a long way to go, and some, like using biomass for electrical power generation, do not have much support. Geothermal looks like the perfect renewable energy: consistent, CO_2-free, nonpolluting, with a very small impact on the environment. However, it comes with some problems and limitations. In Chapter 3, we discussed that with current technology there were only a few countries that had ample geothermal energy (that is easily accessible). Geothermal energy is a niche market and even if it doubles in the next twenty years, it would only add up to 30 GW of carbon-free electric power capacity. The world had 3,869 GW of renewable energy capacity as of 2023 (IRENA 2024). Unless new technology makes geothermal less expensive (currently about $2 million per well dug) or a new method of extracting heat from the earth becomes widespread, geothermal will continue to be a powerful form of renewable energy mainly for the few countries that have easy access to it. However, geothermal home heating and cooling is available almost everywhere. Geothermal heat pumps (GHPs) use the constant temperature of the earth (30 ft underground) to both cool and heat a house using heat exchange systems (EERE 2024). These systems can be used in both hot and cold climates and significantly lower the energy needs of a home. They are long-term investments that can last up to twenty-five years, but the upfront costs have prevented them from becoming common. Governments could potentially create incentives for GHPs to decrease energy needs.

Tidal energy is a slow-growth area. There have been successful tidal power plants for decades, but few additional plants have been constructed due to limited areas that are ideal for tidal energy and the fact that many tidal areas are used for other purposes such as commercial shipping, fishing, leisure boating, and recreation. Wave energy is an emerging technology. There have been successful trials in Portugal and Scotland, though there are not any permanent large-scale wave power operations. Wave energy or wave power is a wildcard in the future energy mix. It is likely that some countries will take advantage of this enormous amount of energy. There is enough potential energy in waves to cover 100 percent of the world's energy needs. Until there is a dominant type of wave technology that is scalable, however, wave power will remain a bit player in the larger energy game.

Burning biomass was the cornerstone of human energy systems for millennia. Biomass is almost any organic material that can be burned: wood, plants, agricultural waste, gas from landfills, alcohol (derived from plants), black liquor (derived from trees in the paper-making process), some types of garbage, as well as human and animal waste (EIA 2020). Currently, methane from landfills is gathered and used, black liquor is burned by paper mills to produce their own electricity, and in some places trash is burned to create electricity. (This practice is decreasing due to pollution concerns.) One of the biggest opportunities that is not being seized is the potential to thin fire-prone forests and use that wood to create electricity. In Australia and the United States, fires during the 2019–20 fire season ravaged forests, killing many people and billions of animals. The two most prominent reasons for the severity of the fires are climate change causing droughts and forest management that allowed clear-cutting and suppressed natural fire cycles. This has painted forest managers into a corner because these unnatural forests can no longer handle natural fire cycles. One of the ways that this can be remedied is by thinning forests to the point that they can tolerate a natural fire regime once again. However, this is expensive work and there is currently little market for the wood. Increasing biomass power plants near these forests could change that. Biomass power plants already produce 2 percent of the electricity in the United States and 1 percent of the electricity in Australia.

According to the *2016 Billion-Ton Report* that was jointly written by the US Department of Energy and Oak Ridge National Laboratory, the United States could quadruple the amount of biomass energy produced. Barriers to growth in this area include the following: (1) with business as usual inertia, utilities do not want to convert to biomass plants unless they are forced to do so (Dubay 2020); (2) there is a fear that if the biomass market grows, farmers will convert to biomass/biofuels and grow less food; and (3) there is pushback on how renewable and clean biomass is due to the time needed to regrow forests and the particulate matter that can be released into the atmosphere from the burning of biomass, especially wood. Areas such as the Western United States could benefit from programs that use wood that will likely burn uncontrolled in a forest fire in the near future. Biomass power plants can be built to capture carbon and reduce particulate emissions. Australian fires are estimated to have increased the total CO_2 output of the country by over 50 percent. If some of this wood had been burned for electricity, it could have actually decreased Australia's CO_2 output for the year. However, this is an area that is unlikely to grow without government support. The private sector is already producing electricity with fossil fuels and is unlikely to switch without financial incentives.

Hypothetical Energy Sources

There are other types of energy that have not been proven. These energy types are theoretical, and some of them could be quite powerful. One type of energy that could completely change the energy systems of the earth is the same energy that the sun produces: fusion power. Nuclear fission is the splitting apart of atoms to create energy and is used in nuclear weapons and nuclear power. Fusion is the opposite process, fusing together two atoms. Stars (including our sun) undergo fusion reactions constantly, creating heat and light. The intense pressure and gravity of a star combines hydrogen atoms to create helium. The Holy Grail of renewable energy would be to replicate the process on earth in a controlled situation. As of 2024, few fusion experiments have created more energy than they have used, so it is currently an unlikely source of energy. Billions of dollars are being spent on experiments attempting to create fusion power. While many have shown promise, the technology is still nascent and will unlikely be available in the foreseeable future. If fusion power ever becomes a reality, it could solve all of the earth's energy problems.

Another breakthrough that could potentially completely change the energy field involves graphene. Graphene is a thin sheet of carbon that is only one-atom-thick. It has been shown to ripple and buckle in such a way that it can create limitless low-voltage power (McGowan 2021). Current configurations have not created enough energy to be useful, but if the process can be scaled it may be a pollution-free form of limitless energy.

The first uses will likely be handheld devices and watches. If it can be used for larger devices, this form of graphene may also be revolutionary.

Thoughts on Predictions

Predicting the future is fraught with difficulty, but we can make a series of assumptions based on the past that have a high probability of coming true. However, there are some energy fields that have been underestimated year after year. For example, the IEA predictions for the growth of renewable energy excluding hydropower have been so off target, it is almost comical. Many experts assumed that solar and wind installations would plummet due to the Covid-19 pandemic, but they were completely wrong. Worldwide renewable installations were up 45 percent in 2020 (Chappell 2021) in what looked to be the new normal. Prices and politics have combined to allow for a market that is favorable for renewable energy. As countries across Europe set up interconnectors (energy exchanges across international borders), and many countries start investing in smart grids (discussed in detail later in this chapter), this process will only continue to pick up speed. Renewable energy production will likely continue to outpace the predicted growth by experts due to falling prices and political shifts in supporting renewable energy.

Renewable energy and EV car ownership will likely continue to grow exponentially, and expert opinions will continue to underestimate the rate of the transition to renewable energy. The Biden administration earmarked $15.5 billion for EV production loans and grants, along with monies for EV components. As countries across the world look to securing domestic production of EVs, this will increase innovation and competition, lowering the costs of EV production.

POSSIBLE MAJOR TRANSITIONS

The Second World War–Level Mobilization

The world has seen changes in its economy that are even more dramatic than what would be needed to avert the worst effects of climate change. The biggest and most recent example of this took place during the Second World War, when, for example, the United States stopped producing cars, commercial planes, and leisure vehicles of any kind. Instead, manufacturers were required to build weapons of war. A resulting synergy developed between capitalism and socialism. The means of production (the factories) were still owned by the private sector, but the market was guaranteed by government contracts. Owners of private businesses and stockholders made more money than before the war as government contracts worth billions of dollars greatly increased production. All of the major auto manufacturers had been unionized by 1941, leading to high-paying jobs for all the new employees who flocked to Detroit from Appalachian areas and the Deep South. These new workers were able to support a family on one salary. Those who had two people working (the Second World War changed prohibitions on women in the factories, temporarily) were able to pay their bills and save for their future. Health care was provided by their employers, as well as retirement pensions. This was a real win-win situation. Owners, employees, and the government were able to achieve their objectives. The United States won the war thanks to unprecedented production. Strong union jobs became the backbone of the rising US middle class. This led to a boom in homeownership, car ownership, luxury purchases (at the time) such as washer and dryers, and the family vacation.

However, not all Americans benefited during this period. Black families saw a system that was built on racism exclude them from homeownership and life in the suburbs. Government regulation on newly built communities explicitly banned non-white families from owning homes in these communities. The practice of redlining banned banks from lending to families in traditionally Black neighborhoods. This depressed prices and only allowed for cash buyers. Soon integrated neighborhoods became Black only; the residents were Black, but the homes were owned mostly by whites. The next government investment in its people and its environment can work to fix these wounds from the past.

The idea of investing heavily in a transformation from the carbon economy to a net-zero economy is not new. It has been called the Green New Deal in a nod to the programs of the New Deal in the 1930s that set out to employ jobless Americans and to build great things: hydropower dams, roads, highways, bridges, and trails. The Green New Deal has been referred to as a climate plan for environmental justice. The European Union calls their plan the European Green Deal, while China calls their plan to be carbon-neutral by 2060 the Green Revolution. For simplicity's sake, we will call it the Green Energy Transition (GET). The GET will require the same energy and effort that was necessary to win the Second World War. To unlock the power of enterprise and ensure that all sectors of society will benefit, governments must lead. The first thing that will need to be done is the ending of government subsidies for carbon extraction of all types. The second necessity will be creating an international carbon market that can adapt quickly to changes in demand for carbon credits. All money made in the purchasing of carbon credits must be used to remove carbon from the air and sea. This would be much less invasive than the Second World War–era control over industries and would allow companies to use their creativity and flexibility to adapt to a quickly changing energy market. US government oversight can be used to be sure that new jobs in emerging energy markets will be catered to traditionally underrepresented minority groups. This can be done in order to remedy past programs that destroyed the opportunity to create generational wealth. While this plan is not a panacea that can fix all wounds and completely transform the energy market overnight, it outlines the first three steps that are needed to set the world on a path to the Green Energy Transition.

Home Generation

Solar PV panels have become more affordable than ever before. In most countries and in most parts of the United States, the monthly payment on a loan for a new solar power system is equal to or lower than the monthly savings in electric bills. This means that installing home solar does not cost any "out-of-pocket" money in many cases. The monthly spending of a family may actually go down with solar. This can be true with wind in many areas as well. Increased and continued government subsidies in the area of home generation should push this quickly growing market into the mainstream. Energy-efficient homes produce more energy than it uses. This energy can be used to charge an electric vehicle, stored in a battery system, and/or sent to neighbors who need it. (We will cover this in detail in the next section on smart grids.)

The three biggest obstacles to creating a net-zero home, a home that produces as much electricity (renewable) as it uses, are cost, storage, and integration with a utility. The problem with cost has all but disappeared. Home solar and home wind is affordable (see Chapter 3). The only trouble is it is not always easy to find financing. That too is changing. Many smaller banks have filled the financing gap. Banks from the Pacific Northwest are financing solar home projects all across the United States. More banks may follow as they realize what a safe and profitable investment it is to finance home solar PV power, home wind power, and home battery storage systems. Home wind power and home solar power are cost-effective, with some systems actually paying for themselves in seven years and continuing to provide "free" electricity for decades afterward. Electric battery storage is not yet as affordable. The payback on battery storage can be much longer (6–16 years). That will change in the near future, as battery prices have dropped precipitously and should continue to do so.

As battery storage becomes affordable, many people will opt to have battery storage in their home in case of emergencies. They will know that they will have power even if the power lines to their home go down. These batteries can also be linked to the larger grid and used to balance the energy needs of the entire community. EVs can be used in this way as well. Their large batteries (along with home energy storage batteries) can be linked to the grid and could eventually replace peaker plants (power plants that supply energy during times of peak demand). These batteries could provide energy to the grid during peak power demand and recharge at times of lower demand. Utilities are learning how to do this while still making money. They reduce their overall costs by paying customers who supply energy at lower rates than it costs peaker plants to produce energy.

As with many of the solutions that are covered here, these solutions are available and working. They only need the help of governments to push them into the mainstream. The United States has moved forward on subsidizing renewables and battery storage with the 2021 Bipartisan Infrastructure Bill and the 2022 Inflation Reduction Act (the most significant legislation on climate in US history). Home electricity generation is especially appealing, as it does not have any impact on the natural world. Built homes are ideal for solar PV. Small wind turbines are generally safe for birds and less of a threat than outdoor cats, which kill 2.4 billion birds annually in the United States alone (American Bird Conservancy 2021). Together, home solar, wind, and storage will be fundamental to the Green Energy Transition (GET), and it will only take a little incentive from government agencies to expedite that transition in time to do the most good possible in the struggle to avert the worst effects of climate change.

Smart Grids

"The grid" refers to electrical systems that stretch across cities, states, and sometimes countries. The grid in the United States is made up of thousands of generating units, hundreds of thousands of miles of transmission lines, and millions of homes and businesses. This patchwork system arose from early electrification in the 1890s. A smart grid would be a system that communicates between providers of electricity, users of electricity, as well as to home energy generation units (rooftop solar and backyard wind) and to EVs, which can both store electricity and release it back into the grid.

A smart grid would replace the current grid and make it more efficient and able to handle more dispersed forms of energy production and storage. As the world transitions to more home energy production and more EVs, a smart grid will become a necessity. It could reduce peak demand through smart systems that can charge during off hours, and a smart grid system would be better able to deal with interruptions in services and have local control that would be able to keep vital services running in an emergency. (For more information on smart grids from the US Department of Energy, go to smartgrid.gov.)

Electric Vehicles

Electric vehicles (EVs) are on pace to replace internal combustion engine (ICE) vehicles. The sales of EVs are growing at exponential rates. For example, EV sales in the United States were 17,000 cars in 2011 and 918,500 in 2022. That is a growth of 5,403 percent in eleven years. (Sales in the United States will be well over a million in 2023.) Worldwide growth in EVs has had a similar trajectory. Sales in 2019 were 2.2 million cars even though sales were depressed due to the Covid-19 outbreak. China is the biggest market for EVs and also took the biggest hit to sales in 2019 as the virus was already in play.

However, sales in China are a quickly growing percentage of vehicles with 31 percent in 2023. EV sales are a smaller percentage of total car sales in the United States at 8 percent in 2023. In Europe, that percentage is growing quickly for the entire continent; sales of plug-in electric vehicles (EVs and plugin electric hybrids or PEVs) were 22 percent of all vehicles sold. Individual European countries are doing even better as some countries have turned the corner from ICE vehicles to EVs: in Norway PEVs sales were 90 percent of all vehicles sold in 2023. Sweden is following close behind as ICE vehicle sales slipped below 58 percent in 2023. The trends are obvious. The speed will depend once again on government subsidies of oil extraction and on EV purchases. Government incentives can either slow the transition or accelerate it.

The rate of growth of EVs is many times the rate of growth of car ownership in the twentieth century or smartphones in the twenty-first century. Governments did not subsidize the horse and buggy market with which the new automobiles competed, nor did governments subsidize simple cell phones with which smartphones competed. In both of these comparisons, new technology was superior and quickly took over the markets despite the increased costs of ownership. The EV market differs in that, while upfront costs might be higher than for an ICE vehicle, the long-term ownership costs of the EV are usually less, due to the fact that electric charging is significantly less expensive than gasoline/petrol.

EVs have many advantages over their ICE competition. They accelerate faster because all power is available to an EV immediately. ICE vehicles are lighting gasoline on fire to produce motion, and that process takes a little longer. EVs are simpler machines. They have fewer parts, including parts that need constant replacement like oil, belts, fuel filters, spark plugs, and mufflers. The biggest expense on an EV would be the loss of a battery, as that can be very expensive. However, batteries are built to last longer each year with most warrantied for 8–10 years or up to 150,000 miles.

EVs used to take a long time to charge and charging stations were not as widely available, leading to a condition called range anxiety. People were afraid that their car would run out of charge and that they would be stranded on the road. That is no longer the case. Rapid chargers can charge up to 80 percent of a battery in less than 20 minutes. In this case, a person can stop at a charging station, go to the bathroom, pick up a snack, and be back on the road. The number of charging stations is also increasing across the world. For most people, charging is done mainly at home and at work. For daily commutes, charging stations are not needed, but they are needed for longer road trips.

Norway is an example of how a country that has profited from oil sales can turn around and transform into an EV leader, with most cars running on renewable energy. Norway had a state-owned oil company named Statoil, now merged with Norsk Hydro and called Equinor, and the Norwegian government still owns a 67 percent stake. Funds from this source help Norway to offer generous subsidies to EV customers. Norway is also powered by hydropower with some wind power and a little gas power. Norway could be the example for all petrol nations to replicate. If nations that produce a lot of oil can use some of that revenue to create incentives for EV purchases, countries may be able to ease the transition before the inevitable decline and crash of the petroleum extraction and export market.

Carbon Markets

The idea of carbon markets has existed for a long time. There are many successful examples of carbon markets. Most use the cap-and-trade system. In this system, there are a set amount of carbon credits that are sold on auction each year. They can be traded and sold on an open market. Each year the cap limits decrease. The difficulty is setting the correct cap limits on CO_2. If the limits are set too high, it can limit business's ability to thrive and provide vital services like electricity. If the limits are set too low, the auction can be a failure by raising less money than anticipated and by not decreasing CO_2 emissions as much as possible. Despite the tricky analysis and decision regarding cap limits, the cap-and-trade method has been extremely successful.

The Regional Greenhouse Gas Initiative (known as RGGI, pronounced Reggie) is a market-based initiative that includes twelve states in the Northeast part of the United States: Maine, Vermont, New Hampshire, New York, Connecticut, Massachusetts, Rhode Island, New Jersey, Delaware, Maryland, Virginia, and Pennsylvania. There is a regional cap set on CO_2 emissions and the cap decreases by 2.5 percent annually. According to the Acadia Center's *The Regional Greenhouse Gas Initiative: 10 Years in Review*, 90 percent of these allowances (or carbon credits) are distributed through quarterly auctions. CO_2 emissions from the power plants that RGGI applies to have decreased by 47 percent, which is 90 percent better than the rest of the United States. Electricity prices have actually decreased by 5.7 percent, while the rest of the country (not including California, which has its own program) saw increases on average of 8.6 percent. The GDP of RGGI states has increased by 47 percent, which is 31 percent more than the rest of the country. Auctions have raised more than $3.2 billion, which is then invested in renewable energy and energy efficiency. The reduction in co-pollutant emissions has produced over $5.7 billion in productivity and health benefits (Acadia Center 2020). The RGGI system is clearly producing real world results. It demonstrates the power of political will and using coalitions to create functional cap-and-trade systems to improve not only the environment but public health as well.

California has its own cap-and-trade system for selling carbon credits. It has been enormously successful. The three main goals of the program are (1) to reduce CO_2 emissions to 1990 levels by 2020, to be 40 percent below 1990 levels by 2030, and to be 80 percent below 1990 levels by 2050 (overlapping with the state's goal of being carbon neutral by 2045); (2) to raise money to reduce GHG emissions through California's Greenhouse Gas Reduction Fund; and (3) to allocate 35 percent of the funds raised to low-income communities and environmentally disadvantaged communities. California achieved its "CO_2 emissions to 1990 levels" goal four years early, in 2016 (C2ES 2021). Figure 4.1 demonstrates that California has completely decoupled GHG emissions and GDP (gross domestic product). This means that California is growing its economy as it decreases its emissions. This is important because there has been a (now debunked) claim that an economy cannot grow without increasing emissions. California is excelling at both economic growth and GHG decline.

California's system is different from the RGGI program in that it includes multiple sectors of the economy, not just electricity production. California also works with the Canadian province of Québec to jointly auction emissions credits. (There are also a number

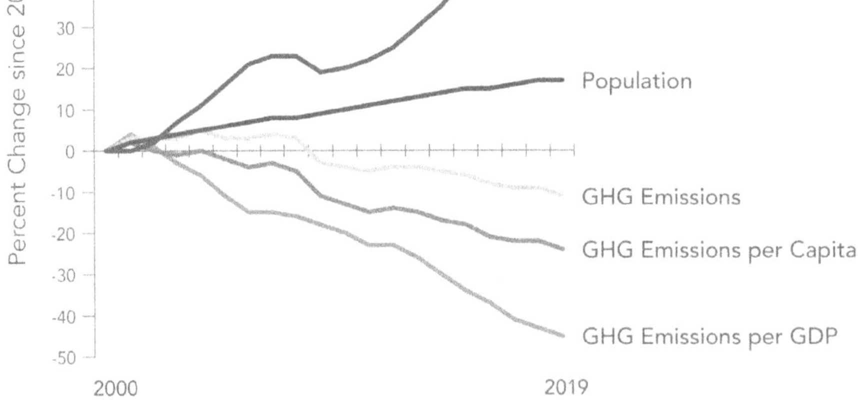

FIGURE 4.1 Change in California GDP, population, and GHG per capita emissions, 2000–19. *Source*: California Air Resources Board, 2020, https://ww2.arb.ca.gov/ghg-inventory-graphs.

of free credits provided for certain sectors that vary depending on the size of the business.) A goal of this joint effort is to continue to add other states and provinces to the program, allowing states and provinces that are ahead of their respective federal governments on addressing climate change to move forward with the process of reducing and eventually ending the majority of GHG emissions.

The European Union Emissions Trading System (EU ETS) has also been tremendously successful. The EU ETS achieved its 2020 goal of a 21 percent reduction of 2005 emissions six years early in 2014. It operates similarly to the California-Québec plan in that it is a mix of free credits (to grandfathered businesses) and credits that are auctioned to future projects. While this has led to criticism, the overall success of the program demonstrates its effectiveness. The EU ETS is the largest GHG emission trading program in the world. It includes all twenty-seven EU member states as well as Norway, the United Kingdom, Iceland, and Liechtenstein. The program regulates 45 percent of all the European Union's emissions and covers over 11,000 heavy-energy-using installations (European Commission 2021). Cap and trade works on the idea that each company must surrender the number of credits needed to cover their emissions for the year. If they use less, they can sell those credits to those that need more. This process allows for reductions of GHGs where they are least expensive. This also invites innovative solutions into the market. If a manufacturer can reduce GHG emissions at less than the market rate of the credits, it can actually turn a profit selling its excess credits. Many companies have done so, and while that in turn can lead to criticism, the overall success of the system cannot be denied.

The EU ETS provides funding for the Innovation Fund through the auctioning of carbon credits. The Innovation Fund supplies funding for programs that create and prove low-carbon technologies. It is the largest such fund in the world. NER300, the predecessor of the Innovation Fund, saw mixed success with carbon capture technologies. There were technological advances demonstrated by patents approved, but much of the new technology has not yet been applied to reducing emissions. Overall, the money from this cap-and-trade program is likely to lead to more innovation in an area (such as carbon capture) that is not yet profitable.

Any carbon market works better than no carbon market. There is inherent difficulty in creating a carbon market in that if it is too strict it can stifle business growth and lead to the loss of jobs. There is also the possibility that a carbon market that is too lenient might lead to companies polluting more than they would in a tighter market. Many of these carbon markets are created through state, national, or international governing bodies, and governments are not usually nimble. A market that can quickly adjust to new technologies and dropping demand for carbon credits will be the most successful. Every carbon market has been criticized by all sides, and new markets will be as well. Pro-business groups believe that carbon markets might suffocate businesses and add unnecessary burdens on companies. Pro-environment groups argue that there should be no free credits and that the number of credits should decrease precipitously every year. This creates a difficult situation, but these markets have succeeded regardless. They have demonstrated not only that it is possible to regulate carbon, but that goals should be higher than they are. In every case, the demand for carbon credits has decreased faster than estimated. This demonstrates that companies are able to use other energy sources; all they need is the proper motivation.

Market Disruptions and Transitions

Transitions are a messy business. There are winners and losers. There are those that make the transition from one technology to the next, but many times it is the upstart businesses that are the big winners. A classic example is the shift from horse-drawn carriages to automobiles. Some companies transitioned with varying levels of success; the Studebaker company is an example of one that survived (until 1967). The company had flourished as a builder of wagons and carriages, and it quickly pivoted to successfully producing cars. This is an exception to the rule. Most companies turn to the government to slow or stop transition to new technologies. Today, coal companies ask the government for more financial support to prop up an energy system that is quickly becoming obsolete. Oil companies asked for loans and grants to continue operation as oil prices collapsed in 2020. ICE (internal combustion engine) automobile manufacturers ask for countries to roll back miles per gallon (mpg) regulations because they fear (rightfully) that they will have a disadvantage when competing with EVs. Renewables will replace fossil fuels, and EVs and fuel cell vehicles will replace ICE vehicles. It is just a matter of when. Cell phones have generally replaced landlines, but landlines still exist. It may take another fifteen years until they are about as common as a horse and buggy. Let us take a look at four market transitions that have already begun.

Energy is changing rapidly. Coal is leveling out and will soon be diminishing. There is a slight uptick in nuclear energy worldwide. Oil has likely reached peak demand. (That could change temporarily, but long-term demand will continue to decrease.) Natural gas has been increasing, hailed as a gap fuel for the renewable energy transition. Solar and wind continue to grow exponentially. There are also nascent energy technologies, such as fusion, that could one day upend all these systems. What is clear today is that fossil fuels are on a path to obsolescence, while renewables are set to eventually dominate energy for both electricity production and transportation.

Coal is going away in most Western countries. Both the European Union and the United States are on track to use no coal by 2040. Climate Analytics estimates that all European coal plants will be economically unfeasible by 2031. Germany and Poland will have the hardest time making this transition. The United States has halved its use of coal since 2000. It is likely that coal will become a niche fuel by 2040. Without government intervention, it is unlikely that any coal plants will be able to produce electricity using coal past 2040. Coal is generally more expensive than solar, wind, and gas plants. Once carbon markets are put into play, coal will become ridiculously expensive compared to the other options. Countries that rely on coal exports need to be ready for this change. The large decreases in coal use in 2019 and 2020 were countered by a rebound after the Covid-19 pandemic receded. However, it is unlikely that coal will ever again regain its position, producing the majority of electricity.

Oil companies are still producing incredible amounts of crude oil, 2019 appeared to be the peak year of consumption. Oil demand increased to record numbers in 2023, but demand appears to be plateauing. The growth of EVs is likely to blunt any increased demand for oil. ExxonMobil has long been the most valuable energy company in the United States, but that changed in 2020 when NextEra Energy, an electricity power company that is transitioning to renewable energy, became the most valuable energy company in the country. This is part of an ongoing trend. Shell, Total, and BP are three traditional oil companies based in the Netherlands, France, and Great Britain, respectively. They are all buying renewable energy companies in an effort to survive the transition from fossil fuels to renewables.

EVs are part of this transition to renewable energy. Their sales are growing exponentially, as discussed earlier in this book. European sales have crossed tipping points in many countries and have become the majority of new car sales in some countries already. This is especially true in Norway. EV market share of new car sales doubled in Europe and then doubled again. This is a process that will repeat itself because EVs are better technology. Most new technologies hit a tipping point in which the old technology no longer appears competitive to the consumer and sales catapult from the old tech to the new tech. This process is likely to occur across most markets by 2030.

Transportation outside of personal vehicles is likely to transition at a slower rate. Cities buy whole fleets and create public transportation that takes decades to build and decades to replace. Despite the pace, change is still coming. Electric trains have already replaced diesel trains in many countries. Cities across the globe are moving toward hybrid, electric, or fuel cell buses. China is way ahead of every other nation on electric buses with over 400,000, which is 99 percent of all electric buses on the planet. China's high adoption rate, due to government incentives, means that the rest of the world will be able to pay much less for electric buses now that they are made on such a large scale. It is estimated that up to half of the world's bus fleets will be electric by 2025.

Heating and cooling is an area that is ripe for big shifts, but it will depend on how governments regulate and incentivize. The coolants used in some refrigerators and air conditioners are hydrofluorocarbons (HFCs). A remarkable international deal was struck in October 2016 (built on the Montreal Protocol) that will phase out all uses of HFCs by 2030. High-income countries began phasing them out in 2019. It is imperative that countries find ways to properly dispose of these old systems so that HFCs do not end up in the atmosphere, where they are extremely potent GHGs. If this is done correctly, governments have a fairly simple way of reducing the equivalent of 90 gigatons of CO_2 (CCAC 2024).

Heating and cooling through home geothermal systems can significantly decrease the need for both air conditioning and heating. Governments have the opportunity to incentivize homebuyers and homebuilders to install these systems. Geothermal systems are especially applicable to large public or private buildings and multifamily structures. Currently a small niche market, these systems could become mainstream with a nudge from government incentive programs, potentially reducing heating and cooling costs by over 50 percent.

Landfill Mining and Thermal Plasma Gasification

Landfill mining is a process in which old landfills are mined for recyclable materials and the remaining contents burned in the thermal plasma gasification process. This process can convert up to 99.99 percent of trash into energy (after removing recyclables). By using a plasma arc that can reach temperatures of over 25,000°F, most matter is broken down into its basic elemental form. It is split into syngas, which can be burned to create electricity, and slag, which can be used in construction and manufacturing. Metals can also be recovered in this process, depending on the temperature used. A number of countries around the world successfully use this system. However, as it is expensive and complicated to correctly implement the system, many other attempted plants have failed.

Burning trash to create electricity has existed for decades, but most of those types of plants have been closed down due to the heavy pollution created, not only CO_2 but also dangerous pollutants like lead, mercury, dioxin, and small particles that can become lodged in the lungs. Some studies have demonstrated that more modern gasification plants are not able to provide the energy promised, they pollute more than they should, and they cannot dispose of the amount of waste that they claim (Tangri and Wilson 2017). Other research has focused on the improvements over classic trash-to-fuel technology and the long-term possible benefits of waste gasification using thermal plasma (Fabry et al. 2013). The cost and benefits of these programs are going to depend on a number of factors: How much land does a country have to continue to bury waste? Can a country decrease waste through reusing, recycling, and composting to a point at which the remaining material would not be sufficient to run a thermal plasma gasification plant? Are the potential costs of this newer technology worth the possible tradeoffs: less landfill but potentially more pollution?

Thermal plasma gasification is a technology that is still moving forward in fits and starts. If it can be perfected, there is almost limitless fuel in the form of buried trash all over the world. Landfill mining can also greatly decrease the need to mine virgin lands. In a world that values the benefits of undisturbed ecosystems, landfill mining may be a more practical approach compared to strip mining. However, like most new technologies, it will take substantial investment by private and public entities to get it to the point in which landfill mining combined with thermal plasma gasification is both efficient and clean.

Hydrogen Fuel Cells

Hydrogen fuel cells are promising technology that has been hampered by the fact that hydrogen is abundant yet difficult to obtain in a pure state. Hydrogen exists in both air and water (H_2O). Separating water into hydrogen and oxygen is difficult and takes energy. Most hydrogen is actually derived from natural gas because that is the easiest and the cheapest way to obtain it. This process makes hydrogen and hydrogen fuel cells not renewable at all, from a climate change perspective.

However, new technology is changing this. Advances in electrolysis to split water into hydrogen and oxygen using renewable energy could prove to catapult hydrogen into becoming the renewable energy source of the future. If the price of this technology can be lowered to the point that it is competitive with gasoline and/or EVs it may become extremely popular. The advantage in a hydrogen fuel cell–run vehicle is that it can run like a combustion engine and be refilled at a station. Hydrogen is also very light, so it is easy to store enough fuel on a truck, or a boat, or a plane. Hydrogen fuel cells can also be used to fuel large ships, which are currently one of the most polluting forms of transportation;

they use the dirtiest form of oil because there are no air pollution rules on the open sea and because these heavy oils are very cheap. Hydrogen fuel cells could also be used to fuel airplanes. That technology is newer, and no fuel cell planes bigger than six-seaters have been built yet.

Hydrogen fuel cells are less efficient than EVs in a well-to-wheels process comparison. EVs use about 60 percent of the energy they receive from the grid; fuel cell vehicles use about 40 percent of the energy needed to produce, store, and convert hydrogen; and gasoline or petrol engines use about 20 percent of the energy from the well-to-wheel process (CAFCP 2016). This puts fuel cells right between gasoline vehicles and EVs. However, fuel cells face the classic problems for new technologies: infrastructure is not yet completed and economies of scale have not begun to decrease costs. So, fuel cells remain rare and expensive. If governments invest in hydrogen infrastructure, those costs will drop significantly. Advances in hydrogen technology will also improve efficiency while dropping costs.

Hydrogen fuel cell vehicles that are run with hydrogen harvested from water using renewable energy are the cleanest vehicles in the world. Their only output is clean water. It will take serious investment by both government and business to make hydrogen fuel cell technology available to the common person. It does appear that hydrogen will be part of the renewable energy economy of the future. Hydrogen will be able to fill in the gaps that EV technology cannot.

Nuclear- or Renewable-Powered Carbon Capture

Nuclear-powered carbon capture would be a low-level form of geoengineering. One advantage of using nuclear energy for this purpose is that the nuclear plant would not need to be near any large population centers, thereby potentially overcoming a big obstacle to new nuclear plants: fear of a disaster that contaminates a heavily populated area. The success of nuclear plants that pull carbon out of the atmosphere would depend on lower nuclear prices and international funding of carbon capture.

It is more likely that renewable energy would be used for this process at current prices. Renewables are much cheaper and using them in places that are far from population centers is a good use of renewable energy that would be expensive to bring to city centers. There are many sunny and/or windy areas that are inhospitable to human settlement. These areas (if there are no imperiled species) can be used to produce electricity for capturing carbon. The intermittency of wind and solar would be irrelevant in this situation, as a carbon capture plant does not need to run continuously. Once again, establishment of this kind of plant will depend on international funding of carbon capture. As it becomes clear that the world needs to take more extreme measures to avoid catastrophic temperature increases, governments may pursue more of these relatively low-risk geoengineering projects.

True Valuations for Ecosystem Services

In order to transition from a business-as-usual framework toward an economic system that accounts for the true value of ecosystem services, major changes to how a government manages its land must be made. To better understand this idea, it is important to think about what a properly functioning ecosystem provides: clean air and oxygen through trees and plants; clean water that is filtered through forests, streams, creeks, and rivers; lumber and game or land for crops and cattle. Private landowners value some of these products depending on what they are currently using their land for, and they are often dependent on the surrounding forests and mountains for the water they need. Government lands are not always treated with the same respect that people would treat their own lands.

This is because governments do not always value ecosystem benefits. When these benefits disappear due to overgrazing, clear-cutting, or mountaintop removal mining, other people pay the costs. Farmers down river may lose clean water. Cities may have to pay for new systems to clean dirty water or build new wells. People may have to deal with dirtier air and resulting detrimental health effects. Clear-cutting in some areas will eventually lead to deadly and intense forest fires decades later. (Many drier forests need a structure that has large old trees mixed in with smaller newer growth as well as periodic low-intensity fires in order to renew the forests and keep them healthy.) The desire to promote business opportunities on public lands has had deleterious consequences. Mismanagement of these areas is largely due to the fact that the lands are valued only for what can be extracted from them. A system that truly values ecosystem services can change this.

Private landowners understand the immediate value of their lands. Grazing fees are a classic example. An Animal Unit Month (AUM) is a measurement for estimating the carrying capacity of grazing lands based (roughly speaking) on the amount of food one animal would eat in a month in that area. Private landowners charge an average of $21.30 per AUM for grazing land in the Western United States, with a range of $10–$46 depending on the area. The US federal government, however, via the Bureau of Land Management (BLM), charges only $1.35 per AUM. BLM lands also tend to be overgrazed because the BLM does not have the money to monitor grazing and keep land up to its own standards. This is because politicians want to appease rural districts. The ranchers might do well in this scenario, but the public does not. These damaged lands can no longer provide clean water, the correct quantity and quality of vegetation, nor healthy soils. If the value of ecosystem services can be included in future BLM grazing leases, this can be remedied.

Hard rock mining in the United States produces more toxic waste than any other industry in the country according to the Environmental Protection Agency (EPA). The General Mining Act of 1872 governs all mining on federal lands. This act allows companies to mine for free. When commodity prices drop or mining companies stop turning a profit, the US taxpayer is stuck paying the bill to clean up thousands of abandoned mines. The original act of 1872 was written when much mining was done by hand. Both Democrats and Republicans have fought against any update to this law. Companies both foreign and domestic profit as the environment is destroyed. The projected costs of cleanup of just sixty-one older mines are between $15 and $20 billion. The EPA is responsible for between $465 and $613 billion in liabilities overall. Many of these from companies that have dissolved or gone bankrupt rather than pay the cost of the clean-up (GAO 2023). Presumably, cleaning up all hard rock mines would cost in the hundreds of billions of dollars. The communities that thrived off of the ecosystem services of clean water, clean land, and clean air must find a way to pay for remediation or the areas may be left uninhabitable. This is a problem all over the world. The poorer the affected community, the less likely the people will be able to either remedy the situation or move, which can lead to people continuing to live in places with increased cancer rates and diseases due to dirty water, land, and air. A functional system would require mining companies to pay up-front for the expected costs of remediation. The US administration of President Barack Obama did begin to require companies to guarantee cleanup with funds or bonds. This requirement was removed by the Trump administration. The Biden administration has a number of proposals to overhaul the mining laws in the United States, but they are unlikely to pass the Republican-controlled US House of Representatives.

True valuation for ecosystem services must be included in the future if any country hopes to have functioning ecosystem services to provide for their people. As the word population grows and wilderness shrinks, this will become even more of an imperative. It will once again depend on political will to avoid these troubles becoming a complete catastrophe. There are areas that have already begun to value these services and have spent the money necessary to preserve them for the future. Once countries codify the value of ecosystem services through laws and regulations, it will be easier to preserve these systems.

Internalization of Environmental Externalities

When a company makes a business plan, they look at all their costs: labor, buildings, supplies, and the like. They do not calculate the cost of the carbon they will produce or the trash they will generate. In some places, they do not pay for pollution put directly into the ground or into waterways. These costs are described as externalities, as they are external (outside) to the company's scope and finances. These externalities are paid for by citizens through taxes to their government or by suffering the effects caused by these externalities.

For example, when a coal power plant opens up in a neighborhood, it can cause up to fifty additional deaths each year due to air pollution, depending on its size (Penny, Bell, and Balbus 2009). It can also cause asthma, neurological problems, cancer, and heart and lung ailments. The community bears these costs directly in the form of lost wages, medical expenses, and the loss of loved ones. The closer a person lives to a coal power plant, the more likely that they will pay with their health or their life. The government may choose to clean up pollutants or remove CO_2 from the air. These costs are paid by the greater community through taxes raised to pay for remediation. Externalities are not part of the economic decision to build another coal power plant, but they are costs that have to be paid by someone, costs caused directly by the coal plant. So, unless a society forces business to internalize these negative environmental externalities, businesses will continue to be incentivized to pollute and operate unsustainably. Internalizing these costs would quickly change corporate behavior in a manner that benefited the environment.

If an economic system internalizes environmental costs into the coal power plant, those costs have to be addressed in the business plan. Some of the questions that have to be answered in calculating the cost of externalities include: How much will the additional medical care needed by citizens that live near the plant cost? How much will the plant have to pay to relatives of those who die? How much will it cost to remove particulates from the air? How much will it cost to remove the equivalent amount of CO_2 from the atmosphere? These environmental costs will most likely lead companies toward moving the plant away from populations, as it would be less expensive to add transmission lines than pay for health care and human life lost. This system would create a fair playing field for alternative energy. It would end an invisible subsidy to polluting industries. The only downside is that it would raise the upfront costs of polluting products (in this example, energy). However, citizens, through their taxes, are already paying these costs. In countries that have socialized medicine (all advanced economies except the United States), the costs are equally distributed monetarily. In the United States, both the physical and financial costs will fall heavily on those that live closer to the plant.

The internalization of environmental externalities is something that all energy providers need to consider. For example, a hydropower dam might have to consider the costs of lost salmon runs or the cost to create systems to allow a salmon bypass. Wind power would have to consider the costs of birds lost to the ecosystem. Solar power plant would have to consider the costs of lost habitat. These evaluations would most likely change the behavior of companies producing all types of energy. Coal and gas plants would spend more to clean up emissions and build away from population areas. Hydropower dams would include salmon ladders and other devices to allow rivers to continue to produce fish in similar numbers. Wind power would locate wind turbines away from known bird migration routes. Some of these things are already happening, due to environmental regulation, but others are not. The internalization of environmental externalities will take care of problems as they are created, instead of the current practice of allowing environmental costs to go unpaid until governments are forced to address them all at once.

Benefit Corporations and Corporations for Renewable Energy

Benefit corporations or public-benefit corporations (PBC) are simply corporations that balance the best interest of the corporation (i.e., profits) with the best interest of workers, the environment, the community where they are located, and society at large. This is a legal definition, which is important. If a CEO or a board of directors makes a decision in a traditional C corporation (most major corporations are C corporations) that loses money for the corporation but benefits the environment, they could be sued by shareholders for not working for the best interest of the corporation. Benefit corporations (sometimes called B corporations) are able to make decisions that balance the whole needs of the people, the community, the environment, and the corporation.

PBCs are able to incorporate what is sometimes called the triple bottom line of sustainability: people, planet, and profit. While doing right to the environment and to workers usually also helps profit, it does not always. Transforming a C corporation to PBC creates the legal cover for corporate leaders to work toward creating a better world. It is also a way to hold corporate leaders accountable. PBCs make up a very small percentage of corporations, but laws in thirty-seven states and the District of Columbia allow companies to become designated as a benefit corporation or PBC (Stracqualursi 2017). Patagonia, Kickstarter, and King Arthur's Flour are among the more well-known PBCs. Certified B corporations are also interested in the same balance of people, planet, and profit, but their status is based on an annual certification through a private company instead of a legal definition through registration with a state government. Some famous B corporations are Ben and Jerry's, Kleen Kanteen, and New Belgium Brewing. These new types of corporations are working to benefit more than just stockholders. They could become a model for capitalist power being used for the good of society. These companies are also more likely to use renewable energy and embrace employee ownership programs. (For a long-standing example of an employee-owned corporation, look to the Spanish corporation Mondragon.)

RE100 is a program through which corporations pledge to obtain 100 percent of their energy from renewable resources. This program has seen huge success, in large part because many corporations have discovered that using renewable energy is both a public relations bonanza and a money saver. Corporations as large as GM, 3M, Adobe, Budweiser, and Ikea are a part of this pledge. Many companies have already achieved 100 percent renewable energy. As of 2024, there were 426 members of RE100 accounting over 500 TWh (terawatt hours) a year. (For current membership and compliance, see https://www.there100.org/.)

Corporations are sometimes blamed for the worst and most destructive practices in the battle over energy resources. Corporations are chartered by governments and controlled by people. Governments can put the guardrails on corporations, or, in the case of PBCs, corporations can put the guardrails on themselves. In these ways, corporations can be a positive force in transitions toward renewable energy as the world faces the existential crises of climate change.

New Food Systems

A discussion of energy resource conflicts would not be complete without the inclusion of the human body's sources of energy: food and the systems that produce that food. While energy of all types accounts for 72 percent of GHG emissions, agriculture is responsible for 11 percent and land use change and forestry accounts for 6 percent. Most of land use change involves forests that are cut down to create grazing pastures for cattle.

Approximately 14.5 percent of all GHGs are from the livestock sector (Gerber et al. 2013), as livestock requires both land and agricultural products. The term "livestock" includes all animals raised for meat, milk, and eggs. Some of the most common livestock animals are cattle, pigs, sheep, goats, and chickens. Cattle raised for meat and milk are the largest contributor of GHGs, accounting for 77 percent of all livestock emissions. For comparison, pigs and poultry combine for only 10 percent of all livestock emissions. In poorer countries, GHG emissions from cattle can be 100 times greater due to poorer feedstock (Herrero et al. 2013). Producing 1 kilogram of beef in the United States and Europe creates 10 kilograms of CO_2 emission, but producing one kilogram of beef in the arid lands of Sudan or Ethiopia creates 1,000 kilograms CO_2. Because of this, cattle in developing countries account for 75 percent of all cattle emissions even though they only account for a small fraction of the total cattle across the world. Developed nations consume much more beef but produce it more efficiently.

The primary solution to the cattle issue is simply for humans to eat less beef. The difficulty with this proposal is that beef and other meats can be a status symbol in developing countries. Cattle can be relatively easy to raise, if there is sufficient land, so there is also a monetary incentive to put cattle out to graze until they reach market weight. Another solution is to only raise cattle in areas that make environmental sense. If rangeland exists that cannot support agriculture and there is not a native population of ungulates that are consuming the grass and brush, there could be room in the ecosystem for cattle. The next and most important step is to only graze enough cattle to fill the ecological niche that exists there and to not overgraze the area. In most cases, producing too many cattle in an area leads to overgrazing, which degrades the land, leaving compact, denuded soil that can lead to erosion when rains come. The usual solution to degraded lands is to find new lands, which are sometimes produced by cutting and burning down forests in places like the Amazon. It is also important that cattle are not subsidized by government policies. In the United States, subsidized corn creates an oversupply. The oversupply of corn is so inexpensive that cattle are fed corn instead of grasses. This distorts the true price of beef and thus increases demand. This positive feedback loop increases the amount of beef consumed and increases GHGs. Another way to decrease red meat consumption is for countries to have public service campaigns that demonstrate that increased red meat consumption can lead to increased risk of type-2 diabetes, certain cancers, stroke, and heart disease (Wolk 2017).

There are numerous practices that decrease the amount of GHGs that are produced from agriculture and animal husbandry. Some of these involve replacing annual crops with perennials; others involve growing crops in layers (with fruit trees above coffee plants, for example). Decreased fertilizer use would lead to less runoff and help protect oceans from the negative effects of nitrogen and the hypoxic zones it creates (killing most ocean life in the area). For more information on the many problems and possible solutions related to agriculture in the United States, see *The Farm Bill: A Citizen's Guide* by David Imhoff and Christina Badaracco (2019). For further reading on worldwide solutions, see *Climate Smart Agriculture: Building Resilience to Climate Change* edited by Leslie Lipper, Nancy McCarthy, David Zilberman. The simplest solution is still for humans to have a more plant-based diet (or at least a diet that minimizes the consumption of beef), as that is better for both the health of the planet and the health of humans.

TRANSITIONING COUNTRIES

The transition away from fossil fuels and toward green or renewable energy will vary in difficulty depending on the natural resources available. It will also depend on what percentage of current energy used is from renewable sources. Countries such as Albania, the Central African Republic, Bhutan, Lesotho, Iceland, Nepal, Ethiopia, and Namibia already obtain 100 percent of their electricity from renewable energy. Norway, Costa Rica, Tajikistan, Uruguay, Zambia, and Kenya all obtain over 90 percent of their energy from renewables. In almost all of these examples, hydropower is the main renewable energy used. Kenya, Costa Rica, and Iceland all have ample geothermal electricity as well. Despite the meteoric growth of solar, only a few countries obtain more than 10 percent of their electricity from this source: Spain, Greece, Chile, Jordan, Australia, Netherlands, Honduras, Germany, Israel, Hungary, Lithuania, Japan, and Thailand. That is likely to change in the next five years as many countries (and unions) install enormous amounts annually; the European Union, the United States, and especially China continue to install record amounts of solar every year. Denmark, Ireland, Portugal, Germany, and the UK all produce over 20 percent of their power from wind. China (282,000 GW), the EU (255,000 GW), and the United States (119,000 GW) dwarf all other countries in total capacity. Annual additions in wind continue to be enormous. China tops all others with almost 38,000 MW in added wind power in 2022. Overall, electricity production is transitioning away from coal and oil, but natural gas remains a growth area that still directly competes with renewable energy.

Electricity generation is not the only driver of CO_2 emissions. Transportation (15 percent), manufacturing and construction (12.4 percent), and agriculture (11 percent) remain major CO_2 contributors (C2ES 2021). The transition to a world economy that has net-zero carbon admissions will require drastic changes in those sectors as well. Norway is leading the way in transitioning personal vehicles to EVs. China is leading the way in transitioning from diesel buses to electric buses. In order from 1 to 5, Sweden, Switzerland, Finland, Denmark, and Norway are all leading the way in the energy transition according to the World Economic Forum. The top 10 countries are all in Europe. The UK and France are the only G20 (the twenty countries with the largest economies) countries to make the top 10, but they demonstrate that large economies do not need large amounts of fossil fuels. There are many countries that are moving toward sustainable energy systems. The examples in this section demonstrate how countries can transition toward carbon-neutral economies.

Costa Rica

Costa Rica is known for its beautiful beaches, lush rainforest, and raging rivers. Those raging rivers are the number one reason the country is already on the road towards a sustainable future. Costa Rica runs on hydropower; over 93 percent of all electricity in Costa Rica is generated from hydroelectric dams. A total of 95 percent of all electricity comes from renewables. Costa Rica has safeguarded over 30 percent of its land from development; currently 52 percent of the country is forested. Costa Rica also has a plan to be completely carbon-neutral by 2050. Its biggest hurdle is transportation.

Costa Rica's plan is detailed in the "2018–2050 National Decarbonization Plan" (available online at https://unfccc.int/sites/default/files/resource/NationalDecarbonizationPlan.pdf). This plan calls for the electrification of transportation by first transitioning 70 percent of all buses to EV by 2030. Setting up a system of charging stations will encourage the transition from ICE vehicles to EVs. If Costa Rica can transition toward electric mobility, they will be the leader in sustainability across the Americas (North, South, and Central).

Costa Rica has long been a leader in ecotourism; using tourism and tourism dollars to protect the ecosystems that make visiting Costa Rica so desirable. National parks and protected lands and waters continue to draw people to the country. Many companies have jumped on the ecotourism bandwagon. Offerings include net-zero vacations that are carbon-neutral, trees planted in the rainforest for every visitor, food served from onsite organic farms, and lodging that is made of natural and local resources while being chemical free. Ecotourism looks to continue to fuel the government's quest to reforest the Costa Rican jungle.

Costa Rica ranks thirtieth in the Environmental Performance Index (EPI), a ranking that would be considerably higher if not for Costa Rica's low scores on agriculture. Costa Rican pineapples and bananas require intense amounts of pesticides, resulting in the country being ranked highest in the world for per-hectare pesticide use. Costa Rica is a leader in renewable energy, wild lands preservation, and ecotourism. If the country can electrify its transportation industry and find more natural ways to grow pineapples and bananas, it will both continue to lead the renewable energy transition and be an example of how to transform a country sustainably.

Kenya

Kenya creates 75 percent of its energy through renewable sources and is on track to achieve 100 percent according to government objectives. It has hydroelectric power as well as geothermal power, which combined produce the vast majority of the country's electricity (IEA 2020d). Kenya plans to expand all renewable sources of energy with a specific focus on geothermal energy. The government is also spending money on wind and solar energy. Kenya provides an example of a developing country that is "skipping" the fossil fuel electrification process and heading straight toward a green economy.

Kenya will face many of the same problems that other developing nations do, such as creating the infrastructure that a developing nation needs without having the funds of a developed nation. If it continues to focus on green energy and green transportation, it is likely to emerge as a leader in Africa. Transportation, home heating, and electrification are all possible problems but also opportunities.

Currently, rooftop solar is growing in popularity due to the high cost of connecting to the grid in Kenya. As the cost of battery systems continue to drop, many of these solar-powered homes will likely be able to have power all night long as well. There are also seven small (under 100 MW) commercial solar power plants, six of which are under construction. When all seven are combined, they will produce 327 MW of solar power. Currently, Kenya is producing 170 MW of energy. Wind power is growing at a much faster rate. With the addition of the 310 MW Lake Turkana Wind Power Project (LTWP) in 2019 (Dahir 2019), Kenya receives over 11 percent of its electricity from wind sources.

While Kenya had been ambitious with its 2020 goal of 100 percent green electricity, it had been hampered by the push to develop a coalmine at Kitui and a coal power plant at Lamu off Kenya's east coast. Legal wrangling and protests continued, and the plans for the plant were eventually canceled. This particular project demonstrates the pressures developing nations have to endure. The Industrial and Commercial Bank of China was providing 1.2 billion in loans before they pulled out of the project. Outside funding can easily guide a nation toward fossil fuels or toward renewables. China is investing in coal power plants in other countries even as it is closing some of its own plants (while also opening new ones). These coal plants provide opportunity for Chinese workers and Chinese business, as no other country has more coal power plants than China. Coal-mining companies in other countries, like China, also encourage new coal power plants in developing nations because they can continue to sell coal even as their own nations move away from coal due to pollution and CO_2 issues. As technology advances and renewables continue to become less and less expensive than coal, many developing nations will be on the hook to pay back expensive loans for outdated and polluting plants. It is important for countries like Kenya to question why foreign countries want to fund fossil fuel projects.

If Kenya can continue its march toward renewables, it will continue to lead not only Africa but also the world in the race to achieve 100 percent renewable energy and carbon neutrality. Kenya has the opportunity to build a transportation system that is tapping into the abundant renewable energy of the country. It is also on the cutting edge of creating solar-powered desalination plants to meet the water needs of the country's driest areas. Kenya may already be building the energy systems of the future. Its green energy plan will keep Kenya independent, protect its people from pollution, and help keep the world from further warming.

Japan (and China)

Japan is covered in lieu of China in this section for three reasons: (1) Japan is an example of the difficulties of using nuclear energy for clean energy; (2) Japan is a mature economy that is actually shrinking its use of electricity; and (3) Japan has plans for transitioning to a hydrogen-based energy system. China is a much bigger player as the number one emitter of CO_2 and the world's second-largest economy. It is also a leader in renewable energy in both production and utilization. Japan is a mature economy that is able to decrease CO_2 by both switching to renewable resources and decreasing demand for electricity. This puts Japan in a position to transition to carbon-free energy at a much faster pace.

China is the undisputed leader in coal power plants as well as in solar power and wind power. If China moves quickly away from coal, it will be the most impressive transition by any country on earth. As of 2022 China produced between 56.2 percent of its electricity using coal. It also produces up to 28 percent of its electricity from renewables. Nuclear and gas produce 3–5 percent. China has the goal of reaching peak emissions by 2030 and zero emissions by 2060. China faces an uphill battle as there are internal obstacles from groups that prefer coal and will try to keep coal a large part of the energy mix.

Japan's renewable energy sector had always been second to its nuclear energy program, but that all changed with the Fukushima disaster in 2011. Despite this, Japan already obtains 17 percent of its electricity from renewable sources and plans to have up to 24 percent of its power from renewable sources by 2030.

Japan's nuclear energy program was producing 30 percent of all electricity for the country before the earthquake and tsunami that took out the Fukushima plant. Japan was expanding nuclear energy, with plans to be 40 percent nuclear-powered by 2030. The Tohoku earthquake did not damage any of Japan's fifty-four nuclear reactors, but the ensuing tsunami caused the failure of the cooling systems of the Fukushima 1 Nuclear Power Plant. The accident did not kill anyone directly, but it did release three times as much radiation as the Three Mile Island accident (but significantly less than the Chernobyl accident). Approximately 140,000 people had to evacuate, and many were never able to return to their homes. The Japanese people no longer trusted the nuclear industry or the government to run it. Japan shut down all its nuclear reactors and began to reassess each one for safety concerns. The country is slowing and carefully restarting its nuclear energy program, with ten reactors have comeback online. There are plans to restart another seventeen reactors by 2030 in order to meet Japan's commitments to the Paris Climate Accords. Japan's situation demonstrates the dangers with nuclear power. One accident can destroy public support for nuclear energy despite decades of successful and safe use. Nuclear power was replaced by more coal and gas power with some renewable power as well.

As mentioned earlier in this book, the Japanese Fukushima disaster turned the German people against nuclear energy as well, and Germany now uses more lignite (brown coal) to replace its shuttered nuclear plants. Until there is nuclear power that is completely accident proof, it will always remain a hard sell with the people of many countries. Nondemocratic countries, either autocratic dictatorships like Russia or communist rule like China, do not need the support of the public because those that display dissent from stated government positions can end up in prison or worse. Countries without democracy are the ones most likely to continue to grow their nuclear power.

Japan has reduced per capita energy use significantly, currently consuming less energy per person than it did at the end of the 1980s. This decrease is due to multiple reasons, some unique to Japan. Japan has a shrinking economy, an aging workforce, and a shrinking population. The country has also invested in both nuclear energy and renewable energies such as solar and wind. Public transportation has grown in popularity, as Japan's population has become 92 percent urban. Japanese policy has encouraged the development of high-speed trains, and both public and private trainlines proliferate across the islands. Subway systems make travel in the city efficient and affordable. Expensive tolls on roadways discourage car ownership and miles driven. For those that do drive, the transition to EVs continues to decrease oil use and oil imports annually. Decreases of 4 percent annually are becoming the norm in Japan (EIA 2020). Japan's oil imports declined precipitously during the Covid-19 pandemic and are predicted to continue to decline at a 5–10 percent annual rate. Despite the transition away from oil, Japan may be adding coal power plants. There were twenty-two new plants approved as of 2020 (Tabuchi 2020). These coal plants are proposed to replace the energy capacity lost when the country decreased its nuclear power after the Fukushima disaster. As of 2023, only one new coal plant has been built. There is significant pushback on these plans as coal is generally expensive and polluting. It is yet to be seen if there will be a coal resurgence in a country that is looking to move toward a hydrogen-based economy.

Japan plans to create a hydrogen-based economy. The idea is fairly simple, but the implementation may prove extremely difficult. Japan would like to be able to separate out the hydrogen from water using electrolysis that is powered by renewable energy or hydrogen-based energy. If hydrogen can be obtained without the need for any fossil fuel, then Japan could run all transportation, heating, and electricity in a manner that creates no CO_2 or other GHGs. In this scenario, Japan's only exhaust would be water because that is all that remains from hydrogen fuel cells or the burning of hydrogen gas. As mentioned in the section on hydrogen fuel cells, the problem with current hydrogen technology is that in some cases more energy is used than created in the process of obtaining pure hydrogen. Japan is hoping to scale up emerging technology so that hydrogen could be obtained with less energy. Using excess renewable energy to separate out hydrogen will further improve the efficiency of using inconsistent renewables such as wind and solar. Instead of needing batteries or pumped hydro systems to store energy, excess energy could be turned into hydrogen for future energy needs.

If Japan sets up a system of hydrogen fuel stations across the country, transportation could quickly transition to hydrogen fuel cells and the country could become the leader of this new technology. Japan will then be able to sell their technology to countries across the planet. Japan's plan is a big gamble, however, as it is currently more expensive to use hydrogen than most other energy sources. If Japan can transition to a hydrogen-based economy, the rest of the world will soon be trying to catch up.

Norway (and Iceland)

There are a number of Scandinavian or Western European countries that could be highlighted here depending on the metric used. Northwestern Europe is leading the world in the energy transition. Switzerland was ranked number one in 2018 by the EPI and Denmark was first in 2020. Robeco ranked Sweden number one in 2018, Norway first in 2019, Sweden first again in 2021, and Finland first in 2023. In this book, the focus is on energy resources, so metrics involving the percentage of power derived from renewable energy sources are appropriate. Norway is both close to using 100 percent renewable energy and transitioning its transportation system from petrol to electric. Iceland is also important to mention, as it uses 100 percent renewable energy and continues to find new ways to export its renewable energy.

Iceland's electricity is provided by 73 percent hydropower and 27 percent geothermal power (Government of Iceland 2020). Overall energy use in Iceland is 85 percent renewable, due to the fact that most Icelanders heat their homes using geothermal heat. The only energy that is not renewable is fossil fuel energy for vehicles. This too is changing. Sales of EVs in 2022 surpassed 41 percent market share. Iceland is encouraging this transition by adding charging stations all around the country. It plans to ban all new diesel and petrol vehicles by 2030. Iceland has a small population (fewer than 400,000) and that allows the government to more easily control the transition. Iceland provides an example of using a country's abundant natural resources to free itself from fossil fuel dependency.

Norway is a small country that has abundant oil resources, abundant hydroelectric resources, and also abundant wind. The well-managed Statoil energy company (now called Equinor), owned by the Norwegian government, has provided the money needed to transition the nation off of oil. It is an ingenious twist to use oil profits to create a country that is no longer dependent on those oil markets for its transportation systems.

Norway uses incentives to increase EV ownership. EV owners pay no sales tax, they get free or discounted parking fees, and do not have to pay for toll roads. Norway has a goal that all new cars sold by 2025 will be zero-emissions vehicles. Rather than banning gas-powered vehicles; the government is incentivizing the purchase of zero-emissions vehicles by removing all the taxes on them except for the scrapping fee. The idea is that only those who pollute will pay. By allowing citizens to choose, the government avoids pushback from those who want the freedom of choice. Norway has also installed over 10,000 charging points, including many fast chargers (Norsk elbilforening 2021). The country's transition is particularly impressive because Norway is both rural and cold. Norwegians buy more EVs, PHEVs, and hydrogen fuel cell cars, combined, than any other people. Norway demonstrates how countries can transition from big polluters to green in a very short time, through smart policy and targeted incentives.

The United States

While this book has covered energy policy in the United States fairly extensively, we would be remiss if we did not touch upon a few additional major and internationally important points. The general trend is that the United States lurches more toward renewable energy with national Democratic administrations and the process slows with Republican administrations. However, that pattern does not hold true on the state level of government. States that have deregulated their energy systems and allowed for a truly unencumbered free market approach have seen a large increase in renewable energy, specifically wind energy. The top three wind energy-producing states in the United States are all "red" states (states that tend to vote for Republicans for statewide offices). Texas is number one in wind energy, producing almost three times as much energy as its nearest competitor. One could argue that a combination of open-energy markets and a price on CO_2 could quickly transform both "red" and "blue" states into renewable energy powerhouses.

US renewable energy production continues to grow at an exponential rate. Solar power generation has increased 100 times from 2010 to 2020. Wind power has increased four times as much in the same time period, and because it is more established, it is now producing over 10 percent of the nation's energy compared to 6 percent from solar. The only barriers to the growth of renewables are political. It is less expensive to build renewable energy commercial power plants than it is to operate coal power plants. United States has therefore shut down or announced plans to shut down over 900 coal-burning units. (Coal plants can be made of a number of separate units that operate independently.) The United States still has 157 coal-powered plants, but the number continues to decline. If these plants were required to remove the amount of carbon they put into the atmosphere, it is unlikely there would be any.

EVs are finally becoming mainstream, and sales are increasing dramatically. Sales of plugin EVs and hybrid EVs are a combined 14 percent of the world market, and that percentage continues to grow quickly year over year. EV manufacturers like Tesla are unable to keep up with demand. Many would-be EV owners are waiting for over a year to receive their car. If governments encourage this transition through financial incentives and charging infrastructure, the United States could transition even faster. Unfortunately, EVs are also sometimes treated like a political football, and the US policy toward EVs can vary dramatically from one administration to the next. There are also huge disparities between states. Some states have little infrastructure for charging while others have incentivized companies to create it or created it themselves. Disparities between cities and rural areas also exist, for more practical reasons, as early-generation EVs did not have the necessary range to attract rural customers. While this is no longer the case, perception takes a long time to catch up with new realities.

The federal system of the United States can also hamper the transition from fossil fuels to renewables. Some states, mainly clustered in the Southeastern part of the country, have passed laws that make it almost impossible to install renewable energy. Some of the sunniest states produce the least solar energy. This is due to politicians who view renewable energy negatively and/or who are financially connected to fossil fuel interests. The financial realities of the affordability of renewable energy are slowly forcing states to be more open to it. These states are decades behind other states that incentivized renewables or mandated public utilities to incorporate them.

The United States produces the second highest levels of CO_2 of any country, with only China producing more. The United States also produces the second highest amount of renewable electricity, again with only China producing more. If it can muster the political will to do so, the United States has the chance to lead the renewable energy transition.

Final Questions

The human race has been in conflict over energy resources since humans first learned how to throw a spear. The conflict, over time, moved from hunting grounds and grain fields to wood, coal, and oil. The current conflict is no longer tribe against tribe or country against country but humanity against itself—its better angels against its worst impulses. Can national governments reign in those that profit from the processes that emit carbon? Will humans find a way to continue to thrive alongside other species? Will a warming earth limit the amount of livable land and decrease the number of species that can survive? Humanity has the power, the knowledge, and the tools for success in restoring the health of the planet; we must use them to overcome energy resource conflict.

FURTHER READING AND REFERENCES

Ahmad, Reaz. 2020. "Higher-Magnitude Flood Frequency Increases in Bangladesh." *Dhaka Tribune*. December 25. https://www.dhakatribune.com/bangladesh/agriculture/2020/12/25/higher-magnitude-flood-frequency-increases-in-bangladesh.

American Bird Conservancy. 2021. *Cats and Birds*. October 4. https://abcbirds.org/program/cats-indoors/cats-and-birds/.

Anderson, R. S. 2015. "Citing Open-Source Images and Formatting References for Presentations [PowerPoint Slides] in Historical Ecology." April 15. Northern Arizona University. Course lecture. Files no longer available online.

Bar-On, Yinon M., Rob Phillips, and Ron Milo. 2018. "The Biomass Distribution on Earth." *Proceedings of the National Academy of Sciences* 115, no. 25 (May 25): 6506–11. https://doi.org/10.1073/pnas.1711842115.

Blackburn, T., G. H. Edwards, S. Tulaczyk, M. Scudder, G. Piccione, B. Hallet, N. McLean, J. C. Zachos, B. Cheney, and J. T. Babbe. 2020. "Ice Retreat in Wilkes Basin of East Antarctica during a Warm Interglacial." *Nature* 583, no. 7817 (July 22): 554–9. https://doi.org/10.1038/s41586-020-2484-5.

Buis, Alan. 2020. "Milankovitch (Orbital) Cycles and Their Role in Earth's Climate." *NASA Global Climate Change*. February 27. https://climate.nasa.gov/news/2948/milankovitch-orbital-cycles-and-their-role-in-earths-climate/.

California Fuel Cell Partnership (CAFPC). 2016. *Air Climate Energy Water Security: A Guide to Understanding the Well-to-Wheels Impact of Fuel Cell Electric Vehicles*. Frontier Energy Inc. https://cafcp.org/sites/default/files/W2W-2016.pdf.

Center for Climate and Energy Solutions (C2ES). 2021a. *California Cap and Trade*. July. https://www.c2es.org/content/california-cap-and-trade/.

Center for Climate and Energy Solutions (C2ES). 2021b. *Extreme Weather and Climate Change*. July. https://www.c2es.org/content/extreme-weather-and-climate-change/.

Center for Climate and Energy Solutions (C2ES). 2021c. *Global Emissions*. July. https://www.c2es.org/content/international-emissions/.

Chappell, Bill. 2021. *Renewable Energy Growth Rate Up 45% Worldwide In 2020; IEA Sees "New Normal."* NPR. May 11. https://www.npr.org/2021/05/11/995849954/renewable-energy-capacity-jumped-45-worldwide-in-2020-iea-sees-new-normal.

Climate & Clean Air Coalition (CCAC). 2024. "Hydrofluorocarbons (HFCS)." https://www.ccacoalition.org/short-lived-climate-pollutants/hydrofluorocarbons-hfcs.

Cox, Peter. 2020. *Guest Post: Could Climate Change and Deforestation Spark Amazon 'Dieback'?* CarbonBrief. February 13. https://www.carbonbrief.org/guest-post-could-climate-change-and-deforestation-spark-amazon-dieback.

Dahir, Abdi Latif. 2019. "Africa's Largest Wind Power Project Is Now Open in Kenya." *Quartz Africa Weekly*. July 22. https://qz.com/africa/1671484/kenya-opens-africas-largest-wind-power-project-in-turkana/.

Deryabina, T. G., S. V. Kuchmel, L. L. Nagorskaya, T. G. Hinton, J. C. Beasley, A. Lerebours, and J. T. Smith. 2015. "Long-Term Census Data Reveal Abundant Wildlife Populations at Chernobyl." *Current Biology* 25, no. 19 (October 5): R824–R826. https://doi.org/10.1016/j.cub.2015.08.017.

Doo, Stephen S., Peter J. Edmunds, and Rober C. Carpenter. 2019. "Ocean Acidification Effects on In Situ Coral Reef Metabolism." *Scientific Reports* 9, no. 1 (August 19): 1–8. https://doi.org/10.1038/s41598-019-48407-7.

Dubay, Taylo. 2020. Interview with Taylo Dubay of the Ecological Restoration Institute. October 2.

Energy Efficiency & Renewable Energy (EERE), Office of. n.d. "Geothermal Heat Pumps." Department of Energy. Accessed June 25, 2024. https://www.energy.gov/eere/geothermal/geothermal-heat-pumps.

European Commission on Climate Action. 2021. *EU Emissions Trading System (EU ETS)*. European Union. May. https://ec.europa.eu/clima/policies/ets_en.

European Nuclear Society (ENS). 2020. *Fuel Comparison*. European Nuclear Society. December. https://www.euronuclear.org/glossary/fuel-comparison/.

Fabry, Frédéric, Cristophe Rehmet, Vandad Rohani, and Laurent Fulcheri. 2013. "Waste Gasification by Thermal Plasma: A Review." *Waste and Biomass Valorization* 4, no. 3 (February 5): 421–39. https://doi.org/10.1007/s12649-013-9201-7.

Fawcett, P., J. Werne, R. S. Anderson, Jeffrey M. Heikoop, Erik T. Brown, Melissa A. Berke, Susan J. Smith, Fraser Goff, Linda Donohoo-Hurley, Luz M. Cisneros-Dozal, Stefan Schouten, Jaap

S. Sinninghe Damsté, Yongsong Huang, Jaime Toney, Julianna Fessenden, Giday WoldeGabriel, Viorel Atudorei, John W. Geissman, and Craig D. Allen. 2011. "Extended Megadroughts in the Southwestern United States during Pleistocene Interglacials." *Nature* 470 (February 23): 518–21. https://doi.org/10.1038/nature09839.

Gerber, P. J., H. Steinfeld, B. Henderson, A. Mottet, C. Opio, J. Dijkman, and G. Tempio. 2013. *Tackling Climate Change through Livestock: A Global Assessment of Emissions and Mitigation Opportunities*. Food and Agriculture Organization of the United Nations (FAO). https://www.fao.org/3/i3437e/i3437e.pdf.

Government of Costa Rica. 2017. "National Decarbonization Plan: 2018–2050." *Decarbonizing Costa Rica, Country Commitment, 2018–2050.* https://unfccc.int/documents/204474

Government of Iceland. 2020. "Energy." Ministry of the Environment, Energy, and Climate. November. https://www.government.is/topics/business-and-industry/energy/.

Greeley, Michael N. 1999. "National Reclamation of Abandoned Mine Lands." Presentation to Small Business and Enforcement (SME), Environmental Protection Agency (EPA). March. https://www3.epa.gov/npdes/pubs/amlpaper.htm.

Gruber, Nicholas, Charles D. Keeling, and Nicholas R. Bates. 2002. "Interannual Variability in the North Atlantic Ocean Carbon Sink." *Science* 298, no. 5602 (December 20): 2374–8. https://doi.org/10.1126/science.1077077.

Haskins, Rosalind K., Kevin I. C. Oliver, Laura Jackson, Richard A. Wood, and Sybren Drijfhout. 2019. "Temperature Domination of AMOC Weakening Due to Freshwater Hosing in Two GCMs." *Climate Dynamics* 54, no. 23 (October 4): 273–86. https://link.springer.com/article/10.1007/s00382-019-04998-5

Herrero, Mario, Petr Havlík, Hugo Valin, An Notenbaert, Mariana C. Rufino, Philip K. Thornton, Michael Blümmel, Franz Weiss, Delia Grace, and Michael Obersteiner. 2013. "Biomass Use, Production, Feed Efficiencies, and Greenhouse Gas Emissions from Global Livestock Systems." *Proceedings of the National Academy of Sciences* 110, no. 52: 20888–93.

Holland, M. 2020. "All Posts Tagged 'EV Sales.'" *Clean Technica*. October. https://cleantechnica.com/tag/ev-sales/.

Imhoff, Daniel, and Christina Badaracoo. 2019. *The Farm Bill: A Citizen's Guide*. Washington, DC: Island Press/Center for Resource Economics.

International Energy Agency (IEA). 2020a. *Hydropower Roadmap Foldout*. October. https://www.ieahydro.org/media/1ea1287f/Hydropower_Roadmap_FoldOut_WEB.pdf.

International Energy Agency (IEA). 2020b. *Snapshot 2020: IEA Technology Collaboration Programme*. November. https://iea-pvps.org/snapshot-reports/snapshot-2020/.

International Energy Agency (IEA). 2020c. "Global Total Coal Demand, 1971–2020." IEA World Energy Outlook. November. https://www.iea.org/data-and-statistics/charts/global-total-coal-demand-1971-2020.

International Energy Agency (IEA). 2020d. "Kenya Electricity Generation by Technology in the Stated Policies Scenario, 2010–2040." IEA World Energy Outlook. December. https://www.iea.org/data-and-statistics/charts/kenya-electricity-generation-by-technology-in-the-stated-policies-scenario-2010-2040.

International Renewable Energy Agency (IRENA). 2019a. "Future of Photovoltaic: Deployment, Investment, Technology, Grid Integration and Socio-economic Aspects: A Global Energy

Transformation Paper." https://www.irena.org/-/media/Files/IRENA/Agency/Publication/2019/Nov/IRENA_Future_of_Solar_PV_2019.pdf.

International Renewable Energy Agency (IRENA). 2019b. "Future of Wind: Deployment, Investment, Technology, Grid Integration and Socio-economic Aspects: A Global Energy Transformation Paper." International Renewable Energy Agency. October. https://www.irena.org/-/media/Files/IRENA/Agency/Publication/2019/Oct/IRENA_Future_of_wind_2019.pdf.

International Renewable Energy Agency (IRENA). 2019c. "Renewable Energy Now Accounts for a Third of Global Power Capacity." April 2. https://www.irena.org/newsroom/pressreleases/2019/Apr/Renewable-Energy-Now-Accounts-for-a-Third-of-Global-Power-Capacity.

International Renewable Energy Agency (IRENA). 2024. "Renewable Capacity Statistics 2024." *IRENA*, March 1. https://www.irena.org/Publications/2024/Mar/Renewable-capacity-statistics-2024.

Irle, Roland. 2021. "Global EV Sales for 2020 H1." *EV-Volumes, The Electric Vehicle World Sales Database*. https://www.ev-volumes.com/.

Johnson, N. C., D. R. Zak, D. Tilman, and F. L. Pfleger. 1991. "Dynamics of Vesicular-Arbuscular Mycorrhizae during Old Field Succession." *Oecologia* 86 (May): 349–58. https://doi.org/10.1007/BF00317600.

Kane, Mark. 2020. "Close to 1.18 Million Plug-In Electric Cars Were Sold in China in 2019." *Inside EVs*. January 22. https://insideevs.com/news/394229/plugin-electric-car-sales-china-2019.

King, Michalea D., Ian M. Howat, Salvatore G. Candela, Myoung J. Noh, Seongsu Jeong, Brice P. Y. Noël, Michiel R. van den Broeke, Bert Wouters, and Adelaide Negrete. 2020. "Dynamic Ice Loss from the Greenland Ice Sheet Driven by Sustained Glacier Retreat." *Communications Earth & Environment* 1, no. 1 (August 13): 1–7. https://doi.org/10.1038/s43247-020-0001-2.

Kulp, Scott A., and Benjamin H. Strauss. 2019. "New Elevation Data Triple Estimates of Global Vulnerability to Sea-Level Rise and Coastal Flooding." *Nature Communications* 10, no. 1: 1–12. https://www.nature.com/articles/s41467-019-12808-z.

Liebowitz, Jon. 2023. "Rewilding in an Unlikely Landscape [Northeast USA]." *Rewilding Successes*, October 4. https://rewilding.ecologicalcitizen.net/2023/02/15/rewilding-in-an-unlikely-landscape-northeast-usa/.

Lindsey, Rebecca. 2022. "Climate Change: Global Sea Level: National Defense Authorization Act (NDAA)." April 19. https://www.climate.gov/news-features/understanding-climate/climate-change-global-sea-level.

Lipper, L., N. McCarthy, D. Zilberman, S. Asfaw, and G. Branca. 2017. *Climate Smart Agriculture: Building Resilience to Climate Change*. Springer Nature. https://link.springer.com/book/10.1007/978-3-319-61194-5.

McGowan, Matt. 2021. "Circuit Generates Clean, Limitless Power from Graphene." *Tech Briefs, Engineering Solutions for Design & Manufacturing*. January 1. https://www.techbriefs.com/component/content/article/tb/pub/techbriefs/electronics-and-computers/38330.

Monzón, Javier, Roland Kays, and D. E. Dykhuizen. 2014. "Assessment of Coyote–Wolf–Dog Admixture Using Ancestry-Informative Diagnostic SNP s." *Molecular Ecology* 23, no. 1: 182–97.

National Centers for Environmental Information. 2020. *A Warming Earth Is Also a Wetter Earth.* NOAA National Oceanic and Atmospheric Administration. July 2020. https://www.ncei.noaa.gov/news/warming-earth-also-wetter-earth.

National Oceanic and Atmospheric Administration. 2019. *Coral Reef Ecosystems.* February 1. https://www.noaa.gov/education/resource-collections/marine-life/coral-reef-ecosystems.

Norsk Elbilforening. 2021. *Norwegian Environmental Policy: Norway Is Leading the Way for a Transition to Zero Emission in Transport.* Norwegian EV Association. May. https://elbil.no/english/norwegian-ev-policy/.

Penney, Sarah, Jacob Bell, and John Balbus. 2009. *Estimating the Health Impacts of Coal-Fired Power Plants Receiving International Financing.* Environmental Defense Fund. https://www.edf.org/sites/default/files/9553_coal-plants-health-impacts.pdf.

Perera, Frederica, David Cooley, Alique Berberian, David Mills, and Patrick Kinney. 2020. "Co-benefits to Children's Health of the US Regional Greenhouse Gas Initiative." *Environmental Health Perspectives* 128, no. 7 (July 29). https://doi.org/10.1289/EHP6706.

Pérez-Izquierdo, Leticia, Karina E. Clemmensen, Joachim Strengbom, Gustaf Granath, David A. Wardle, Marie-Charlotte Nilsson, and Björn D. Lindahl. 2020. "Crown-Fire Severity Is More Important Than Ground-Fire Severity in Determining Soil Fungal Community Development in the Boreal Forest." *Journal of Ecology* 109, no. 1 (October 21): 504–18.

Pitteloud, J. 2021. *Global Wind Installations: Total Installed Capacity [MW].* Wind Energy International. January. https://library.wwindea.org/global-statistics/.

Ray, Deepak K., Paul C. West, Michael Clark, James S. Gerber, Alexander V. Prishchepov, and Snigdhansu Chatterjee. 2019. "Climate Change Has Likely Already Affected Global Food Production." *PloS One* 14, no. 5 (May 31). https://doi.org/10.1371/journal.pone.0217148.

Renforth, P., and G. Henderson. 2017. "Assessing Ocean Alkalinity for Carbon Sequestration." *Reviews of Geophysics* 55, no. 3 (June 15): 636–74. https://doi.org/10.1002/2016RG000533.

Rignot, Eric, Jérémie Mouginot, Bernd Scheuchl, and Mathieu Morlighem. 2019 . "Four Decades of Antarctic Ice Sheet Mass Balance from 1979–2017." *Proceedings of the National Academy of Sciences* 116, no. 4 (January 14): 1095–103. https://doi.org/10.1073/pnas.1812883116.

Stracqualursi, Maria. 2017. "The Rise of the Public Benefit Corporation: Considerations for Start-ups." *Boston College Law Lab.* March 21. http://bclawlab.org/eicblog/2017/3/21/the-rise-of-the-public-benefit-corporation-considerations-for-start-ups.

Stutt, Jordan. 2019. "The Regional Greenhouse Gas Initiative: 10 Years in Review." *Acadia Center.* September 17. https://acadiacenter.org/wp-content/uploads/2019/09/Acadia-Center_RGGI_10-Years-in-Review_2019-09-17.pdf.

Swetnam, Thomas W., Craig D. Allen, and Julio L. Betancourt. 1999. "Applied Historical Ecology: Using the Past to Manage for the Future." *Ecological Applications* 9, no. 4: 1189–206.

Tabuchi, Hiroko. 2020. "Japan Races to Build New Coal-Burning Power Plants, Despite the Climate Risks." *New York Times.* February 3. https://www.nytimes.com/2020/02/03/climate/japan-coal-fukushima.html.

Tangri, Neil, and Monica Wilson. 2017. *Waste Gasification & Pyrolysis: High Risk, Low Yield Processes for Waste Management. Report: A Technology Risk Analysis.* GAIA. https://www.no-burn.org/wp-content/uploads/Waste-Gasification-and-Pyrolysis-high-risk-low-yield-processes-march-2017.pdf.

United Nations. 2020. "Food." *UN Global Issues*. March 10. https://www.un.org/en/sections/issues-depth/food/index.html.

USA for UNHCR. 2020. "Refugee Crisis in Europe." *The UN Refugee Agency*. December 8. https://www.unrefugees.org/emergencies/refugee-crisis-in-europe/.

US Energy Information Administration. 2020. *Japan*. November 2. https://www.eia.gov/international/analysis/country/JPN.

US Energy Information Administration. 2022. *Biomass Explained*. June 2. https://www.eia.gov/energyexplained/biomass/.

US Department of Agriculture. 2022a. "17 State Grazing Fees: Animal Units." National Agricultural Statistics Service. January. https://www.nass.usda.gov/Charts_and_Maps/Grazing_Fees/gf_am.php.

US Department of Agriculture. 2022b. "USDA Forest Services Announces 2022 Grazing Fees on the National Forests and Grasslands in Western States. January 31. https://www.fs.usda.gov/about-agency/newsroom/releases/usda-forest-service-announces-2022-grazing-fees-national-forests-and

US Government Accountability Office (US GAO). 2023. "From Gold Rush to Rot—The Lasting Environmental Costs and Financial Liabilities of Hardrock Mining." https://www.gao.gov/blog/gold-rush-rot-lasting-environmental-costs-and-financial-liabilities-hardrock-mining.

Wheeler, Everett. 2020. "Electric Vehicles to Set New Market Share Record in 2020." *S&P Market Intelligence*. June 15. https://www.spglobal.com/marketintelligence/en/news-insights/latest-news-headlines/electric-vehicles-to-set-new-market-share-record-in-2020-59050766.

Wolk, A. 2017. "Potential Health Hazards of Eating Red Meat." *Journal of Internal Medicine* 281, no. 2: 106–22.

CHAPTER 5

Contributed Essays

INTRODUCTION

The following essays look at both the bad (Ike) and the good (Anderson) that can be done through corporate power in the energy sector and in ameliorating the effects of climate change. We look at both the end of coal (Munir) and the limits of solar and wind (Solman). The promise of nuclear energy, which has proved elusive at times, is revisited (Sadiqa). In all of these essays, we look toward the future as we try to find a path toward obtaining energy without creating new conflicts.

The next conflict over energy is truly humanity against itself. As global temperatures rise, we must still meet the demands of hundreds of millions of people longing for electrification and the modern conveniences that come with it. The question of how to do that remains. We know it will not work when corporations interfere with government attempts to limit GHGs (Ike). We also know that corporations can unleash innovation and accelerate the energy transition (Anderson). Coal as an energy source needs to be put behind us (Munir). Will nuclear (Sadiqa) or wind and solar (Solman) lead us toward a carbon-free energy future? The authors in this section give us insight into these ideas and will leave us with new ideas about what our shared future will look like.

THE DANGERS OF CORPORATE POWER CONCERNING CLIMATE CHANGE

Vivian Ike

Corporations use their economic power to influence countries like the United States and those in Europe by making them prioritize policies that further their interests above climate change. Historically and to date, corporations have greatly contributed to disapproval for, and shaped the agenda of, climate change policy (Levy and Egan 1998; Kolk and Pinkse 2007). The major strategies used to achieve this include financing climate science propaganda that has since evolved into developing research and technology in market-based solutions and then lobbying policymakers to promote and formulate market-based policy.

To better understand the dangers of corporate power, we must ask what is climate change and what is its relationship with corporations? Climate change refers to alteration in climate patterns due to global warming, which is largely caused by fossil fuel energy–intensive activities that emit tons of greenhouse gases (GHGs). As this became a serious problem in the 1990s, countries from around the world formed an international alliance in 1992 called the United Nations Framework Convention for Climate Change (UNFCCC), thereby creating policies such as the 1997 Kyoto Protocol and 2015 Paris Agreement. Countries have also formulated domestic policies to regulate and reduce GHG emissions.

Fossil fuel-heavy corporations from industries such as oil and gas, electricity, chemicals, mining, metals, vehicles, and transport, however, did not welcome these developments because their profits are gained from energy-intensive activities and consumer products that emit much GHG. Such corporations combined contribute about one-third of global GHG emissions, posing one of the greatest dangers for climate change (Taylor and Watts 2019). In the late 1980s, when countries began deliberating an alliance, powerful corporations such as DuPont, ExxonMobil, Chrysler, and British Petroleum formed the largest industry alliance against climate change, called Global Climate Coalition (GCC).

The GCC's early strategy was the financing of climate science to propagate misinformation. It spent millions of US dollars on scientific research throughout the 1990s, promoting the claim that uncertainty regarding the effect of GHGs on climate cannot justify that fossil fuel energy–intensive activities cause climate change (Climate Investigation Centre 2019). Millions were also spent to spread such misinformation through mediums such as pamphlets and ad campaigns. The GCC sent science experts to promote misinformation about climate change at international meetings where countries were negotiating the UNFCCC and Kyoto Protocol (Savage 2019). This strategy was used to counter and discredit findings of the International Panel of Scientists on Climate Change (IPCC) formed by countries in 1988 to help investigate climate change and provide policy recommendations.

Also during the 1990s, some GCC members spent millions in lobbying and financing the campaigns of political candidates within the United States, thus forming close ties with Congress. This resulted in the 1997 unanimous Byrd–Hagel decision, co-sponsored by a majority in the Senate, to delay and eventually remove US support for the Kyoto Protocol on the grounds of costly economic impacts, hence also blocking any domestic policy that could have been subsequently implemented. GCC members financed the campaign of Byrd–Hagel's lead sponsor, Senator Chuck Hagel, as well as many other political candidates during the 1994–6 elections (Fisher 2006; Levy and Egan 1998). Then GCC chairman William O'Keefe gave a statement at the Senate about the negative economic impacts of the protocol, and by 2001, President George W. Bush credited the GCC for its part in the government's decision to reject the protocol (MIT 2014).

Similar strategies have also been employed in other major countries such as Canada, which as a result withdrew from the Kyoto Protocol in 2012, and Australia, which struggles to domestically implement the protocol (Young and Coutinho 2013). This strategy persists in the United States where, for instance, Exxon Mobil—a former GCC member and one of the country's largest corporations—contributed millions to Donald Trump's 2015–16 presidential election campaign. In 2017–18, former Exxon employees Rex Tillerson and Scott Pruitt were appointed secretary of state and head of the Environmental Protection Agency (EPA), respectively (Influence Map 2016). Thus, the president's decision in 2017 to withdraw from the 2015 Paris Agreement and refusal to implement domestic policy comes as no surprise. This has had negative consequences for climate change, as the United States is home to the world's largest corporations and emitters of GHGs.

Within Europe, however, corporations have employed less antagonistic strategies but have been particularly influential in reaching a compromise with European Union (EU) countries to adopt market-based solutions for climate change (Kolk and Levy 2004; Pinkse 2007). EU-based corporations such as BP and Shell were among the first to depart from the GCC in the late 1990s and adopt a more open stance toward climate change science and policy. Throughout the 1990s, EU countries such as Germany, France, and Denmark proposed a carbon tax on fossil fuel energy both regionally and internationally, which corporations and countries like the United States, Canada, and Australia opposed (CorpWatch 2001). They instead pushed for voluntary and flexible market-based solutions that were eventually included in the Kyoto Protocol, such as carbon emissions trading, carbon offset projects and sinks, and clean energy research and development, as opposed to strict solutions that require direct regulation of corporations and activities that emit much GHG.

It wasn't until after much lobbying from and talks with EU-based corporations that EU countries settled on a compromise. As a result, by the early 2000s, EU countries drafted an Emissions Trading Scheme (ETS), the largest carbon trading market, in place of the stringent carbon tax. To ensure its success, the final ETS policy was designed to accommodate the interests of EU-based corporations, especially the largest GHG emitters that refused a mandatory ETS and a cap on the amount of carbon that could be emitted (Markussen and Svendsen 2005). Furthermore, EU-based corporations, in order to gain an advantage in the global economy, have spent millions in developing research on market-based solutions, thus dominating the market for renewable technologies (European Commission 2015).

Non-EU corporations like those in the United States have begun in recent years, albeit slowly, to adopt similar strategies given the undisputed advances in climate science. Corporations including ExxonMobil and Chrysler, for example, invest millions into energy-efficient technology such as biofuels and hybrid cars, which nonetheless require fossil fuels to function (ExxonMobil 2018; Schoenberger 2019). Such solutions, therefore, do not significantly alter business as usual nor produce transformations necessary to address climate change when compared to other renewable alternatives like solar energy and electric cars. The new strategy poses a danger to limiting climate change and is strengthened by the fact that corporations lobby and finance political candidates so that their market-based solutions are supported and promoted. Despite, for instance, the limitations of biofuels as evidenced by research and hesitation by the EU to adopt them, the IPCC in its recent report has openly recommended biofuels as a supplementary addition to renewable energy solutions (Fortuna 2018). This is only one of the many examples of the dangers of corporate power concerning climate change.

FURTHER READING AND REFERENCES

Climate Investigations Centre. n.d. "Global Climate Coalition Document Index." Accessed April 22, 2020. https://climateinvestigations.org/global-climate-coalition-documents-index/.

Climate Investigations Centre. 2019. "Global Climate Coalition: Climate Denial Legacy Follows Corporations." April 25. Accessed July 14, 2024. ttps://climateinvestigations.org/global-climate-coalition-industry-climate-denial/.

CorpWatch. 2001. "Corporate Campaign to Corrupt the Kyoto Protocol Continues after COP-6." April 1. Accessed July 14, 2024. https://archive.corporateeurope.org/observer8/cop6.html

ExxonMobil. 2018. "Advanced Biofuels and Algae Research." September 17. Accessed July 14, 2024. https://corporate.exxonmobil.com/what-we-do/transforming-transportation/advanced-biofuels/advanced-biofuels-and-algae-research

European Commission. 2015. "The European Union Leading in Renewables." Accessed July 14, 2024. https://ec.europa.eu/energy/sites/ener/files/documents/cop21-brochure-web.pdf.

Fisher, Dana. 2006. "Bringing the Material Back In: Understanding U.S. Position on Climate Change." *Sociological Forum* 21, no. 3: 467–94.

Fortuna, Gerardo. 2018. "Electricity and Biofuels Needed in Tandem to Meet Climate Goals, UN Report Says." *Euroactive*, November 18. Accessed April 12, 2019. https://www.euractiv.com/section/agriculture-food/news/electricity-and-biofuels-needed-in-tandem-to-meet-climate-goals-un-report-says/.

Influence Map. 2016. *An Investor Enquiry: How Much Big Oil Spends on Climate Lobbying.* March. Accessed July 14, 2024. https://content.influencemap.org//site/data/000/287/Lobby_Spend_Report_March_2016.pdf

Levy, David, and Daniel Egan. 1998. "Capital Contests: National and Transnational Channels of Corporate Influence on the Climate Change Negotiations." *Politics and Society* 26, no. 3: 337–61.

Kolk, Ans, and David Levy. 2004. "Multinationals, Environment and Global Competition Climate Change: Issues for the Automotive and Oil Industries." *Research in Global Strategic Management* 9: 171–93.

Kolk, Ans, and Jonatan Pinkse. 2007. "Multinationals' Political Activities on Climate Change." *Business and Society* 46, no. 2: 201–28.

Markussena, Peter, and Gert Tinggaard Svendsen. 2005. "Industry Lobbying and the Political Economy of GHG Trade in the European Union." *Energy Policy* 33, no. 2: 245–55.

Massachusetts Institute of Technology (MIT). 2014. *The Fossil Fuel Industry's Role in Hindering Climate Change Action.* April 8. https://www.fossilfreemit.org/wp-content/uploads/2014/08/FossilFreeMIT-Lobbying-Disinformation.pdf.

Pinkse, Jonatan. 2007. "Corporate Intentions to Participate in Emission Trading." *Business Strategy and the Environment* 16, no. 1: 12–25.

Savage, Karen. 2019. "Global Climate Coalition: Fighting Global Climate Action in Favor of Fossil Fuels." *Climate Liability News,* April 25. https://www.climateliabilitynews.org/2019/04/25/gcc-global-climate-coalition-un-fossil-fuels/.

Schoenberger Robert. 2019. "Fiat Chrysler to Focus on Hybrids, Not Full EVs." December 9. https://www.todaysmotorvehicles.com/article/fca-uaw-contract-light-evs/

Taylor, Matthew, and Jonathan Watts. 2019. "Revealed: The 20 Firms Behind a Third of All Carbon Emissions." *Guardian*, October 9. https://www.theguardian.com/environment/2019/oct/09/revealed-20-firms-third-carbon-emissions.

Young, Nathan, and Aline Coutinho. 2013. "Government, Anti-reflexivity, and the Construction of Public Ignorance about Climate Change: Australia and Canada Compared." *Global Environmental Politics* 13, no. 2: 89–108.

THE POWER OF BUSINESS CORPORATIONS TO DRIVE ENERGY TRANSFORMATION: THE RE100 CASE

Haley N. Anderson

Introduction

In 1988, human-induced climate change was officially recognized through the establishment of the Intergovernmental Panel on Climate Change (IPCC). In 2017, Carbon Disposal Project (CDP) released a report that stated that over 70 percent of global industrial greenhouse emissions have been linked to just 100 active fossil fuel-producing companies (CDP Carbon Majors Report 2017). According to CDP (2017), "This group is the source of 635 billion tonnes of GHGs emitted since 1988." In 2018, a report from the IPCC warned that a global temperature increase of 1.5°C would be catastrophic for the planet and that the original goal of 2°C established during the Conference of the Parties to the United Nations Framework Convention on Climate Change would not be enough to avoid disaster (IPCC 2018). While these figures are alarming, business corporations are responding by creating their own climate action plans, purchasing renewable energy at record pace (Scott 2019), investing in renewable facilities and generating their own renewable energy (Warren 2018), and publicly committing to 100 percent renewable energy by 2050 (RE100 Overview 2019).

The Case

An example of a commitment that some corporations are making to alleviate their impact on global warming is the RE100 initiative. The RE100 initiative, led by The Climate Group in conjunction with CDP, has set the goal of getting the world's most influential businesses to publicly commit to 100 percent renewable energy by 2050. Since RE100's launch at Climate Week NYC in 2014, more than 400 companies have joined the initiative (RE100 2024).

At the initial Climate Week NYC and at each annual meeting since, leading businesses have stressed their desire for strong climate action from governments, including increased regulations and emission standards on corporations like themselves (The Climate Group 2015). Momentum behind the RE100 initially continues to grow annually. In the United States 500 companies represent "$13.7 trillion in revenues, $1.1 trillion in profits, $22.6 trillion in market value, and employ 28.7 million people worldwide" (*Fortune*

2019). Nearly 20 percent of the Fortune 500's top 100 companies have signed onto the RE100 initiative (Fortune 2019; RE100 Companies 2020), including Walmart and Apple, ranked number 1 and 3, respectively.

RE100 companies can hit their targets through the production of renewable energy from their own facilities, or the purchase of renewable energy from market suppliers, and upon joining are responsible for reporting their progress for publication in the RE100 Annual Report. The 2019 Annual Report (RE100 2019), demonstrates some of the influence that this initiative has had on the private sector, including the following statistics: One in three members are already at least 75 percent renewable and one in two have experienced cost savings as a result of their increased consumption of renewables. Additionally, and perhaps most importantly, these companies are driving the demand for renewable energy and using their influence to guide stakeholders to "overcome policy barriers and push for a faster transition to renewables" (RE100 2019).

Theory

While there are many theories as to why a business corporation would invest in renewable energy, two that have gained traction among business and political scholars alike are the theories of Corporate Social Responsibility (CSR) and Corporate Environmental Responsibility (CER). At their core, CSR and CER are based on the idea that corporations have an ethical obligation to society and to the environment to mitigate the adverse impacts of their operations.

CSR has had a long history in the literature and has varied definitions (Carrol 1999), dimensions (Dahlsrud 2006; Garriga and Mele 2004), and organizational clusters (Lindgreen, Swaen, and Johnson 2009). The notion of CSR is related to ethical decision-making behavior by corporations (Branco and Rodrigues 2006). Consistent with McWilliams and Siegel (2001), here we define CSR as "actions that appear to further some social good, beyond the interests of the firm and that which is required by law" (117). Corporations can interpret and relate CSR to complex issues such as community relations, public health responsibilities, and environmental protections (Branco and Rodrigues 2006). The same concept can be applied to CER, by looking at companies' relationship to the environment and what companies are doing to further environmental good, beyond what is required by law. In the case of the companies that have joined the RE100, they are recognizing their responsibility to act on global climate change.

CSR and CER initiatives, like the RE100, are an important way that companies communicate what they are doing to give back to society, and numerous studies have shown that consumers are more likely to want to affiliate themselves with companies that act in socially responsible ways (McGuire, Sundgren, and Schneeweis 1988; Creyer 1997; Stanwick and Stanwick 1998) and environmentally (Hart 1995; Russo and Fouts 1997; Ambec and Lanoie 2008; Wahba 2008; Gunningham 2009). Becker-Olsen, Cudmore, and Hill (2006) found in two studies conducted on consumers, that "greater than 80% of respondents believed that firms should engage in social initiatives and 76% felt those initiatives would benefit the firms" (52), showing that these responsible actions can affect customer decision-making and impact a corporation's bottom line.

Conclusion

In the case of climate change, many companies recognize that they have an ethical obligation to mitigate their impact on climate change and the environment and are taking steps to act in the best interest of their stakeholders and communities. Through initiatives like the RE100, the public is able to recognize work that these corporations are doing to move to be more climate conscious and influence policy and political stakeholders to push for more aggressive climate action plans. Though corporations and the private sector have not historically held a positive reputation when it comes to their relationship with the environment, it is clear that through their efforts to drive energy transformation, that reputation is changing.

FURTHER READING AND REFERENCES

Ambec, S., and P. Lanoie. 2008. "Does It Pay to Be Green? A Systematic Overview." *Academy of Management Perspectives* 22, no. 4: 45–62.

Becker-Olsen, K. L., B. A. Cudmore, and R. P. Hill. 2006. "The Impact of Perceived Corporate Social Responsibility on Consumer Behavior." *Journal of Business Research* 59: 46–53.

Branco, M. C., and L. L. Rodrigues. 2006. "Corporate Social Responsibility and Resource-Based Perspectives." *Journal of Business Ethics* 69: 111–32.

Carroll, A. B. 1999. "Corporate Social Responsibility: Evolution of a Definitional Construct." *Business & Society* 38, no. 3: 268–95.

CDP Carbon Majors Report. 2017. *CDP Worldwide.* https://cdn.cdp.net/cdp-production/cms/reports/documents/000/002/327/original/Carbon-Majors-Report-2017.pdf.

The Climate Group. 2015. "Fortune 500 Companies Pledge to Use 100% Renewable Electricity." Accessed July 14, 2024. https://www.there100.org/our-work/news/fortune-500-listed-companies-pledge-use-100-renewable-electricity-news.

Creyer, E. H. 1997. "The Influence of Firm Behavior on Purchase Intention: Do Consumers Really Care about Business Ethics?" *Journal of Consumer Marketing* 14, no. 6: 421–8.

Dahlsrud, Alexander. 2006. "How Corporate Social Responsibility Is Defined: An Analysis of 37 Definitions." *Corporate Social Responsibility and Environmental Management* 15: 1–13.

Fortune. 2019. "The Top 10." https://fortune.com/fortune500/.

Garriga, E., and D. Mele. 2004. "Corporate Social Responsibility Theories: Mapping the Territory." *Journal of Business Ethics* 53: 51–71.

Gunningham, N. 2009. "Shaping Corporate Environmental Performance: A Review." *Environmental Policy and Governance* 19: 215–31.

Hart, S. L. 1995. "A Natural Resource-Based View of the Firm." *Academy of Management Review* 20, no. 4: 986–1014.

Intergovernmental Panel on Climate Change (IPCC). 2018. *Global Warming of 1.5°C*. Geneva, Switzerland: IPCC. https://report.ipcc.ch/sr15/pdf/sr15_spm_final.pdf.

Lindgreen, A., V. Swaen, and W. J. Johnston. 2009. "Corporate Social Responsibility: An Empirical Investigation of U.S. Organizations." *Journal of Business Ethics* 85: 303–23.

McGuire, J. B., A. Sundgren, and T. Schneeweis. 1988. "Corporate Social Responsibility and Firm Financial Performance." *Academy of Management Journal* 31, no. 4: 854–72.

McWilliams, A., and D. Siegel. 2001. "Corporate Social Responsibility: A Theory of the Firm Perspective." *Academy of Management Review* 26, no. 1: 117–27.
RE100 Companies. 2024. *RE100*. http://there100.org/companies.
RE100 Overview. 2020. *RE100*. http://there100.org/re100.
Russo, M. V., and P. A. Fouts. 1997. "A Resource-Based Perspective on Corporate Environmental Performance and Profitability." *Academy of Management Journal* 40, no. 3: 534–59.
Scott, M. 2019. "Companies Continue to Drive Demand for Clean Energy." *Forbes*. September 9. https://www.forbes.com/sites/mikescott/2019/09/09/companies-continue-to-drive-demand-for-clean-energy/.
Stanwick, P. A., and S. D. Stanwick. 1998. "The Relationship between Corporate Social Performance and Organizational Size, Financial Performance, and Environmental Performance: An Empirical Examination." *Journal of Business Ethics* 17, no. 2: 195–204.
Wahba, H. 2008. "Does the Market Value Corporate Environmental Responsibility? An Empirical Examination." *Corporate Social Responsibility and Environmental Management* 15: 89–99.
Warren, G. S. 2018. "1-Click Energy: Managing Corporate Demand for Clean Power." *Maryland Law Review* 78: 73–104.

THE DIRTY FUTURE OF COAL

Laine Munir

Coal is the self-destructive energy source we just cannot seem to quit, despite the decades of research demonstrating its unsustainable environmental, health, social, and financial costs. In 2019, surface and underground coal extraction expanded, world consumption increased, and combustion still accounts for 40 percent of global CO_2 emissions from energy use (Jakob et al. 2020).

Coal's damaging effects begin before it is taken out of the ground. It causes biodiversity loss and environmental damage as forests are felled and mountains disfigured to clear surfaces. This land can never be returned to its exact prior ecological balance or used for planting crops. The mining process then releases numerous environmental toxins, including high amounts of mercury that can poison waterways, wildlife, and human populations. Both extracting and burning coal for use discharge carbon monoxide, carbon and sulfur dioxides, nitrogen oxides, lead, arsenic, and toxic heavy metals and soot that cause global warming and acid rain. Methane is particularly concerning because it is eighty-six times more potent than carbon dioxide at trapping heat over twenty years (Jackson et al. 2020). To keep global warming within 1.5 °C of pre-industrial levels, researchers argue that in most scenarios, there needs to be a complete cessation of coal use by 2050 (Dessler 2016, 72). The causal link between coal and climate change is clear.

Some argue that coal use may continue more sustainably with the increased use of Carbon Capture and Storage (CCS) systems. This process involves capturing carbon dioxide from industrial processes, such as fossil fuel power stations, and then storing it securely underground. However, CCS is an emerging technology that remains inefficient, especially when compared with cleaner forms of energy generation, and it is still unproven at the scale necessary to meaningfully address climate change. CCS also does not reduce other harmful pollutants produced in the fuel cycle of coal.

Meanwhile, coal continues to imperil public health. In addition to coal mines poisoning the land communities rely on, air pollution from coal-fired power plants causes asthma, cancer, heart and lung ailments, and neurological problems under certain conditions. A recent study indicates that the continuous inhalation of coal-related hazardous substances triggers cardiovascular disease, brain degeneration, DNA damage, and autoimmune and reproductive disorders (Gasparotto and Martinello 2020). Furthermore, coal mining and processing remain one of the most dangerous professions. In addition to continual exposure to these toxins, miners may endure underground gassings, incinerations, or explosions from three deadly gases (methane, carbon monoxide, and carbon dioxide) (Bradley 2020, 9). There is also the danger of cave-ins, floods, and falling rocks (Hamby 2020, 32). Although coal mining labor is highly regulated in the developed world, it is hugely unmonitored, and thus perilous, in the Global South.

There is less attention on the social conflicts arising from the coal industry. Grievances occur when communities are displaced from land to make way for extraction sites or left uncompensated for mining-related losses. Particularly in developing countries, coal mining is linked to environmental injustice, gender-based violence, and labor exploitation, including that of children (Cardoso 2015; Cock 2019). In India alone, an estimated 1/5 of all mineworkers are under the age of seventeen (Chandran 2016). As a whole, mining often has unequal gender outcomes, with women less likely than men to find employment opportunities or leadership positions in the sector (Buss et al. 2017).

Coal's popularity persists despite the indications that it is increasingly more expensive than alternative forms of energy. Most of the world's surface-level "easy" coal has been mined already, forcing extraction from deeper underground, which increases costs. There are now cheaper fuels, including natural gas, which is why 2019 was the first year in which renewables surpassed coal in energy production (Rowland 2020). There are no expectations that coal will fall out of favor any time soon, however.

It would seem that if the environmental and health impacts of coal were not enough to end its use, at least the supposed laissez-faire principles of energy pricing would. Yet, the question remains in the face of its economic inefficiency, "Why is coal so hard to quit?" Because, as Sengupta (2018) describes it, coal is a "powerful incumbent." It exists underground in a finite amount and eager companies—backed by powerful governments and esurient banks—are in a rush to access it as soon as possible. National electricity grids were designed for it and coal plants help politicians to deliver cheap electricity to voters. The coal mining industry employs about eight million people globally and creates revenues of more than US$900 billion a year. Coal use is unlikely to decline in the immediate future, and reductions in the Global North are offset by growth in China, India, and other Asian countries, ensuring future demand (Sengupta 2018).

The principle solution to the continued demand for coal in the face of its apparent harm is an international ban. With any maneuvering room at all in law, policy, or regulation, those who profit from it will find a way to keep coal use ongoing at an unsustainable rate for the planet and people. The urgency of climate change, public health, and socioeconomic well-being call for a global embargo gradually rolled out over the next decade. It is crucial that the phase-out be fair and that the process dovetails with political realities. Law and policymakers must understand in more detail who "will be affected by a transition away from coal, how these societal groups can be effectively compensated, and how powerful vested interests can be counterbalanced" (Jakob et al. 2020). Despite the challenges leaders will face, to ignore the pressing need for a ban on coal would be environmentally and socially careless.

FURTHER READING AND REFERENCES

Bradley, M. 2020. *Blood Runs Coal: The Yablonski Murders and the Battle for the United Mine Workers of America*. New York: W. W. Norton.

Buss, D., B. A. Rutherford, J. Hinton, J. M. Stewart, J. Lebert, G. E. Côté, A. Sebina-Zziwa, R. Kibombo, and F. Kisekka. 2017. *Gender and Artisanal and Small-Scale Mining in Central and East Africa: Barriers and Benefits*. Montreal: McGill University and the Institute for the Study of International Development.

Cardoso, A. 2015. "Behind the Life Cycle of Coal: Socio-environmental Liabilities of Coal Mining in Cesar, Colombia." *Ecological Economics* 120: 71–82. https://doi.org/10.1016/j.ecolecon.2015.10.004.

Chandran, R. 2016. "Children Working in India's Coal Mines Came as 'Complete Shock', Filmmaker Says." *Thomson Reuters Foundation*, July 6. Accessed July 14, 2024. https://www.reuters.com/article/lifestyle/children-working-in-india-s-coal-mines-came-as-complete-shock-filmmaker-says-idUSKCN0ZM1DY/

Cock, J. 2019. "Resistance to Coal Inequalities and the Possibilities of a Just Transition in South Africa." *Null* 36, no. 6: 860–73. 10.1080/0376835X.2019.1660859.

Dessler, A. E. 2016. *Introduction to Modern Climate Change*. Cambridge: Cambridge University Press.

Gasparotto, J., and K. Da Boit Martinello. 2020. "Coal as an Energy Source and Its Impacts on Human Health." *Energy Geoscience*. https://doi.org/10.1016/j.engeos.2020.07.003.

Hamby, M. 2020. *Soul Full of Dust: A Fight for Breath and Justice in Appalachia*. New York: Little, Brown.

Jackson, R. B., M. Saunois, P. Bousquet, J. G. Canadell, B. Poulter, A. R. Stavert, and A. Tsuruta. 2020. "Increasing Anthropogenic Methane Emissions Arise Equally from Agricultural and Fossil Fuel Sources." *Environmental Research Letters* 15, no. 7: 071002.

Jakob, M., J. C. Steckel, F. Jotzo, B. K. Sovacool, L. Cornelsen, R. Chandra, O. Edenhofer, C. Holden, A. Löschel, and T. Nace. 2020. "The Future of Coal in a Carbon-constrained Climate." *Nature Climate Change* 10, no. 8: 704–7.

Rowland, T. 2020. "Economics Wins the War on Coal." *Regulation* 43: 72.

Sengupta, S. 2018. "The World Needs to Quit Coal: Why Is It So Hard? *New York Times*, November 24. Accessed July 14, 2024. https://www.nytimes.com/2018/11/24/climate/coal-global-warming.html.

LIMITS OF SOLAR AND WIND

Helena Solman

Introduction

Wind and solar energy are two of the key renewable energy resources in the world, accounting for approximately 27 percent of all the renewable electricity produced in 2018 (International Renewable Energy Agency 2018). Wind turbines can be located on land (called onshore wind) and on the sea (called offshore wind). Solar energy is generated by photovoltaic (PV) modules, commonly referred to as solar panels (see Figures 5.1 and 5.2).

FIGURE 5.1 Solar park. Image by Sebastian Ganso from Pixabay.

FIGURE 5.2 Wind park. Image by Mar from Pixabay.

Current Trends: Upscaling Technologies

In the last decades, technologies of solar panels and wind turbines have been gradually improving (Karnøe and Garud 2012; Jacoby 2016). However, in order to meet a growing demand for electricity generated by renewable sources, diffusion of solar and wind energy is needed at a large scale. Wind turbines have already moved from relatively small to very high turbines. For example, GE Renewable Energy recently developed at 13 MW turbine that is 260 m high (GE Renewable Energy 2020). While such turbines are mostly used offshore, the marine environment is harsh, making it more expensive to mount turbines, transport electricity, maintain the infrastructure, and decommission wind turbines after their lifespan (Van Bussel et al. 2001). In some places around the world, the sea is too deep to mount the turbines on the bottom of the sea. To address this challenge, an alternative design solution was proposed: floating wind turbines.

Challenges of Developing Solar and Wind Energy

Finding Space The challenge of finding space is a growing problem, especially for onshore wind energy. Development of wind turbines is commonly regulated by governmental legislation, for example, on the maximum noise level (Koppen and Fowler 2015). Such restrictions are necessary to assure safe and livable energy landscapes; however, in countries where land is scarce it is difficult to find appropriate locations for development (Rudolph and Kirkegaard 2019). One of the possible ways to make more efficient use of space is to develop multifunctional energy landscapes (Stremke 2010). For example, by developing wind energy or solar parks on agricultural land (in a way that allows crops to grow under and around the installations), food and energy production are combined.

Energy Variability and Storage The key advantage of wind and solar energy is that they rely on naturally occurring resources of sun and wind. However, the amount of wind and sunshine differs significantly across different geographical regions. Variability in energy supply is a great challenge to traditional energy systems as they were built to operate on constant flows of power (Boyle 2009; Ueckerdt, Brecha, and Luderer 2015). In other words, the challenge is to manage the supply of renewable energy so that it is dependable. On the other hand, there is also a need to improve the energy infrastructure so that it can deal with varying amounts of energy. One of the potential solutions for the grid includes development of units for energy storage—batteries absorbing surplus of energy. Thus far, batteries remain expensive and are often considered a nonoptimal solution, mainly due to a negative environmental impact of hazardous material contained in them (Dehghani-Sanij et al. 2019).

Public Acceptance and Environmental Impacts Apart from these technical challenges related to the exploitation of wind and solar energy, there is a range of social and environmental issues related to implementation of solar and wind energy. For example, the public has voiced concerns about the visual impact of wind turbines on landscapes (Möller 2006), bird strikes (Horbaty, Huber, and Ellis 2012), or noise (Wolsink et al. 1993). While large solar parks tend to also create concerns related to landscape preservation (Zoellner, Schweizer-Ries, and Wemheuer 2008), little is known about their impact on fauna and flora (Gasparatos et al. 2017).

In addition to issues with possible negative impacts of these technologies, there are also concerns related to how wind and solar energy projects are planned and managed (Solman et al. 2021). For example, when wind or solar parks are being developed in areas where people live or work, those people might want to be involved in decisions about how, where, and by whom these projects are going to be developed and managed over time. When participation is missing or is perceived as unfair, opposition tends to emerge (Haggett 2014). Opposition can be avoided by assuring a fair distribution of benefits from energy generation (Lienhoop 2018). One of the ways of doing that is to invite citizens to invest in local projects or to create a community benefit scheme (Li et al. 2013).

Conclusions

Both wind and solar energy have a large potential to become a part of sustainable and green energy infrastructure; however, there is a range of social, environmental, and technical challenges that need to be considered while planning and managing such projects. While many of the technical challenges require innovating the design of both solar and wind technologies, there are also challenges linked to grid integration and maintenance. Success of both wind and solar energy projects is also linked to the extent to which project developers meaningfully engage local communities and other stakeholders.

While there are clear environmental benefits of using wind and solar energy, both sources of renewable energy also have impacts on nature. For this reason, energy efficiency remains an important factor in increasing the sustainability of the current energy systems.

FURTHER READING AND REFERENCES

Boyle, G. 2009. *Renewable Electricity and the Grid: The Challenge of Variability.* London: Routledge.

Dehghani-Sanij, A. R., E. Tharumalingam, M. B. Dusseault, and R. Fraser. 2019. "Study of Energy Storage Systems and Environmental Challenges of Batteries." *Renewable and Sustainable Energy Reviews* 104: 192–208.

Gasparatos, A., C. N. Doll, M. Esteban, A. Ahmed, and T. A. Olang. 2017. "Renewable Energy and Biodiversity: Implications for Transitioning to a Green Economy." *Renewable and Sustainable Energy Reviews* 70: 161–84.

GE Renewable Energy. 2020. "An Industry First. Haliade-X Offshore Wind Turbine. The World's Most Powerful Offshore Wind Turbine in Operation." Accessed July 14, 2024. https://www.ge.com/renewableenergy/wind-energy/offshore-wind/haliade-x-offshore-turbine.

Haggett, C. 2014. "'Planning and Persuasion': Public Engagement in Renewable Energy Decision-Making." In *Renewable Energy and the Public*, edited by P. Devine-Wright, 47–60. London: Routledge.

Horbaty, R., S. Huber, and G. Ellis. 2012. "Large-scale Wind Deployment, Social Acceptance." *Wiley Interdisciplinary Reviews: Energy and Environment* 1, no. 2: 194–205.

International Renewable Energy Agency. 2018. "Renewable Energy Technologies." Accessed November 7, 2020. https://www.irena.org/Statistics/View-Data-by-Topic/Capacity-and-enerat ion/Technologies.
Jacoby, M. 2016. "The Future of Low-cost Solar Cells." *Chemical & Engineering News* 94, no. 18: 30–5.
Koppen, E., and K. Fowler. 2015. "International Legislation for Wind Turbine Noise." In *Proceedings of the Euronoise*, edited by C. Glorieux, 321–6. Maastricht, The Netherlands: NAG, ABAV. Accessed July 14, 2024. https://www.conforg.fr/euronoise2015/proc eedings/data/articles/000225.pdf.
Li, L. W., J. Birmele, H. Schaich, and W. Konold. 2013. "Transitioning to Community-Owned Renewable Energy: Lessons from Germany." *Procedia Environmental Sciences* 17: 719–28.
Lienhoop, N. 2018. "Acceptance of Wind Energy and the Role of Financial and Procedural Participation: An Investigation with Focus Groups and Choice Experiments." *Energy Policy* 118: 97–105.
Möller, B. 2006. "Changing Wind-Power Landscapes: Regional Assessment of Visual Impact on Land Use and Population in Northern Jutland, Denmark." *Applied Energy* 83, no. 5: 477–94.
Rudolph, D., and J. K. Kirkegaard. 2019. "Making Space for Wind Farms: Practices of Territorial Stigmatisation in Rural Denmark." *Antipode* 51, no. 2: 642–63.
Solman, H., M. Smits, B. van Vliet, and S. Bush. 2021. "Co-production in the Wind Energy Sector: A Systematic Literature Review of Public Engagement beyond Invited Stakeholder Participation." *Energy Research & Social Science* 72: 101876.
Stremke, S. 2010. "Designing Sustainable Energy Landscapes: Concepts, Principles and Procedures." Dissertation. Wageningen University, the Netherlands.
Ueckerdt, F., R. Brecha, and G. Luderer. 2015. "Analyzing Major Challenges of Wind and Solar Variability in Power Systems." *Renewable Energy* 81: 1–10.
Van Bussel, G. J. W., and A. R. Henderson. 2001. "State of the Art and Technology Trends for Offshore Wind Energy: Operation and Maintenance Issues." *In Proceedings Offshore Wind Energy Special Topic Conference*, December 10–12, pp. 1–4. Brussels, Belgium: European Wind Energy Association.
Wolsink, M., M. Sprengers, A. Keuper, T. H. Pedersen, and C. A. Westra. 1993. "Annoyance from Wind Turbine Noise on Sixteen Sites in Three Countries." *European Community Wind Energy Conference* 271 (March): 276.
Zoellner, J., P. Schweizer-Ries, and C. Wemheuer. 2008. "Public Acceptance of Renewable Energies: Results from Case Studies in Germany." *Energy Policy* 36, no. 11: 4136–41.

THE PROMISE OF NUCLEAR ENERGY

Ayesha Sadiqa

The urgency to deal with looming climate crises has revived local and international interest in nuclear energy. Many commentators and policymakers are convinced that nuclear energy is the answer to climate change and soaring energy demand around the world. Environmentalist James Lovelock (2004) argued that the advancement of nuclear power is the only alternative energy option that can meet growing electricity demand and that associated risks with nuclear energy are insignificant compared to the real danger posed by climate change–related intolerable and lethal heatwaves. Supported by some strong institutes—the US Department of Energy, the International Energy Agency, and the International Atomic Energy Agency—nuclear energy is ready to rapidly grow around the globe and is making its way back into energy policy roundtables. Ambitious plans for increasing nuclear capacity to generate electricity exist in many countries, especially Russia, China, and India. Around 50 new nuclear plants are being constructed in 16 different countries, 100 new plants are planned, and 300 have been proposed. China expects to build twenty-seven new reactors in next fifteen years while Japan is trying to increase its nuclear electricity share to 40 percent by 2040.

Each stage of the fuel cycle in nuclear plants has high capital cost—from production, fuel recycling, waste storage, and decommissioning. Nuclear reactors suffer cost and time overruns regardless of their size. A study found that almost 97 percent of the nuclear reactors constructed between 1936 and 2014 had some type of overrun (Sovacool, Gilbert, and Nugent 2014). Nuclear power plants have long construction time and abundant uncertainties during the construction process. These projects are often state-funded but still have huge subsidies in the form of assurances and limited responsibility for major accidents.

Nuclear energy is not carbon neutral despite the fact that the heat and energy generation stages do not emit greenhouse gases. Fuel processing upstream and downstream in nuclear plants is mainly responsible for the greenhouse gas emissions. Mining, enrichment, milling, fuel fabrication, and waste management all involve intensive capital and energy inputs and hence produce carbon emissions. Dependence of nuclear plants on imported uranium can also increase a country's reliance on other nations, with the risk of price spikes and volatility (Lenzen 2008). Historically, uranium prices are somewhat volatile, increasing six times from 1973 to 1976, declining throughout 2002, then dramatically soaring until 2007 before dropped again in 2008 (Senate Energy, Utilities and Communications Committee 2007). Uncertainty and volatility in uranium prices can have considerable effects on the operation of nuclear power plants. According to the Nuclear Energy Agency, every 1,000-MW reactor needs 200 metric tonnes of uranium; fuel is responsible for 15 percent of the cost of the power plant.

During fission, there is no burning and oxidation process, so almost all of a plant's fuel is converted into waste with slight reduction in mass. Typically, a standard nuclear reactor generates 30 tonnes of waste that could be radioactive for a very long period of time (250,000 years). Assuming the current annual storage cost of US$35,000 for 1 ton of nuclear waste, every nuclear reactor has an additional cost of US$875 million. Only 15 percent of the nuclear waste is recycled, and the recycling process is not as efficient (only 1 percent) as the recycling process for non-reprocessed fuel (Sovacool 2010). Nuclear plants and waste can pollute and degrade land in five different areas (Fleming, Harvey, and Bowers 2022: (1) plants generate nuclear waste at the plant site; (2) plants need uranium mines and mills; (3) during operation, plants regularly discharge small quantities of radioactive isotopes; (4) during an accident, plants can discharge huge amounts of pollution; and (5) plants produce plutonium waste.

Different stages of nuclear plant operation and fuel cycle, milling, mining, and waste storage not only withdraw and consume water but also contaminate water supplies. Since the concentration of the uranium is extremely low, the process is extremely water intensive. These practices can contaminate and pollute local land and water supplies directly. Massive supplies of water are usually required to cool down a reactor core during operation, and some of that water is lost as it turns into steam. Nuclear power generation uses more water than any other electricity generation facility (Bajwa et al. 2017; Okano et al. 2022).

Nuclear power plants have questionable safety records at best. Human error is considered responsible for the mishap at Three-Mile Island in the United States and design flaws were blamed in the case of Chernobyl in the Soviet Union (now Ukraine). No fewer than ninety-nine nuclear accidents, accounting for US$20.5 billion in damage, occurred between 1952 and 2009 around the world. (An accident is defined, according to US federal government criteria, as any incident that resulted in either the loss of human life or more than US$50,000 of property damage.) When the damage and fatalities of nuclear accidents are compared with other energy generation technologies, nuclear power is the second most deadly source of electricity generation after hydroelectric dams. The nuclear industry claims that it has worked very hard to ensure the safety of nuclear plants. Newly designed reactors have safety features that can overcome the issues that led to Chernobyl and Fukushima (in Japan). New reactors still face criticism and questions about earthquake handling, and human error still has the potential to lead to a nuclear accident (Cooper 2014; Sovacool 2008).

In summary, significant resources (financial and material) are required for nuclear plant manufacturing, fuel availability, recycling, waste storage, recovering, and decommissioning. New nuclear reactors can still lead to the risk of serious accidents and breakdowns. The decline in availability of high-quality uranium ore is a looming threat, and the fuel cycle of nuclear plants can degrade and contaminate water resources and supplies. Significant amounts of greenhouse gas emissions are associated with the nuclear lifecycle. Nuclear waste storage sites invariably have a huge environment cost. When all of these criteria are considered, renewable energy technologies offer superior alternatives to nuclear energy for policymakers. They can reduce the possibility of fuel disturbances and scarcity while improving transmission systems. These technologies are ecologically friendly and less expensive to build and maintain for a longer period. Small renewable generators produce electricity in small steps and do not need to depend on continuous government subsidies. The waste and greenhouse gas emissions are negligible as compared to other conventional electricity generation technologies and do not have significant risks of accidents.

FURTHER READING AND REFERENCES

Bajwa, B. S., S. Kumar, S. Singh, S. K. Sahoo, and R. M. Tripathi. 2017. "Uranium and Other Heavy Toxic Elements Distribution in the Drinking Water Samples of SW-Punjab, India." *Journal of Radiation Research and Applied Sciences* 10, no. 1: 13–19. https://doi.org/10.1016/j.jrras.2015.01.002.

Cooper, M. 2014. "The Economic Failure of Nuclear Power and the Development of a Low Carbon Electricity Future: Why Small Modular Reactors Are Part of the Problem, Not the

Solution." NIRS, May. Accessed July 14, 2024. https://www.nirs.org/wp-content/uploads/reactorwatch/newreactors/cooper-smrsaretheproblemnotthesolution.pdf.

Fleming, M., K. Harvey, and K. C. Bowers. 2022. "Development and Testing of a Nuclear Regulator Safety Culture Perception Survey." *Safety Science* 153: 105792. https://doi.org/10.1016/j.ssci.2022.105792.

"James Lovelock: Nuclear Power Is the Only Green Solution." 2004. *The Independent*, May 23. Accessed July 14, 2024. https://www.independent.co.uk/voices/commentators/james-lovelock-nuclear-power-is-the-only-green-solution-564446.html.

Lenzen, M. 2008. "Life Cycle Energy and Greenhouse Gas Emissions of Nuclear Energy: A Review." *Energy Conversion and Management* 49, no. 8: 2178–99. https://doi.org/10.1016/j.enconman.2008.01.033.

Okano, I., A. Rosenberg, M. Dworkin, V. Murthy, S. Jayaraman, and K. Takabe. 2022. "Direct and Indirect Health Effects of the Nuclear Power Plant Disasters: A Review for Health Care Professionals." *IJS Global Health* 5, no. 5: e71. https://doi.org/10.1097/GH9.0000000000000071.

Senate Energy, Utilities and Communications Committee. 2007. "Hearing before the Senate Energy, Utilities and Communications." https://seuc.senate.ca.gov/jimhardingintroduction.

Sovacool, B. K. 2008. "Valuing the Greenhouse Gas Emissions from Nuclear Power: A Critical Survey." *Energy Policy* 36, no. 8: 2950–63. https://doi.org/10.1016/j.enpol.2008.04.017.

Sovacool, B. K. 2010. "A Critical Evaluation of Nuclear Power and Renewable Electricity in Asia." *Journal of Contemporary Asia* 40, no. 3: 369–400. https://doi.org/10.1080/00472331003798350.

Sovacool, B. K., A. Gilbert, and D. Nugent. 2014. "Risk, Innovation, Electricity Infrastructure and Construction Cost Overruns: Testing Six Hypotheses." *Energy* 74: 906–17. https://doi.org/10.1016/j.energy.2014.07.070.

PART II

A–Z Entries

INTRODUCTION

Part II aims to consolidate much of the information that can be found throughout this book into relatively concise yet comprehensive, alphabetically arranged entries that each address a particular topic. For example, wind power has been discussed in three different chapters, which delve into the history of wind power, current wind power capabilities, and the future of wind power based on growth projections and intentional commitments to attempting to stop the worst effects of climate change by transitioning to renewable energy. All of that information is summarized in one entry on "Wind Power" in this section. Additional technical information that may not have been appropriate for other sections of the book is included here as well. This section allows the researcher to delve deeper into a particular subject and understand the technical aspects as well as some of the nuances that were not addressed in other parts of the book. Entries are arranged alphabetically, although some entry titles contain multiple terms. (For example, one entry covers carbon tax, carbon markets, carbon credits, and cap and trade; it is alphabetized under "Carbon Tax.") Cross-references between entries appear in the text where appropriate. While no entry is exhaustive, this section aims to cover in detail all the major subjects of *Energy Resource Conflict.*

AGRICULTURAL REVOLUTION

The agricultural revolution was the slow process of groups of humans transitioning from gathering and hunting to growing crops for food. This gradual process occurred during prehistoric times, which means there are no written records of the events. Archaeologists are able to observe the change through ancient sites that demonstrate the creation of tools for growing food crops as well as the equipment needed to turn crops into something easily edible and/or able to be stored for long periods of time.

We intentionally refer to "gathering and hunting" instead of "hunting and gathering" because, despite the fact that hunting inspired the movement of humans out of Africa and across all the continents except Antarctica, both hunters and gathers obtained the majority of their calories from gathering. This is due to the intermittent nature of hunting; without a guaranteed alternative source of calories, a few failed hunting trips could sound the death knell for a tribe.

There are no shortages of theories on how and why humans transitioned from gatherers to growers. Most likely gatherers had learned millennia before the agricultural revolution that there must be seed left behind if they were to gather the food source again, regardless of the food source. Some nuts must remain in order to grow trees, seeds from the fruits must remain for new plants to sprout, and some grass seeds (grain) must remain in order for there to be grain in the future. Ancient humans were enmeshed in their natural environment; they were fully a part of it. They knew the cycles of the moon, the patterns of the tides, the migration of animals, and the life cycles of plants. These keen observations no doubt enticed humans to encourage the growth of the plants and trees that most benefited their livelihood. This truism can be observed in many forests and savannas where people gathered and hunted. The trees that exist in areas people frequented have a greater percentage of fruit- and nut-bearing trees and plants. Trees that are less useful for humans are used for firewood. Trees that provide excellent building material also tend to remain at higher percentages in the forests, as they are allowed to grow until they are large enough to serve the desired purpose. Conversely, in areas where game is to be increased, humans will set fire to the dry grasses in the fall in order to destroy young saplings. This increases the amount of grassland, which in turn increases the size of the animal herds that live off of that grassland. During this early period, humans altered the natural environment in order to increase local food resources, but it was still a long way from agriculture and animal husbandry.

As this process continued, humans spent more and more time encouraging the growth of grains. A grain is basically the seed of a tall grass-like plant. Why did ancient civilizations across the planet tend to settle on grains like barley, rice, wheat, quinoa, amaranth, sorghum, teff, and millet? (Other cultures also grew root vegetables such as potato, cassava, and sweet potatoes. Humans have been digging up and roasting root vegetables for at least 170,000 years.) These grains were easily stored for long periods of time in two ways: (1) as dried grains in built storage areas and (2) as fermented beverages. It is a distinct possibility that early agriculture and the agricultural revolution in particular owe their success to humans' drive to make beer and wine. Grain can be both the main dish and the accompanying beverage. The argument that remains is what came first: Did humans accidentally overferment some dough and start making beer? Did early gatherers and hunters ferment grains to create seasonal beer and then decide it would be better to grow enough to have beer available all year long? These are questions that cannot be definitively answered. The Chinese were brewing beer by 7000 BCE. The ancient Mesopotamian people of the Godin Tepe settlement in Modern Iran were brewing between 3500 and 2900 BCE (Nelson 2005) according to archeological evidence or, as some claim, as early as 10,000 BCE, before settled agriculture appeared in the area. The ancient people of Sumer were likely brewing some time in between these dates. Grain for bread and grain for beer clearly drove the agricultural revolution.

A bigger change that accompanied the agricultural revolution is that people became settled for the first time. Before agriculture most people moved to where the food was during each season. People built habitats where they spent time in the winter, for example, or summer camps. They built religious and ceremonial sites, but they did not live in one place all year. There were no cities. The agricultural revolution was the beginning of walled villages, which became walled cities. These cities became the base for empires to be built. These fortified towns and cities held food and other riches. As farming techniques and technology improved, people were able to specialize. Sculptors, politicians, bakers, religious leaders, and soldiers were able to eat without hunting, gathering, or growing their own food. This changed human society in myriad ways. It led to an explosion of arts and crafts. It led to writing and huge advances in human knowledge. It also led to bigger, longer, and more devastating conflicts over resources.

At the same time people who grew food brought in semi-domesticated animals and made them truly domesticated, dependent on humans for all their needs. The animals were no longer able to roam freely and became an important part of the agricultural process. Humans fed their animals (chickens, goats, pigs, sheep, cows, horses, donkeys, etc.) the chaff-— the part of the plant that was left over after the harvest of grains. Animals in turn created waste that worked as fertilizer for the next season's crops. This beneficial closed-loop cycle allowed people to farm in areas indefinitely without totally exhausting the land. This cycle is the basis for permanent human settlement.

Gatherer and hunting groups would attack villages for their foodstuff: animals, fruits, nuts, and most importantly grain and beer. Nomadic herders would also attack villages looking for food, wealth, and women. The conflict over energy resources intensified as successful farming created a larger population that then convert more wildlands into farmland. This created conflict with the people who already were living off the land by gathering, hunting, and herding. This conflict continues today, all over the world. In South America, the native people of the Amazon River basin are continually confronted by slash and burn farmers, cattle ranchers, and gold miners who all want access to the vast natural resources of the Amazon. One way of life is not compatible with the other, and much blood has been and will be spilt as the agricultural revolution continues to spread into some of the last wild places on earth.

BIOFUELS

Biofuels are, most simply, fuels made from something that can be grown. The source can be corn, sugarcane, farm by-products, or algae, among other agricultural products. The advantage of a biofuel over a fossil fuel is that if the input (energy put into growing the fuel, including fertilizers, pesticides, and power used in machinery) is less than the output (usable fuel after being harvested and processed) then biofuel can be a renewable resource. Biofuels are looked at as a sustainable alternative to fossil fuels; however, this is not always the case.

When corn is used to create ethanol, the total amount of energy used in the process is equal to and sometimes greater than the amount of energy produced. The only reason that corn-to-ethanol fuel is in existence is that federal programs in the United States artificially lower the price of corn through guaranteed minimum floor prices (through subsidized insurance programs) and the government creates a market for corn ethanol by mandating ethanol use in gasoline. Most states mandate a certain percentage of gasoline sold at gas stations be made of ethanol during specific times of the year. This creates a market that cannot use a less expensive product. The hope behind the creation of these programs was that the United States could grow their way to energy independence and create a system that is carbon-neutral or at least less carbon intensive. The data demonstrates that corn to ethanol is not a path to energy independence nor is it a path to less carbon-intensive transportation. If the agricultural waste that remains after the cobs of corn are harvested could be converted into fuel, then it would be a renewable fuel. Currently, that is not the case.

Brazil has the largest fleet of cars that can run on ethanol derived from sugar cane. All cars sold in Brazil since 1975 are required to have an engine that runs on an ethanol-gasoline blend (Briguet Bastos 2007). The program, named Proácool was instituted to use home-grown sugarcane to create a local fuel. The drive for this came out of the need to address the fuel crises of the 1970s. This allows Brazil to be more resilient in the face of volatile oil prices. Brazil is able to grow more of its fuel each year and therefore needs to import less oil. Despite its large population, Brazil has become a net exporter of oil due to the production of sugarcane-based biofuels.

Algae-based biofuels have received a lot of research money over the years. The idea has been around since the 1940s. During the Second World War there was a spike in interest in creating oil through growing algae, especially in countries that did not have a lot of natural oil resources. The difficulty in algae-to-oil systems is that it is usually more expensive to create oil from algae than drilling for oil. Algae-based biofuels have been and will continue to be sold when market conditions are right. Algae can also be used to sequester CO_2 and is a renewable energy source that is net zero. If there is a market for carbon that includes a carbon tax, that tax would not apply to algae-based biofuels. In that type of market algae-based fuels would remain competitive and profitable.

The future of biofuels will depend on a number of factors. The two most important are the ability of companies to continue to lower the cost of biofuels and the ability of governments to create demand for biofuels by regulating CO_2. If the world hopes to reach the goals stated in the Paris Climate Accords, certain biofuels should have a place in the market. It should be noted that biofuels derived from foodstuff are less desirable than those that are created from agricultural waste and algae. It could become problematic to have edible crops being used for fuel, as that could lead to rising food prices and food insecurity for many people. However, a properly focused biofuel program could be used to create a carbon-neutral fuel source that could be part of the effort to move the planet toward a carbon-neutral future.

BIOMASS ENERGY

Biomass energy is using organic materials that come from plants or trees directly for fuel. This is different from biofuels as there is no process to transform the material to fuel. Biomass, such as wood or agricultural chaff (materials remaining after edible parts are removed, such as wheat), is burned and the heat produced creates steam. The steam is used to turn a turbine that generates an electric current, just as would be done in a coal-burning power plant. Biomass can be controversial for a couple of reasons. Some NGOs argue that the use of forest products (primarily trees) for power plants will lead to deforestation. There is also an argument that rightly ascertains that burning agricultural products creates more pollution than using renewable energy.

Some argue that biomass should be allowed to compost and/or decompose back into the soil in the natural carbon cycle that has always existed. The counter argument is that the burning of biomass can be beneficial in some places—for example, there are many forests in the western half of the United States and across many other areas of the world that have been poorly managed and require thinning in order to avoid the catastrophic fires that have grown in severity over the last decade due to climate change. An additional fear is that if governments create a market for wood to be used as energy, bad actors in the forestry space would eventually find ways to clear-cut forests and abuse the system. Forest management is a "sticky wicket" or a "wicked" problem (a problem that has so many variables and possible solutions that it is extremely difficult to find an answer) in that there are a great number of factors that vary from forest to forest. To create a functioning system, professional forest experts would need to be consulted in each distinct forest so that they could inform the work done by the companies extracting forest products. This would require an investment by each country's federal government to be sure that they had enough experts who were also empowered to make and enforce their decisions. In this scenario, it would be possible to manage each forest according to its needs. Some forests thrive with low-intensity fires while others are not equipped to survive fires well at all. Some forests thrive with a dense forest canopy and high number of trees per acre, while other forests must be spread out with a very low number of trees per acre. These forests require extremely different management. There are also some virgin forests that do not require any human intervention at all; these forests simply must be protected from any human interference. Virgin forests are only a small fraction of all the forests in the world. The vast majority of forests have already been modified and changed by humans. With correct management, these forests could be used for biomass energy without harming the species that live in them.

Agricultural chaff is already used in many places to create electricity. Farmers want to use their fields the most efficient way possible. That can mean growing crops up to three to four times a year, depending on the climate. They do not want to take up space on their farm for agricultural chaff. Agricultural chaff is defined as an agricultural product that is left over after the harvest. For example, this can be cornstalks from corn or the pea plants that remain after the peas have been harvested. If this chaff is incorrectly disposed of (put into a landfill) it can create methane gas, which is one of the most powerful GHGs that humans create. If the government provides a market for agricultural chaff, these GHG emissions can be avoided and renewable energy can be created. Burning any biomass creates CO_2, particulate pollutants, and other GHGs. It is important that new biomass power plants also include CO_2 recovery and intense pollution controls. If these power plants do not run as cleanly as possible, the negative effects could outweigh the positive. Air pollution is responsible for an estimated 7 million deaths per year (4.2 million are due to outdoor air pollutants) according to the World Health Organization (WHO 2021), so it is important that biomass does not contribute to that. Most of the deaths from pollution are connected to fossil fuels such as pollution from coal power plants and exhaust fumes from gasoline (petrol) and diesel vehicles.

There is space for biomass plants in net-zero energy plans. It is another situation in which it will take experts and scientists to determine the right quantities of biomass and create the best power plants to use the biomass. They must also create the least amount of CO_2 possible. This will likely require carbon capture and in this scenario could actually be carbon negative. Plants take up CO_2 in the growing process, so if the CO_2 that is released from the burning process is captured, biomass power plants can be removing CO_2 from the atmosphere. Removing CO_2 from the atmosphere may become necessary if humanity hopes to achieve the goals of the Paris Climate Accords. Biomass power plants, done correctly, may be one opportunity to do just that.

CARBON CAPTURE AND STORAGE

Carbon capture and storage (CCS) is the process of capturing carbon either directly from the atmosphere or from a power plant that burns fossil fuels, biofuels, or biomass. The processes vary, but the end goal remains the same: take CO_2 that would otherwise be released into the atmosphere and find a way to store it in the earth as permanently as possible.

Carbon storage is important for two reasons: it can be used on site in conjunction with any power plant that burns fossil fuels or biomass, and it can be used in conjunction with devices that capture carbon from the air and the ocean. (This information is also covered in Chapter 3 under the heading "Carbon Storage.") The carbon that is captured is usually in the form of CO_2. The gas is then injected into the earth. Ideally, the CO_2 will be injected into a site that will permanently store it and not eventually release it into the atmosphere. The two most common types of underground storage are oil and gas wells and geological storage. Underground carbon storage is most effective when geologists can identify formations in which CO_2 will chemically react with minerals in a formation to create a new but permanent geological formation. However, the most common current use of CO_2 is in the oil and gas industry. CO_2 can be injected into old oil and gas wells to push out the remaining oil and gas and may also be useful as part of the hydraulic fracking process. CO_2 is needed commercially for a number of applications, including the creation of carbonated beverages such as Coke or Pepsi products. There is also a growing demand (as of 2024) for carbonated water.

Carbon capture has not gained widespread use for two reasons: it is very expensive, and there are very few places that have created a carbon market by putting a price on releasing carbon into the atmosphere. There is however new research that has demonstrated a process that not only captures CO_2 but also converts it into a fuel. Cutting-edge companies like LanzaTech are already converting captured carbon emissions into jet fuel. (These emissions are captured at a fossil fuel power plant.) Carbon capture can also be done anywhere on the planet. Public sentiment might be more inclined to approve nuclear-powered carbon capture facilities if they are located far from large population centers.

Remote nuclear-powered CCS plants may become necessary in order to remove carbon on a scale that can actually change the likely detrimental rise of atmospheric CO_2 levels. There is another possibility as well: using renewable energy during times of overproduction to remove carbon from the atmosphere. If remote nuclear and renewable CCS can be combined with nascent technology that converts CO_2 into fuel, two problems could be solved at once: the continued need for liquid fuels in certain industries and the need to remove carbon from the atmosphere. With proper investment, CCS and CO_2 to fuel technology could revolutionize efforts to control CO_2 concentrations.

While CCS is still an emerging technology, it has an important role to play in the transition from a carbon-based energy system to a carbon-free energy system. Some industries will likely continue to need to use carbon-based fuels. If those fuels could be sourced from the atmosphere, the world would no longer be adding to already high atmospheric CO_2 concentrations.

CARBON DIOXIDE

Carbon dioxide (CO_2) consists of a carbon atom double-bonded to two oxygen atoms. It makes up a relatively small percentage of the atmosphere, 0.04 percent or 412 parts per million. Historically, over the past 800,000 years, CO_2 levels never crossed 300 parts per million. Pre-industrial levels were at 280 parts per million. Human activities—specifically the burning of fossil fuels to heat homes, create electricity, and power transportation over the past 200 years—have caused this significant jump in the amount of CO_2 in the atmosphere. Without a doubt, the scientific consensus is that this human-caused acceleration of atmospheric carbon concentration has led to a warmer climate. This climate change has had a wide variety of effects. Some areas are becoming wetter and more prone to flooding. Other areas are becoming drier and more prone to drought and catastrophic wildfires. Almost all areas are becoming warmer, and glaciers are melting over most of the world.

Millions of years ago the planet was much warmer and CO_2 levels were much higher than they are today. This was a world that had a tropical Antarctica. There are very few plants and animals from the present that could live in that world and vice versa. Most of major cities would be under water if sea levels were as high now as they were then. A return to these CO_2 levels would prove disastrous to all life on this planet. Millions of years of CO_2 sequestering by rocks slowly brought down the atmospheric CO_2. The past 800,000 years established a rhythm that included a number of ice ages and kept CO_2 levels below 300. Most of the life on the planet evolved to live in this climate.

Today's CO_2 levels are higher than they have ever been in human history. The effects of these high levels (424 parts per million) as of 2023 are already changing the world in dramatic ways. There are many more natural disasters due to more extreme weather brought on by a warmer climate. There are unexpected or counterintuitive effects as well. For example, warmer winter temperatures have allowed cold Arctic air to escape the area centered around the North Pole, resulting in devastatingly cold temperatures in the European Union and North America at times. Scientists use climate models to develop likely scenarios and to estimate fairly accurately what temperatures will be and where to expect some of the worst effects. However, the processes that create these scenarios could accelerate faster than predicted, which could result in acute changes that do not leave humanity time to prepare. Some feedback loops may accelerate at unpredictable rates. The goal, then, is prepare for likely changes and create contingency plans for less likely, yet possible, extreme changes.

The path forward is clear. Through a concentrated effort that includes thousands of different ways to limit CO_2 produced and to capture and sequester it, humanity can, and most likely will, bring down CO_2 levels. The two most important and effective methods for decreasing CO_2 emissions into the atmosphere are by transitioning from fossil fuel-based electricity generation to renewable energy generation and to transition away from fossil fuel-based transportation.

If nothing is done, in a few generations, cities that have been full of people for millennia will be reclaimed by the sea. Changing climate will cause a huge shift in human settlement patterns. In a very full world that will mean conflict over land and water. Severe storms may make some areas unlivable. While the resulting world is unlikely to mean the end of humanity (unless conflict becomes nuclear), there will be far fewer of our fellow species remaining.

CARBON FOOTPRINT

Your carbon footprint is the total amount of carbon dioxide (CO_2) that is released into the earth's atmosphere as a result of your personal actions. Humans produce CO_2 every time they exhale, but that is a miniscule amount compared to how much CO_2 is released every time someone drives a car or takes a plane or buys something that is shipped across the planet. Analyzing and understanding an individual carbon footprint can empower someone to make personal decisions that can have a positive effect on something as large as climate change.

While much effort has gone into teaching people how they can lower their carbon footprint (and it is true, they can), the biggest change in CO_2 per person comes from government policy. If a government creates a transit system that makes using electric buses easy and affordable, then making a smart decision to lower one's carbon footprint is easy. If the only green transportation available is a bike, there will be few people having a green commute on a winter day in Chicago, for example. Green infrastructure is necessary for people to make decisions that lower their carbon footprint.

There are still a number of fairly simple and straightforward things an individual can do to lower their carbon footprint. Diet is a simple change that can have huge effects on the amount of CO_2 a person is responsible for. Beef is by far the worst emitter of GHGs, especially methane. A pound of beef is responsible for ten times the amount of GHGs created when producing a pound of chicken and twenty times the amount created when producing a pound of a wild caught fish. Beef creates 180 times more CO_2 equivalent than nuts, and nuts also grow on trees that sequester CO_2. (Cows create methane in their digestion process and 1 kg of methane is equivalent to 25 kg of CO_2 emissions.)

Some studies have argued that grass-fed beef that is raised in a sustainable way can be closer to carbon-neutral, due to the carbon stored in perennial grass roots and cow waste that is absorbed into the soil and used by plants as fertilizer. That is, of course, a best-case scenario. If cattle are raised in dry and barren areas, they must eat much more food to receive the needed nutrients. In these cases, cattle can be responsible for 10–20 times more CO_2 equivalent emissions.

A simple solution is to reduce the amount of beef in your diet. It is important to note that small actions can still make big changes. For some people, a change in diet to reduce beef consumption may take the form of a meatless Monday. Chicken or fish are good substitutes, and some people will choose to switch from eating beef to these alternatives. Others will choose to completely change their eating habits toward a more planet-friendly diet by focusing on plant-based foods. If families simply eat less beef and more plants, grains, and nuts, CO_2 emissions from food would significantly decrease. Beef consumption has been trending lower in the United States over the past twenty years, but overall meat consumption (beef, lamb, pork, and chicken) has remained steady with a slight increase. Changes to government policies that artificially depress the cost of beef by creating an overproduction of corn may help to reduce the demand for beef.

Transportation can sometimes be the biggest share of a person's carbon footprint if they commute to work. Government-funded green transportation options are the gold standard for reducing transportation CO_2, but without those options, individuals can still find ways to lessen their transportation carbon footprint. Walking or biking to work is a great alternative if a person is close to their workplace. In cities, electric scooters, skateboards, bikes, and one-wheels are great options. Solar panels for those who own their home are an affordable solution that can be used to charge EV cars, trucks, bikes, and more. Carpooling is another way people can decrease their carbon footprint while saving money. In areas where it is available, public transportation is an excellent option as well. As more companies produce EVs and more people realize that solar panels can save money in most markets, transportation will have a smaller and smaller carbon footprint.

Electricity for many people is the biggest part of their carbon footprint, especially in areas that are powered by coal power plants. Coal is the most polluting form of electricity and releases the most CO_2 into the atmosphere. For those who live in an area that supplies coal-powered electricity, there are a number of options. Homeowners can have solar panels or wind turbines installed (depending on the area). Many utilities offer an option to buy green electricity, usually at a slight upcharge that is used to build more renewable energy. Companies are legally required to be able to produce as much renewable energy as they sell. If enough people sign up for these programs, utilities will be forced to build more renewable energy power plants. Another option is to be more efficient with energy use at home. This can be as simple as changing an incandescent light bulb to an LED bulb or unplugging "energy vampires" (electronics that draw a charge even when not in use). People can insulate their homes, assessing them for leaks where heated or cooled air can escape. Another way to save energy is to get a smart thermostat that can be set to use less energy for heating and cooling when people are gone for the day. Lower a thermostat in the winter and wearing a sweater or raising a thermostat in the summer to use less air-conditioning can also save energy.

All of these options help lower an individual's carbon footprint. If these measures are in widespread use, they can significantly lower the aggregate electricity demand. Nonetheless, the most important thing that an individual can do is to put political pressure on governments and utilities to move to carbon-free energy sources so that turning on a light will not be warming the planet.

CARBON OFFSETS

Carbon offsets are used to counter the CO_2 that a person, company, or municipality creates. The idea is to remove the exact amount of carbon from the atmosphere that was put into it. There are a number of companies that allow individuals to purchase carbon offsets. These companies will also allow the purchaser to choose which carbon offset they are interested in buying. These projects can spread the gamut from collecting methane as it is released from old landfills to creating carbon-negative farms. Larger companies, cities, state/provinces, and countries create their own carbon offsets by foresting land, purchasing timber lands and preserving them, converting areas back to their natural state so that they sequester more carbon, supplying people with more efficient heat and cooking stoves, replacing carbon-intensive energy production with renewable energy production, or other similar projects. Regardless of how the CO_2/carbon is offset, the effect should be the same: CO_2 released should be less than or equal to the CO_2 captured or sequestered. The advantage of carbon offsets is that they allow people, corporations, and governments to engage in behavior that requires fossil fuels without adding to the problem of global warming and climate change. Critics of carbon offsets argue that some actions, such as planting a new forest, would not actually offset the CO_2 put into the atmosphere for up to another fifty years. This is because a tree does not reach its full potential until many years after it is planted. A sapling will not sequester much carbon due to its small size. Another argument is that carbon offsetting is not enough and that humans should be reducing and/or stopping all activity that is carbon intensive *and* working to take carbon out of the atmosphere. Regardless of these complaints, carbon offsets are rising in popularity with corporations, individuals, and governments alike. They are one way to encourage action in the near term to positively affect the atmosphere and the planet.

For individuals, there are basically three ways to offset the carbon they put into the atmosphere. The first option is to create their own carbon offsets by buying solar panels and electric vehicles and by planting trees (either on their own property or on lands that have suffered from fires). The second option is to purchase carbon offsets (through one of many carbon offset sellers, such as TerraPass, Pachama, or the RainForest Alliance) anytime they do something that is particularly carbon intensive. The third option is in many ways the simplest, but it could be more expensive over the long term. A person could compute their annual CO_2 production, counting everything from electricity to transportation to consumer goods and food. The person could then subscribe to a monthly or annual carbon offset purchase through a carbon offset seller. This type of purchase would ensure that the person was not contributing to additional carbon in the atmosphere.

While some companies are willing to use carbon offset sellers, many large corporations have taken a more hands on approach. Even companies that make a living selling carbon intensive products like gasoline are getting involved in what are sometimes called voluntary carbon offsets. Shell Corporation has been selling oil for well over a hundred years. It has recently begun a journey toward net zero, although it has not made any hard promises on when that would happen. Currently, Shell has a number of offset projects that they are running. The company breaks them down into four general categories: conservation projects (also called REDD+), reforestation projects, cookstove projects, and renewable energy projects. A cynic might say that these projects pale in comparison to the detrimental damage the company does every day by drilling, shipping, and delivering gasoline. An optimist might say that one of the dirtiest companies (biggest polluter and CO_2 emitter) of all time is starting to look at ways to curb its carbon footprint. Both these sentiments are true. Shell is not the only company working to net zero using a combination of carbon offsets and lowering its carbon footprint. The RE100 (as in renewable energy 100 percent) is an organization that asks that these corporations commit to getting 100 percent of their energy from renewable energy. Companies are asked to set hard target dates for increasing their percentage of renewable energy. This is another way that some large corporations are looking at changing the way they see their responsibility for the CO_2 they produce.

Many cities have also committed to getting to net zero by using carbon offsets. These cities use a combination of purchases from carbon offset sellers, policy changes, and actual investment in carbon-neutral infrastructure. An example of a city that is working to do that is Flagstaff, Arizona. The city is blessed with ample sun, and the area around it has ample wind. So, the city has built solar panels all across their properties and multiple wind projects are going up in the area. Flagstaff has also committed to the goal of being carbon-neutral by 2030. How does a city of 80,000 people offset their carbon? One method is by creating more open space. If land that was slated for development is covered in mature Ponderosa pine trees, some of the carbon that the citizens release is offset. Another method is using biochar to sequester carbon. Organic material, such as yard waste, branches, plants, and trees, is converted to biochar using the process of pyrolysis (heating in the absence of oxygen). The biochar can then be used for landscaping or gardening, and the carbon is stored underground. It can also be used as bioenergy and to increase biomass (growing plants, bushes, and trees on city lands). This is the example of one small city in high dry western forests, but each city will have a very different set of tools at their disposal to achieve the same goal.

States/provinces are larger, covering a much greater land area. For that reason, they are able to do some things to offset carbon that cities might not be able to do. For example, a state might be able to restore a bog or swamp area that has been degraded by water diversion or dams. Bringing a wetland to its natural condition can create a system that naturally stores CO_2 below the murky waters. A city built on a swamp like Chicago cannot restore the swamp because the area is too built up, but the state of Illinois may know of an area that can be restored at minimal cost to the state.

Countries, of course, are able to do things on an even grander scale. For example, China has mandated that massive amounts of green power be built in the form of photovoltaic solar panels and wind turbines. Although China leads the world in coal use, it also leads the world in both solar energy and wind energy produced. While its renewable energy may not offset all of the CO_2 that the country produces, it does at least limit the amount of coal that is being burned. Countries can also regulate entire industries, create mandatory carbon markets, protect huge swaths of land from deforestation (which is an even better carbon offset than growing a new forest), and create new more sustainable energy systems. For this reason, countries can foment change at a level and scope that is incomparable to lower-level government modalities. The Paris Agreement requires countries to act in a manner that drives their citizens to implement necessary measures.

Carbon offsets help to frame the issue of climate change in a manner that is more manageable for the human mind. They encourage people, companies, states, and countries to work on their own carbon balance sheet in a tangible way. With a concrete number in mind, people can get a hold of the issues and more easily solve them.

CARBON TAX/CARBON MARKET/CARBON CREDITS/CAP AND TRADE

A carbon tax is a broad term for putting a cost on anything that releases CO_2. It can be implemented in a variety of ways. For example, with oil, which becomes gasoline/petrol, a tax or fee could be applied as the oil is removed from the ground or at the refinery. It could also be charged to the companies that ship the gasoline/petrol to the station. Finally, it could be applied directly to the consumer at the pump. These monies can then be used to either decrease CO_2 pollution or remove CO_2 from the air or sea.

A carbon tax puts a price on releasing CO_2, and this creates a carbon market. While releasing CO_2 costs money, sequestering carbon can be profitable. A company that releases CO_2 in their operations can purchase carbon credits. If a carbon sequestering company can sequester CO_2 at a lower cost than the carbon tax, a company that produces CO_2 in their operations would be incentivized to purchase sequestered carbon instead of paying to release it into the atmosphere.

Carbon credits are another tool for carbon management. Entities that create systems to decrease or absorb CO_2 (for example, replacing annual crops with a forest or nut trees) can receive a yearly payment. The owners of this new carbon sink (area that can absorb carbon) could receive money from carbon tax funds for each year that they maintain the carbon sink. Companies or individuals that put up solar panels or wind turbines could also qualify for carbon credits, receiving funds for the equivalent of CO_2 that is avoided from the local power plant. This is especially effective in areas that have coal- or gas-powered plants.

Cap and trade programs already in place are discussed in detail in Chapter 4. In this system a government estimates the amount of CO_2 that an area will need and then decreases that amount annually. The government auctions off a set number of carbon allowances, usually on an annual basis. The allowances cap can be bought and sold on an open market, and the funds raised from the auctions are generally used to decrease CO_2 indirectly or directly. Some governments choose to focus on programs that make homes and businesses more energy efficient, thus decreasing the need for fossil fuels to heat or cool buildings. Others create green transportation options such as bike paths, electric trains/trolleys, and electric buses. They may also use the funds to directly build renewable energy infrastructure and power plants. These programs have demonstrated success and effectiveness beyond expectations. One of the bigger problems is that demand for allowances tends to drop faster than expected. Once the externality of CO_2 pollution becomes a financial internality, businesses react quickly with creative ways to decrease this cost as soon as possible. It is one of the easiest, most commonsense ways for governments to fairly spread the costs for CO_2 remediation. The success of cap-and-trade systems in Europe and parts of the United States is likely to spread across the planet as a way to harness the creativity and talent of individuals and businesses to confront the enormous challenge of climate change and global warming.

The pushback against any sort of carbon tax or trade system results from the fear that adding a cost to doing business will hamper companies from being successful. This has been proven to be false. Companies, including those that run fossil fuel power plants, have been able to succeed and thrive under cap-and-trade systems. In Regional Greenhouse Gas Initiative (RGGI) states, located in the Northeastern United States, 60 percent of energy still comes from fossil fuels, but overall emissions have decreased by 67 percent. States in the RGGI program have received almost $8 billion in funds that are used to create green space, further reduce emissions, insulate buildings, protect coastal areas from flooding, and many other environmentally positive programs. By simply providing companies with a financial incentive to reduce CO_2 emissions, they will find a way.

California and Quebec are both part of a fairly straightforward cap-and-trade system that has proven highly successful. They set the amount of CO_2 they are aiming to allow for a given year based on their estimates for demand. Carbon credits are then auctioned off. If demand is high, prices rise. If demand is low, carbon credit prices default to the set minimum price. This process puts economic pressure on businesses in California and Quebec to decrease the amount of CO_2 in their operations.

California's first goal was to decrease 2020 emission to 1990 levels. This goal was achieved four years earlier in 2016. Their next goal is to decrease CO_2 emissions to 40 percent below 1990s levels. The final goal is to be 80 percent below 1990s level by 2050. California also has the goal to become carbon-neutral by 2045. This will be achieved by carbon-offset programs such as replanting forests as well as CCS. Importantly, both the RGGI program and the California/Quebec program have demonstrated the decoupling of CO_2 output and economic output. These economies continue to grow and thrive as emissions plummet.

The largest carbon market is in Europe. The European Union Emissions Trading System (EU ETS) is made up of thirty-one countries: twenty-seven EU members plus Norway, Iceland, Liechtenstein, and the UK. The system is made up of a combination of free credits granted to "grandfathered" companies that are already established CO_2 emitters and any additional credits that are needed by growing or new companies, which are auctioned off to the highest bidder. The criticism of this program is that there is less motivation for a grandfathered company to decrease CO_2 emissions. However, these companies can sell their credits if they use less than they are allotted. Companies across Europe have done this. The complaint from startup companies is that this gives established companies an economic advantage. While that is true, the program has been effective. It is wide ranging and covers over 45 percent of the emissions of the thirty-one member nations. Its first goal of decreasing 2005 emissions by 21 percent was achieved in 2014, six years earlier than expected.

Another added advantage of these programs is that they raise funds that can be used to ameliorate the effects of pollution, to continue to decrease energy use through modernization and insulation of existing building, to help communities disproportionately affected by emissions from power plants and heavy industry, or to fund innovation in carbon capture and renewable energy. RGGI, California/Quebec, and EU ETS all use their funds differently, but the overall goals are to improve societies with these funds. Each of these programs has raised billions of dollars that have been used to continue to decrease emissions. RGGI funds have been used to invest in renewable energy and energy efficiency programs for homes and businesses. California's cap-and-trade has been used to reduce greenhouse gases (GHGs) by reducing energy demand and increasing energy efficiency. California also invests 35 percent of proceeds into environmentally disadvantaged communities. EU ETS has been investing in innovative projects that hope to have breakthroughs in areas such as carbon capture.

A truly international cap-and-trade program would be the best possible solution. The EU is leading the way on this. A combination of continent-wide programs would also be effective. In the United States, Oregon and Washington state are moving forward with programs that will likely be combined with the California/Quebec program. Many cities and municipalities are working on programs that aim to use some of these methods to encourage businesses to find solutions that limit their CO_2 output. As more countries join this program, more great minds will be working toward cutting emissions and capturing carbon. With the whole world working on this problem, it would not take long for innovation and determination to begin the process of decarbonizing the global economy.

CASCADE EFFECT

The cascade effect, also called the trophic cascade effect or the ecological cascade effect, results from the loss or the introduction of a species into an ecosystem. As all species are interdependent in the web of life, removing one can cause the loss of many others. Humans often hunt animals to local extinction (which can become global extinction). In many cases humans remove a predator species to protect livestock or in hopes of increasing a species that both humans and the predator hunt. While it is intuitive to think that fewer predators will create more prey, that is not always the long-term effect. If an herbivore species such as elk or deer have no predators, there will be an explosion in their populations. This in turn will lead to the herbivores exhausting their food sources, which can lead to huge die-outs. The winter months can prove particularly deadly for these overpopulations. In the spring, human hunters are left with fewer, not more, deer or elk. The loss of plant life from overgrazing can cause other herbivore species to decrease in number as well. Additionally, there are many species that depend on the carcasses left behind by predators, to survive. As each species depends on the other, many species will disappear with the removal of one.

Wolves at one time exceeded two million across the United States but were hunted to near extinction as local, state, and federal government programs paid hunters to kill wolves, offering bounties for their pelts and/or skulls. In the 48 contiguous US states, there are three wolf subspecies. The Northern wolf, sometimes called the timber wolf, is the only population that remained after the over 100-year effort to eradicate wolves across the country; their population was down to 300. The remaining population existed only on the northern tips of Michigan and Minnesota. The Mexican gray wolf was completely eradicated from the wild, but captive populations were used for reintroduction in Arizona and New Mexico. Their native range included the Four Corners region between Utah, Colorado, Arizona, and New Mexico on the north end and down into Mexico on the south end. As of 2023, there was a stabilizing population of 247 Mexican gray wolves in the United States (Maestas 2024). The third major wolf population is the red wolf. Like the Mexican gray wolf, red wolves were driven to extinction in the wild. The red wolf originally had a range that extended across all of the southeastern United States. The red wolf reintroduction program, mainly located in North Carolina, has struggled for two important reasons: local people have shot and poisoned many of the wolves, and the wolves have at times intermixed with coyote populations. This has led to fears that a pure red wolf population will become extinct. As of 2024, the reintroduced population that peaked at 130 was down to 20–2 due to human hunting and poisoning. Alaska is home to two subspecies of wolf that have never been threatened or endangered.

Wolves have historically been seen, by Europeans and their descendants who settled across the planet, as danger and a threat. President Theodore Roosevelt famously called wolves beasts of waste and desolation. These strong feelings toward wolves have made reintroduction problematic, especially in ranching areas where there is the fear of losing cattle and sheep. Yellowstone National Park allowed for a reintroduction experiment that was not hindered by local ranchers and private landowners. Wolves in Yellowstone National Park have been studied extensively to understand how both the lack of wolves and the reintroduction of wolves changed the ecosystem and, in some ways, the actual geography of Yellowstone.

Wolves were hunted and poisoned in Yellowstone until the mid-1900s when they were almost completely wiped out in the lower forty-eight states. This caused a trophic cascade that led to larger numbers of elk. Elk and bison behavior changed; they began to stay near the rivers and eat the young sprouts of willow and aspen. This allowed rivers to meander more, as they were no longer held into shape by the trees that once lined the banks. This led to fewer beavers, which in turn led to fewer ponds. The landscape then had more alluvial plains, which were good for ungulates but created less habitat for many other species.

In 1995 and 1996, thirty-one gray wolves imported from Western Canada were released in Yellowstone National Park as part of a reintroduction program (NPS 2022. Wolves changed the behavior of ungulates in Yellowstone, most profoundly elk and deer. With ungulates no longer eating the new shoots of deciduous trees that lined the banks of the rivers and streams of Yellowstone, new trees brought back the beaver population. Beavers, as engineers of ecosystems, built ponds and small lakes. This created areas for more mice and rabbits, which in turn increased the population of animals that predated on mice and rabbits such as raptors and foxes. Trout populations increased as well. The river and pond borders stabilized, and berry bushes increased. This brought in animals that lived off the berries. Habitat improved for bear and mountain lion populations, and otter populations increased. Overall, the biodiversity in Yellowstone increased as wolf numbers peaked and then leveled off. Not all species increased their numbers; elk population decreased, as did coyote populations. However, elk carrion provides food for all kinds of scavengers, from bears, wolverines, and lynx to ravens and eagles. Beetles and other small scavenger populations also increased.

Wolves give us a great example of trophic cascade effects. Another example is the overhunting of sea otters and/or seals that eat sea urchins, which leads to sea urchins eating up all the macroalgae (seaweed and other marine life that attaches to the seafloor) in an area. Macroalgae is the foundation of sea life near the shore. Without the right predator, sea urchins can turn a vibrant ecosystem in a barren area devoid of life. As humans continue to change the planet on the macro level (climate change) and the local level (deforestation and pollution), it will be important to remember how these examples demonstrate the importance of all members of an ecosystem in ensuring thriving biodiversity.

CLIMATE ENGINEERING OR GEOENGINEERING

Climate engineering can be defined as any human attempt to reverse climate change and global warming. Proposed solutions range from science fiction–like almost impossibilities, such as creating a giant sunshade that floats in space, to those that have already been implemented, like machines that pull CO_2 out of the atmosphere. Approaches generally focus on either directly removing CO_2 or cooling the planet. They can either encourage a natural process, such as reforesting areas in which forests have been lost, or they can be simple human processes like carbon capture from the atmosphere combined with underground storage. Geoengineering is any human attempt to change the natural environment to help humanity; it can include climate engineering, for example, cloud seeding, to increase rain in drought-stricken areas. These methods will be needed if humans hope to reverse the accidental climate engineering resulting from the burning of fossil fuels from the dawn of the industrial revolution until today. "Natural" methods, which we will focus on in this entry, are much less controversial and are usually less expensive than more advanced technological methods.

Reforestation is the most obvious and easy way to capture carbon, but it is by no means the fastest. Planting a forest can be done in a day with enough trees and people, however it takes decades for that forest to take in as much carbon as a mature forest does. Although human population is increasing, people are moving from rural communities into cities that offer more opportunities. With fewer people in areas that used to be forests, governments and nonprofits are likely to encounter less pushback when they propose reforestation. Areas available for reforestation are abandoned or unproductive farms, forest areas that have been converted to rangelands, and recent burn areas. These areas can begin to naturally store carbon at a more productive level every year after planting.

Afforestation involves planting a forest in a place where one did not previously exist. As the climate changes, this will partly be a natural process. While some areas are getting drier, others are getting wetter. Warmer weather means that areas that were once frozen tundra can become alpine or boreal forests. Instead of waiting for tree species to move north on their own, in Canada, Alaska, and Siberia, people are moving these species by planting new forests further north. The costs of these efforts are relatively low, although there are risks related to human interference in natural ecosystems. Afforestation will occur with or without people, but the outcomes will be very different (species/locations/rates of change). As with reforestation, it takes a long time for the planted trees to become a large forest capable of pulling significant amounts of carbon out of the atmosphere.

Stopping deforestation is the most effective natural way to use forests to store carbon, but that does not always mean no wood harvesting. In some forests responsible wood harvesting can help to prevent forest fires and keep the forest healthy. Many places in the Northeastern United States have implemented policies that preserve forests in general but allow for the specific harvesting of valuable mature hardwoods. For example, the strong market for maple and oak trees makes it worthwhile for a logging company to remove only specific trees rather than clear-cutting. Most forests in the Western United States have already been clear-cut at least once. This practice has created unnatural forests that are too dense and are prone to intense crown fires that can destroy a forest permanently by not only burning the trees but also burning out the nutrients in the soil and killing the mycorrhizal fungi on which many species of trees depend. Selective cutting can create a healthier forest that is less prone to lethal crown fires yet open for low-intensity burns that are a part of the natural forest cycle. The absolute most effective way to make sure forests are sequestering carbon is to stop slash-and-burn agriculture in virgin forests. This process involves cutting down a forest, letting it dry, and then setting it on fire. The burned forest releases CO_2 into the atmosphere and loses the ability to sequester mass amounts of CO_2 for decades, possibly indefinitely, depending on local policy. Creating policy that preserves existing forest while providing livelihoods for the people who would be most motivated to slash and burn forest to create land to farm or raise livestock is essential.

Urban forestry involves caring for the "forest" of trees that exist across most cities. Urban lumber is a new concept in which logging companies assess well-settled lands that contain mature trees. Many cities are full of valuable tree species, including ash, oak, maple, apple, juniper, redwood, walnut, and cherry, among many others. Traditionally, when these trees near the end of their life due to old age, overgrowth, or conflict with buildings and/or electrical lines, they are cut down, cut up, and either sent to a landfill, made into wood chips for ground cover, or used for firewood in outdoor fire pits. Traditional sawmills are not usually interested in urban lumber because of the risk of encountering a metal nail or wire that could easily destroy a commercial blade that could cost up to $12,000. However, recently, there has been growth in the area of small businesses that salvage these trees for lumber and/or specialty furniture. By viewing trees that need to be removed as a resource and not a nuisance, billions of board feet of wood can be salvaged and used to replace wood that may have been clear-cut from the forest.

Current mainstream agricultural practices add CO_2 to the atmosphere. Tilling the soil brings up organic matter containing carbon, and when this carbon is exposed to oxygen, it combines to become CO_2. Nitrogen-based fertilizers can combine with oxygen to create N_2O (nitrous oxide) which can have a warming effect on the atmosphere that is 298 times as powerful as CO_2 over a 100-year timescale. Most large farms use petroleum-based fertilizers that eventually run off and are concentrated in places where waters drain into the ocean. For example, where the Mississippi River meets the Gulf of Mexico in the United States, agricultural runoff creates a hypoxic zone (low oxygen area) every year that kills many small fish, crabs, mussels, and other sea life that cannot swim fast enough to escape the roughly 5,000-square-mile death zone.

In contrast, traditional farms, which are smaller and simpler, are carbon-neutral. They use animal waste as fertilizer to grow crops that animals and people eat. People also eat the milk, eggs, and meat that the animals provide. All waste is reused, and there is no need for additional inputs from outside the farm. Many of the suggested ways of turning a GHG-producing farm into a carbon-storing farm come from using more traditional methods.

The United States Department of Agriculture (USDA) Natural Resources Conservation Service has created an extensive list of methods to both store CO_2 and/or limit its release. (See Chapter 3 for more information on these methods.) The USDA rates each activity on a nine-point color-coded scale that goes from red for the most carbon-intensive methods to dark green for methods that store the most carbon. A more advanced system from the USDA that can be used to calculate and lower carbon intensity is called COMET-Farm. COMET-Farm analyzes all of the information about a farm and estimates the amount of CO_2 and other GHGs that are being released. It then gives the farmer a list of improved farm management practices and what the future scenario would be if those practices were implemented at the farm. Healthy soils are carbon rich, so farmers have the opportunity to turn their farms into carbon sinks instead of carbon emitters.

Some of the most effective methods of decreasing CO_2 release from the soil are planting cover crops between commercial growing seasons and implementing no-till, strip till, or direct seeding. Next is putting cover over a fallow (vacant) field. If the farm has both animals and plants, animals can eat the plant parts that humans are unable to digest. If there are no animals, plant residue can be collected and composted and the compost used as fertilizer. Another method that accompanies no-till, which still involves a very small and precise amount of tilling before seeds are planted, is leaving crop residue on the ground. This has many benefits including reducing tilling time by two-thirds, cutting water runoff and evaporation by 70 percent, doubling the number of earthworms per acre, and improving habitat for wildlife (USDA 2021).

Multistory cropping, another old technique, is also extremely effective at storing carbon. Coffee, for example, is traditionally grown in the shade of fruit trees, nut trees, or other economically important trees. In this way a farmer can grow a number of crops at the same time; large, medium, and small trees can be grown with shrubs, vegetables, and herbs underneath. Multi-story cropping is effective for fruits, vegetables, and nuts, but it is more difficult with grains, especially at industrial scales. Recent studies have demonstrated that both wheat and barley can have increased yields with less sunlight; barley specifically saw increased yields with 50 percent sunlight (Arenas-Corraliza et al. 2019). Opportunities for farmers to combine silviculture with grain production therefore exist. Another possibility is to grow crops under solar panels; early research has demonstrated that this can be effective with vegetable crops. Multi-story cropping holds promise, especially as the world heats and many crops will need a little shade.

River restoration and riparian forest buffers offer another way for farmers to naturally sequester carbon and protect farmland. Bare soil is threatened by intense rainstorms and windstorms. Many farms have removed all trees so that it is easy to move large tractors in and out of fields. The future of autonomous tractors will create an opportunity for farmers to no longer feel that it is necessary to remove every possible obstacle. Riparian forests stabilize the banks of rivers, creeks, and streams that flow through farmland, reducing erosion and creating windbreaks that protect soil from being blown away. They also store carbon more intensely every year as they grow and thicken, and they create habitat for fish, birds, reptiles, mammals, amphibians, and helpful insect species. In farms without riparian areas, a hedgerow or windbreak made of trees and bushes can protect against the ravages of wind and rain as well as provide habitat for all types of wildlife.

Another way to decrease CO_2 release on a farm is to create an anaerobic digester, which is a type of composter that operates without oxygen and produces both fertilizer and methane gas. The fertilizer can be applied to fields, and the methane gas can be used to run the farm and heat the buildings. Anaerobic digesters are especially useful on farms that produce mostly pigs and/or cows. Using animal waste to create fertilizer and fuel also stops methane from reaching the atmosphere, where it can be fifty times more destructive than CO_2. There are some creative farms that produce useful items from animal waste. For example, a farm in Connecticut creates "CowPots," biodegradable pots that can be planted directly into the soil, where they will provide nutrients for a growing plant.

While efforts to protect land-based forests for their carbon-storing capabilities attract the most attention, coastal forests and ecosystems can store up to ten times the amount of carbon as terrestrial forests. This is sometimes referred to as "blue carbon." The restoration and advancement of mangrove forests, seagrass beds, coastal wetlands, seaweed forests, and shellfish and seaweed aquaculture are important opportunities as humans look to effectively fight climate change with limited resources.

Mangrove forests store significant amounts of carbon, more than terrestrial forests (including the Amazon). Mangroves accumulate carbon in their tissue in the form of roots and trees, but their thick root systems also trap sediments and debris, thereby burying substantial amounts of carbon. These coastal habitats account for 14 percent of the carbon sequestration for the world's oceans (Alongi 2012). Mangrove forests have been cut down for fuel and/or building supplies, removed to clear water ways, and cleared for hotels and beach goers. As climate change increases the intensity and power of ocean storms and hurricanes, coastal cities and nations are realizing that restoring mangrove forests creates protective coastal barriers during these storms in addition to storing carbon.

Seagrass beds consist of underwater flowering plants of many shapes and sizes that live in coastal waters all over the planet, except Antarctica. They are able to sequester an estimated 10 percent of the carbon that is stored in the ocean, or 27.4 Tg of carbon per year (Blue Carbon Initiative). This is striking because seagrass beds only cover 0.2 percent of the ocean floor, meaning they sequester fifty times as much as the average sea floor area. As with mangroves, seagrasses accumulate sediment and other detritus, storing carbon in the soil up to 12-feet (4 m) deep, allowing them to eventually store twice as much carbon as terrestrial forests per hectare. Seagrass beds also act as a buffer against large storms, flooding, and erosion. They are also important habitat for numerous marine animals including multiple species of fish, turtles, and manatees. Currently, seagrass beds are threatened by dredging, climate change, and polluted waters. The world is losing about 1.5 percent of its seagrass beds a year. Almost one-third of seagrass beds have been lost (Blue Carbon Institute 2021).

Coastal wetlands can consist of many different ecosystems, ranging from swamps (sometimes hardwood swamps) to marshes (both salt and freshwater) and mangrove forest. They are dominated by salt-tolerant reeds, grasses, and sedges that anchor the vibrant ecosystems where fresh and saltwater mix, and they store carbon well. Tidal marshes are particularly important as they can sequester up to four times as much CO_2 as tropical forests, with deep soils that accumulate mineral sediment, organic material, and other detritus. Soils can be up to 21 feet (7 m) deep. In addition to sequestering carbon, tidal marshes filter pollutants from the fresh waters that flow through the marshes and into the ocean. They also slow down flooding and protect against storm damage. Like other most other types of coastal wetland, tidal marshes are disappearing at an annual rate of 1–2 percent. They face a triple threat of coastal development, rising sea levels due to climate change, and conversion for agriculture, especially rice.

Seaweed forests, which are dominated by macroalgae such as kelp, are another excellent example of natural CO_2 sequestration and storage. They take up CO_2 and release oxygen, just as a terrestrial forest does, and they provide critical breeding grounds for many sea creatures, both big and small. These forests have seen significant loss in areas near human habitation due to shipping, fishing, and pollution. As they face the additional obstacle of warming waters, the need to replace and/or restore these areas is more urgent than ever. Programs to replant seaweed forests have seen mixed results, as warmer or more polluted waters may not be able to support the same macroalgae as before. In these cases, scientific experimentation is required to determine how and what to replant. With supportive efforts from politicians, scientists, and local populations, seaweed forests can become an even more important method for the sequestration of CO_2.

Man-made coastal "ecosystems" are created by shellfish and seaweed aquaculture, both of which pull CO_2 out of the ocean. Shellfish store carbon in their shells as calcium carbonate, and fast-growing seaweed stores CO_2 in its tissue. When a shellfish or seaweed is removed from the ocean, so is its carbon. Humans have been shellfish farming since ancient times. For example, ancient Romans enjoyed eating raised oysters. It takes about one and a half to two years to grow oysters to the size that they are ready to go to market. The shells can be crushed and used as aggregate in construction. Due to the fact they are 95 percent calcium carbonate (Cimons 2017), they can also be used to replace limestone used in cement. This way they not only sequester CO_2 but also replace materials that are unsustainable. Using the waste from a seafood meal is an extremely sustainable substitution. (See also the Blue Carbon section in Chapter 3.) Seaweed agriculture is another under-used ancient method of removing CO_2 from the world's oceans. Kelp is one of the most well-known types of seaweed, but there are many types of edible seaweed that make up a popular part of local cuisines across the world. Seaweed is also an excellent fertilizer and can replace ammonia-based fertilizers that require enormous amounts of natural gas to produce. If seaweed can be grown and harvested in large quantities, more CO_2 will be removed from the ocean and at a faster rate. Policies that encourage seaweed agriculture could go a long way to further this process.

COAL

Coal has been used for millennia in places that had outcrop coal, which is coal that can be seen sticking out of the ground. After all outcrop coal was exhausted in these places, digging for coal began. Ancient Chinese smiths used coal for all types of metallurgy. Bronze Age peoples used coal, although generally only for particular applications that called for high heat, such as metal work and funeral pyres. Ancient Romans used it extensively when they ruled southern England, and it was not until the High Middle Ages in the thirteenth century that coal's use became common in England again, and then only for specialized procedures. However, in the 1800s, two things happened in England at the same time: steam engines began to replace water wheels as the means for mechanization of factories; and the supply of wood near London used for heating was nearly exhausted. This created a ready market for coal, and coal mining exploded across the island of Britain. This was the beginning of the Industrial Revolution, and coal was the fuel that fired that revolution. Coal became even more important the following century with the electrification of much of the Western world.

Coal is carbon-based life that has been compressed by pressure, heat, and time. So, coal can be ancient forests, ancient ocean floors made up of sea life and plant life, and ancient animals as well. For example, coal formation may have begun in a low, coastal, wet forest that formed peat bogs (in which slow-moving, acidic water stops dead plants from breaking down). When plants or animals are not able to decompose due to acidic or oxygen-free circumstances, the regular carbon cycle is interrupted, and this carbon-based matter begins to accumulate. The added pressure of successive generations/layers pushes the peat further underground, eventually squeezing out liquids such as water and gases such as CO_2 and methane. The peat becomes concentrated carbon, which is coal.

There are many types of coal, which can be thought of as occurring on a spectrum, depending on how much heat, pressure, and time the coal has experienced. The more heat, pressure, and time that are applied to coal the higher the carbon content, and the cleaner it burns. Higher-carbon coal produces less pollution.

Pre-coal is peat, which creates enormous amounts of pollution when burned, as well as the highest levels of CO_2. It has a carbon content of less than 60 percent (after drying). Burning peat is not safe for human health. Unfortunately, it is still burned for heat and electricity in a few places. Ireland is weaning itself off of peat for electricity production, but it is still popular for house heating.

"Young" coal is called lignite or brown coal. Brown coal is the most polluting of actual coal types. It contains 60–70 percent carbon (after drying). Fortunately, it is the least used coal type. Only Greece and Germany use it to provide a large amount of their electricity. While both countries have made strides to meet their commitments to the EU and the Paris Accords, they both still produce nearly a quarter of their electricity from lignite. Lignite is not imported or exported as it is generally low-grade coal and can contain up to 75 percent moisture. Therefore, it is not economical to ship it for sale. It is used only locally. While all coal use is decreasing, lignite is likely to decrease the most and the fastest. It does not have a future in the renewable energy transition.

Sub-bituminous coal (also called black lignite) is less polluting than lignite as it has a dry carbon content of 70–76 percent (Donev n.d.). It is the most used type of coal in the United States. It still produces a lot of particulate pollution, which is especially dangerous for anyone who has asthma or other respiratory illnesses. Its popularity has grown in the United States because legislation meant to address the acid rain problem in the 1980s called for low-sulfur coal, which was created in areas that had freshwater over brackish or saltwater. These areas are in the Wyoming Powder River Basin where strip-mining allows easy access to coal that is 70 feet below the surface. The other US location for low-sulfur coal is the Appalachian Range in the east. The process for obtaining this type of coal is mountaintop removal, which destroys entire mountain ecosystems by using explosives to blow up the top of the mountain. Mountaintop debris that is full of toxins is pushed into valleys adjacent to the mountains, polluting rivers downstream as water works its way through. This is an unintended consequence of a law meant to clean the air. Before this law, most coal in the United States was obtained through underground mining, which is less destructive to the environment but can be more dangerous for coal miners who work inside the mines.

Bituminous coal is less polluting than sub-bituminous coal as it has a dry carbon content of 76–86 percent (Donev n.d.). It is created by the geological process of compression of former peat bogs and is considered to be of higher quality than both lignite and sub-bituminous coal. It burns cleaner than both. There are various levels/types of bituminous coal. Coking coal is used in the steel-making process as it is low in phosphorus, which can be very destructive to steel. Bituminous coal that is low ash is highly valued in electricity production because it creates less slag that needs to be disposed of and it does not require as much effort to clean or scrub the gasses produced by burning the coal to remove particulate matter. Smithing coal has the least amount of impurities including ash and sulfur. Bituminous coal, while less polluting than sub-bituminous coal and lignite (brown coal), is still full of pollutants; burning it releases toxic metals along with GHGs, especially CO_2. If coal is to be used for the production of electricity and the world hopes to stop or slow climate change, there must be a process for collecting CO_2 at the point of production. While most attempts at clean coal have not been successful, there are a few startup companies that are attempting to collect the CO_2 produced and repurpose it into a new form of fuel (see entry on CO_2).

COMMON POOL RESOURCES

Common pool resources are resources that are used by many people, which can make management to preserve these resources difficult. In vast areas such as the ocean and the atmosphere, it is extremely difficult to get all players to equally participate in a set of rules and/or regulations. The phrase "tragedy of the commons" is used to discuss the problem that exists when, due to dwindling resources, all the players involved in the situation feel motivated to extract what's left. This process can cause everyone to lose. For example, if the resource is a fish species, overfishing can cause a crash in the fish population from which they never recover, meaning all parties can no longer fish for that species. A tragedy of the commons is avoided by prudent management of the resources by the group of people who are using it. In England, each village traditionally had an area that was "common" to all the families in the village. This common area was usually a field used to graze animals. The leaders of the village would monitor the common field and decide if more sheep or cows would be allowed to use the green. This self-regulating system allowed the common field to produce the greatest amount of animals and animal products such as milk and cheese while protecting the common pool resource.

There are many examples of these systems working well. Nobel laureate Elinor Ostrom in her study of Maine lobster fisherman described another local self-regulating system. The basic premise is that each lobster fisherman has been allocated a specific region in which to set his lobster traps. If another lobster fisherman attempts to poach lobster in these areas, the lobster fisherman who has the "rights" to the area will put a brightly colored ribbon on the "illegal" lobster trap to let the poacher know that he is caught. If the poacher continues poaching, the community has set up a series of punishments. The first is a visit to the home of the poacher by a group of lobster fisherman. If the "talking to" does not work, the punishments continue until finally the poacher's boat is sunk. This system is not part of a government program; it is simply how a group of people regulate their shared common pool resource so that the most people can survive and thrive without exhausting the resource that is their livelihood. In groups where everyone knows each other and they are part of a community, it is relatively easy to regulate the system. This is the way humans have been sharing common pool resources for millennia.

This system begins to fall apart when the scale of common pool resources is no longer on a human relationship scale. For example, global fisheries no longer consist of local fishermen who know each other. Many fishing boats are part of larger corporate operations from a variety of countries. These companies are looking for short-term profit, and the destruction of one species would not necessarily destroy their businesses. Their international mobility creates a different set of motivations, making self-regulation unlikely. As a species begins to dwindle, each corporation and each country are motivated to get as much as possible before the fish are gone. So, the long-term survival of a species and its long-term ability to support a fishing community are overridden by the short-term profitability of a fishing season. These fishing interests use their influence with national governments to encourage fishing limits that are not scientifically based. These decisions on limits pave the way for the destruction of a valuable fishery.

Two of the most famous fishery collapses are the Atlantic northwest cod fishery and the Pacific Basin sardines. We will focus here on the former, as it is less dynamic and easier to understand. The Atlantic northwest cod fishery provided jobs and food for the First Nations of Canada for millennia and for European settlers and their descendants for hundreds of years. Bigger trawlers and new technology developed during the Second World War, including radar and sonar, allowed for increased catches which eventually were over four times historical numbers. Along with cod, the new larger ships were capturing a lot of by-catch, noncommercial fish, that were vital to the ecosystem. The loss of these vital species exacerbated the decline of the cod fishery. Within 15 years, more cod was harvested than in the previous 100 years. As the problem was becoming apparent, the Canadian government tried to stop the decline of the cod population by banning foreign boats from Canadian shores up to 200 miles off the coast. Despite repeated warnings from scientists, domestic fishing quotas were not decreased significantly. While the decline temporarily slowed, Canadian fishermen continued to overharvest until 1992 when the population completely crashed to 1 percent of its historical spawning population. This crash caused the government of Canada to completely ban cod fishing, and over 35,000 people lost their jobs. After years of a cod-fishing moratorium, northern (Atlantic northwest) cod has not rebounded to a level that would allow for harvesting.

It appears that low levels of cod have created a situation where there are not enough juvenile fish to overcome predatory pressures. Northern cod may not ever recover to pre-crash levels.

The Atlantic northwest cod collapse demonstrates a classic common pool resource dilemma. Politicians in government are loath to introduce measures that "kill" jobs, so they do not protect a resource (lumber, fish, clean air, clean water, etc.) in order to protect "the economy." However, by not dealing with these problems with an eye toward sustainability, long-term collapse becomes inevitable. Two of the biggest common-pool resource problems the world is facing today are (1) GHGs being released into the atmosphere, heating the planet, and causing climate change and (2) plastic and other ocean pollution. The problem with lowering GHGs is that developing countries like India and China do not feel it would be fair to severely limit their emissions as more developed nations have had free reign to pollute and release as much GHGs as necessary for hundreds of years before India and China began their industrialization process. However, if only industrialized nations cut their GHG emissions, it will not be enough to stop the worst effects of climate change.

Without a worldwide government body that has an enforcement mechanism, it is extremely difficult to solve large common-pool resource issues, also known as "wicked" problems. Plastic pollution in the ocean is a problem that is less difficult to solve, but there has been little progress on it. An international agreement that held manufacturers of plastics responsible for recycling or reusing their own products could be effective. The easiest solution is to require a large enough deposit on any plastic that people are motivated to return the plastic for the deposit. Countries such as Denmark have such high deposits on bottles, for example, that they collect 99 percent of all bottles that are sold. The solution is evident, but major worldwide distributors of plastic containers, such as the Coca-Cola and Pepsi corporations, have pushed governments to not pass deposit laws and in some cases to reverse the ones they already have. An international movement similar to the Paris Accords but with more accountability could change this. Governments need to band together to influence corporations that are destructive to the planet.

While the Paris Climate Accords are a step in the right direction for solving our biggest common-pool resource problem, they are insufficient. The next international agreement needs to have an enforcement system in which countries that do not comply face fines and trade sanctions. Political pressure combined with economic incentives have the potential to positively reverse the current situation in regard to the major common-pool resource problems of climate change and plastic pollution.

CONCENTRATED SOLAR POWER

In simplest terms, concentrated solar power (CSP) is using mirrors to concentrate solar energy toward a focal point. The focal point is usually something that can hold heat for hours after the sun sets, such as a central tower that holds molten salts or water. Another setup for a CSP is to have parabolic troughs of mirrors with a central pipe carrying a liquid that can retain heat. Molten salts are used most often because they store heat better and slowly release that heat to create electricity. The advantage that CSP has over photovoltaic panels (solar panels) is that a CSP plant can produce electricity during the peak demand hours that continue for the first few hours after dark. Some CSP plants may only be able to produce power for a few hours after dark, while the most advanced plants can produce electricity from stored heat for up to fifteen hours after the sun sets. The disadvantage of a CSP plant is that they are much more expensive to build and to operate than photovoltaics.

Spain leads all countries in total CSP installed with 2.3 GW. The United States is second with 1.76 GW installed. Morocco has 530 MW of installed capacity. South Africa has 400 MW of installed power. India has a total of 229 MW installed. (NS Energy compiles these numbers.) China is rapidly increasing its CSP, aiming for 600 MW installed; by 2025, it will likely have more installed CSP plants than all other countries besides Spain and the United States.

CSP plants are problematic for a few reasons. As mentioned, they are much more expensive than photovoltaic panels. They also create an intensely bright light that attracts bugs, which in turn attract birds. The intense heat generated from the CSP can instantly cook a bird to death, in some cases incinerating it. While horrible, the thousands of birds that may die in a year at a CSP plant pales in comparison to the two and a half billion birds killed every year by feral cats and pet cats that are allowed outside.

Ivanpah plant in California is the largest CSP plant in the United States and the second-largest in the world. It is located out in the Mojave Desert, where there is less wildlife and birds. However, it has been plagued by problems, including needing a substantial amount of natural gas to get its molten salts heated up in the morning. It also took a number of years before it was producing close to the amount of power its builders promised. Despite these problems, Ivanpah is able to produce electricity when it's needed most and when most solar panels are unable to produce. While it does use natural gas, the Ivanpah plant uses a tiny fraction of the amount of natural gas that a natural gas power plant uses. Noor Power Station (Ouarzazate Solar Power Station) located in Morocco is the largest CSP in the world at 510 MW, and when combined with its 72 MW of photovoltaic solar power, it can produce 582 MW of electricity.

The ability to produce electricity after the sun sets is extremely valuable as the world transitions toward renewable energy to avoid the worst effects of climate change. It is likely that CSP will be a cost-effective way to provide power after dark when compared to photovoltaic technology combined with batteries, but CSP is currently only financially viable in dry, sunny places. Because of this, CSP is a logical choice in desert regions such as the Middle East, parts of Central Asia, the Southwestern United States, and Peru, where there are dry, sunny locations close to metropolitan areas. CSP will likely continue to expand as costs fall.

CRUDE OIL

Oil is often thought of as a thick black liquid, but it actually consists of many different types, spanning from oily rocks, to oil sands, to traditional "pure" oil deposits. The EPA defines four classifications mainly for the purpose of oil spill cleanup operations: Class A—light volatile oils, which are more flammable and are the easiest type of oil to clean up; Class B—nonsticky oils, which are the classic oil fluid, sticky but less toxic than Class A oils; Class C—heavy, sticky oils, which are difficult to remove due to the fact that they adhere to most things; and Class D—non-fluid oils, which are solid, can only become fluid with additional heat, and are usually easier to clean up due to the fact that they do not spread as fluid oil does.

The American Petroleum Institute (API) created a gravity range to classify different types of oil. Creating a classification system allows for easy trade and sales of the various types of oil, which are not interchangeable. The API gravity system measures the density of oil, which ranges from a thick heavy tar used to create asphalt to ultra-light oil that has a similar "thinness" to nail polish remover. The six oil types, with their API ranges, are as follows:

Tar, Bitumen, and Kerogen: 6–10
Extra-Heavy Oil: 10–12
Heavy Oil: 14–22
Intermediate Oil: 25–30
Light Oil: 31–40
Ultra-Light Oil: 41–50+

Density is basically the mass that is held in a specific area, so the higher the number the denser the oil. The "lighter" the oil, which means it is actually denser, the more likely the oil is flammable, as it consists of more pure hydrocarbon and contains more natural gas as well, such as butane and propane. These lighter oils are dangerous to transport as they could explode, possibly killing people and damaging property. On the other end of the spectrum, heavy tar or bitumen burns dirtier but is less flammable so it is safer to transport. The danger with heavier oils is they are particularly difficult to clean up after a spill.

Each type of oil has different uses and can also be processed into other types of oil for alternative uses. Tar, bitumen, and kerogen are not generally burned at all unless they are processed first. They are used in the creation of roads and in manufacturing. Extra-heavy oil and heavy oil are more likely to be used in heavy marine fuel; they are the most toxic of all the crude oils that are burnt for energy or to transport vehicles, planes, and ships. Intermediate crude oil is likely to be used for diesel fuel, which is used in trains, ships, trucks, and passenger vehicles as well as some planes. It is less polluting than heavy marine fuel but creates more particulate matter than any of the lighter types of crude oil. Particulate matter that comes from the burning of these various types of crude petroleum (oil) is particularly dangerous to human health and extremely toxic to both animal and plant life. Light oil is more likely to be used to create gasoline, which is used in transportation as well. Most automobiles sold in the United States run on gasoline. Gasoline-powered combustion engines can be used to power any type of craft as well as for running generators that create electricity. Technically, any form of crude oil could be burned to create steam and turn a turbine, creating electricity. Ultra-light crude oil is used for jet fuel. The lighter the crude the more volatile it is; it ignites faster and evaporates faster when it is spilled. Lighter crude is more dangerous to transport but easier to clean up, and the heavier crude types are the opposite. Crude oil is estimated to reach peak demand between 2019 and 2030. The Covid-19 pandemic significantly decreased the demand for crude oil, but demand has increased ever since, albeit at a slower pace than before the pandemic. Demand will not top out until EVs become a greater mix of the motor vehicle market. (For more information see https://context.capp.ca/energy-matters/2021/og-101-types-of-oil/.)

DEEPWATER DRILLING

Deepwater drilling is drilling that is done in the ocean, sometimes many miles out to sea and at least 150 m deep. Deepwater drilling is full of risk and great rewards for oil companies. The ocean is a dangerous and unpredictable place. Difficulties include storms, hurricanes (or cyclones or typhoons), isolation, technical issues that cannot be fixed for weeks due to remote locations, and extremely high costs to build, repair, and operate. These are all reasons why deepwater drilling is relatively new on the oil recovery scene. The risk of accidents is also high, with oil leaks almost inevitable in these difficult situations, and the ability to stop leaks is very limited. Retaining a team of people who can operate underwater at high pressure is also difficult. With oil prices rising to record highs early in the 2000s, the costs and the risks were overwhelmed by the possible profits, as many potential drilling sites would access the richest oil reserves in the world. The deep-sea reserves off Venezuela, for example, are part of an oil basin that holds more oil than any other on earth. (Political turmoil has made extraction all but impossible in Venezuela as of 2024.) Deepwater drilling continues to grow across the planet as conventional oil sources dry up and oil prices rise. It is yet to be seen if the world will transition from oil to renewables before these massive deepwater oil reserves are explored and exploited.

Deepwater drilling does not consist of one particular process. There are seven major methods for drilling deep in the ocean, with differing depths at which they can be effective: a Fixed Platform (FP) rig, used for projects in seawater up to 1,500 feet deep; a Compliant Tower (CT) rig, used for projects in seawater between 1,500 and 3,000 feet deep; a Sea Star (sStar) rig, also called a Mini-Tension Leg Platform (Mini-TLP), used for projects in seawater between 500 and 3,500 feet deep; a Floating Production System (FPS), used for projects in seawater between 1,500 and 6,000 feet deep; a Tension Leg Platform (TLP), used for projects in seawater between 1,500 and 7,000 feet deep; a Subsea System (SS), used for projects in seawater up to 7,000 feet deep; and a Spar Platform (SP), used for projects in seawater between 2,000 and 10,000 feet deep (Tisserand 2004). Each of these oil platforms or methods have strengths and weaknesses depending on the conditions of the sea in which they are located. A Fixed Platform rig is what most people imagine when they think of ocean-based oil rigs, a flat platform with cranes hanging off the sides. Metal piles driven into the seafloor support the rig, and there is a central drill that brings up the oil from under the seafloor. A few buildings on top support the crew and are used to process the oil for removal to a ship or into a pipeline. The Compliant Tower rig looks similar to a fixed platform but it is specially designed to withstand shearing forces that might destroy a regular fixed platform. This additional strength allows it to survive in deeper and more turbulent waters. The Sea Star rig is smaller and more economical. It also floats with mini-tension legs holding it to the seafloor and is used in situations where expected oil production is small or for exploratory drilling. The Floating Production System oil rig is semi-submersible and held in place by large heavy anchors. It does not need to stay connected to the wellhead as the wellhead is attached to the seafloor when the drilling is done. This offers more flexibility at greater depth than the previous systems we have explored. The Tension Leg Platform is also a floating structure. Tensioned tendons connect it to the seabed, and pile-secured anchors that also connect to the seabed provide structure to limit vertical motion. While TLPs are usually only used as deep as 4,000 feet, they are rated up to 7,000 feet. The Subsea System is very different from the other rigs we have discussed. It connects to the seabed, but it is not a drill. A moveable drill creates a number of relatively small wells which are then connected to each other by the Subsea System, which also connects to a large production platform that can then distribute the oil to ships. The Spar Platform is the largest type of deepwater oil recovery system. The center of the system is a large cylinder that is connected to the platform, but it is not connected by piles like most platforms. It is attached to the seabed by lines and cables. The cylinder is very large, and it is capable of surviving hurricanes due to its unique construction.

The biggest problem with deepwater drilling is not the technology and the costs. Technology continues to improve, and the costs continue to decrease as technology improves. The biggest hurdle facing these deep-sea drilling ventures is the possibility of catastrophic destruction and the resulting oil leaks. The Deepwater Horizon disaster in 2010 illustrates this risk. The Deepwater Horizon was operating in about 5,100 feet of ocean water, drilling an exploratory well that was deeper than 18,000 feet below sea level. An explosion of methane gas destroyed the drill and killed eleven workers. Ninety-four others were rescued, some with serious injuries. The leaking well spewed oil into the Gulf of Mexico for eighty-seven days. The US government estimates that 210 million gallons were leaked, affecting 70,000 square miles of ocean, an area bigger than the state of North Dakota. The damage caused was so extensive that BP (British Petroleum), the majority owners in the rig, had to pay a $5.5-billion fine as well as $8.8 billion in natural resource damages (EPA 2022). This is considered the greatest oil disaster in United States history.

While this is the most extreme example of the risks posed by deep-sea drilling, it is also a cautionary tale. Deepwater drilling and nuclear power are both energy resources bristling with potential but laden with almost incomprehensible dangers. (Nuclear power, however, has the added benefit of being a carbon-neutral form of energy.) Society is left with difficult decisions. Are the potential oil resources and lower prices that can accompany additional oil added to the market worth the existential risks to the ocean and to ocean life near a deepwater rig? The Deepwater Horizon spill alone killed an estimated 7,600 adult sea turtles and 35,000 small turtles and hatchlings. The local dolphin population is not estimated to return to pre-spill numbers for another 39 years. Other species may need up to 105 years to recover. If a storm were to destroy multiple rigs, the damage could be even worse. These are the risks and rewards that must be balanced as each country decides whether to allow deepwater drilling in their waters.

ENHANCED RECOVERY/HYDRAULIC FRACKING/ HORIZONTAL DRILLING

Enhanced recovery refers to techniques for extracting oil by increasing its ability to flow to a well. It is a blanket term that includes both hydraulic fracking and horizontal drilling, along with other methods such pumping CO_2 into the ground. The combined methods used in enhanced recovery have been instrumental in the oil boom in the United States over the past twenty years and have subsequently been adopted by countries all over the world. One of the most important things about enhanced recovery methods is that they allow oil to be recovered from old oil wells. This cuts out the costs of research and exploratory wells, as companies have been able to pull oil out of areas that had not been productive for over a hundred years. The downside of some of these methods is that they have introduced a number of new consequences to oil drilling. For example, hydraulic fracking can alter water tables and create mini earthquakes. Oil companies inject a slurry mixture into rocks to "release" the oil or gas they are looking for. The exact components of this slurry mixture are proprietary information, and companies do not disclose it. Researchers have found a number of chemicals that are harmful to wildlife and humans in areas where the slurry has been injected. As a result, some areas and states have banned hydraulic fracking altogether. It has become a contentious issue that pits neighbor against neighbor because there is also money to be made by people who want to lease their land to oil companies. Some states, such as Oklahoma, now regularly experience thousands of small earthquakes annually due to enhanced oil recovery methods. At the same time, the United States now produces more oil than any other nation on earth. The United States has the tenth-largest oil reserves, but these techniques have allowed for unprecedented production.

Hydraulic fracking (hydraulic fracturing) or "fracking" for short is the process of injecting a mixture of water, sand, and chemicals into tight rock formations, usually shale or sandstone. These rock formations are already known to contain oil and/or natural gas. The oil deposits in these formations are called tight oil (or gas). Fracking involves a lot of water, sand, and undisclosed chemicals. The process can extract oil that was thought of for decades as unrecoverable, and it has allowed the United States to become the number one oil producer in the world. Fracking, along with other enhanced recovery techniques, has kept the price of oil relatively low. Conventional oil production peaked in 2006, so if new techniques had not been implemented, oil prices would have risen to astronomical levels. Fracking has also led to many personal financial windfalls, as land owners profit by leasing out their land and receiving a share of the oil profits.

Fracking also has significant disadvantages, however. It has caused damage to underground aquifers and caused thousands of small earthquakes, making the ground below homes and businesses less stable. Oklahoma and Texas have experienced earthquakes caused both by fracking and by injecting wastewater from the oil-drilling and fracking process. There have been anecdotal videos of people lighting the gas coming out of their water fauce. Many groups fear that fracking will permanently pollute local water supplies with chemicals that are toxic to humans and wildlife. Cheap oil has also slowed down the transition to fossil fuel-free transportation. Higher oil and gas prices may have led to faster adoption of electronic vehicles, for example. Only time will tell if the benefits of fracking will be worth the damage done in the process. New York state has outlawed the practice, but it is legal in the neighboring state of Pennsylvania. Comparing the two states in the future may shed light on how the process has changed the landscape. In the short term, it is unlikely that fracking will stop in most places, as it is cheaper and easier than deep-sea oil exploration and a financial boon to areas that allow it. The demand for oil is just beginning to ebb. It will be many years before oil demand drops dramatically and permanently.

Horizontal drilling involves drilling both vertically and horizontally. The oil rig drills down to the level of oil formation and then turns toward the area that has the oil. This is useful in places in which land restoration is very expensive or in areas where it can be used in conjunction with fracking. In some situations, it may be difficult to drill a vertical well for a variety of reasons: water that is above the oil or gas formation limits where oil rigs can be placed, only drilling directly vertically will not produce much oil or gas, and horizontal drilling will allow one oil rig to produce more and longer. The downside of horizontal drilling is that it is very expensive, up to three times the cost of drilling a traditional oil well. Because it is often used in tandem with fracking, all of the downsides of fracking apply as well. Horizontal drilling is an important part of the oil and gas drilling renaissance that is occurring in the United States and other countries. It will likely continue to be an affordable and efficient option for those trying to get the most out of both new and old oil fields.

Injecting CO_2 is another process that has been used to increase production in old oil wells. CO_2 can be captured at a power station that burns wood, waste, oil, gas, or coal. This captured CO_2 reacts with the remaining oil in these less productive wells. The oil becomes thinner and easier to extract, but the majority of the pumped CO_2 remains sequestered underground indefinitely. It is possible that in some cases more CO_2 will be sequestered underground than is created by burning the oil recovered. Overall, this method looks like a win-win. More oil is produced, and CO_2 is sequestered to offset the CO_2 that is eventually released. There is the argument that it would be better to sequester the CO_2 without producing any oil at all. However, as long as there is demand for oil, it makes sense to balance oil production and CO_2 sequestration.

The big questions around enhanced recovery methods are: Is cheaper oil a good thing? Would the world more easily transition to a carbon-free economy if more localities, states, and countries banned these advanced oil extraction practices? Another aspect to consider is that OPEC+ (the Organization of the Oil Producing Countries plus allies) has the ability to counter excessive oil production by limiting the amount of oil the OPEC+ countries produce. The United States is not a member of OPEC+ and will continue to produce oil using these methods as long as it is profitable to do so. Advanced methods of oil and gas extraction will continue to allow humans to extract fossil fuels from the earth. Natural gas burns cleaner than most other fossil fuels. Many argue that natural gas is needed to transition the electric power industry from coal to carbon-free electricity.

FOREST MANAGEMENT

Forest management is the process of people making decisions about how to use the forests inside their country. This process can be difficult because there are competing needs and desires. An outsider or a Brazilian conservationist may look at Brazil's vast Amazon rainforest and see an enormous nature preserve, but for some of the people of Brazil, it is also an untapped resource full of timber, potential cattle lands, and mining possibilities. This is the classic balancing act that countries must address: how much to protect and save and how much to use to grow an economy and create economic opportunity for more people? Historically, most Western nations have used forests as their economies grew, and it was not until countries had achieved high levels of economic prosperity that forests were looked at as something that should be preserved. The exception to this were forest preserves created to allow for hunting of the animals that lived in these areas. National parks tended to limit or ban both logging and hunting, but hunting these same animals (that benefited from park forests) outside of the parks was usually allowed. Preserving forests for their own sake did not truly take hold until the environmental movement of the 1960s and 1970s. Not until the dissemination of information on global warming in the 1980s and 1990s did countries begin to realize that forests have an intrinsic value as carbon storage. This knowledge has intensified the debate between those who see forests as a renewable resource and those who believe a forest is more valuable as a continual undisturbed ecosystem. Trees will regrow, and unlike coal, oil, or natural gas, they can be replenished within a human lifetime. Many forests can be harvested every twenty-five years and still produce profitable timber, especially if they are hardwood forests.

In a perfect world, most of the "old-growth" forests would still exist. An old-growth forest is one that has never been cut down by people. These forests hold more carbon than any other type of forest. They have tall trees that have a lot of mass. They are populated with diverse plant and animal life. That rich biodiversity creates a balanced system that is more resilient against forest fires and more able to adapt to climate change conditions. Unfortunately, only about a third (34 percent) of old-growth forest still exists (UNEP 2020). The most important forest management practice that people can implement is the continual protection of old-growth forest from logging. A further difficulty arises in how to define old-growth forests for the purposes of management. Some organizations and governments define old-growth forests as those that have never been harvested for timber, but others require that a never-cut forest reach a minimum age to be considered old-growth. Conflict can arise in a situation where, for example, a forest takes hold in an area that was formerly a meadow. This forest may only be 80–100 years old. To an environmental group, this is an old-growth forest, but to a logging company that has set a 150-year minimum on old-growth forests, it is new growth and eligible for logging. A logging company may also require that the majority of a forest be over 150 years old. This can be problematic as a forest is always a mix of older and younger trees. In some cases, trees that are hundreds of years old are surrounded by a secondary forest and are harvested alongside the newer trees. In many countries forested areas are owned by logging companies, who can make their own regulations or choose to have none. In lands owned and/or regulated by governments, regulators and foresters can make policy and local decisions on what will be cut down and how often it will be harvested.

In places where secondary forests dominate the landscape, foresters face a wide variety of problems. In areas that are drying out due to climate change or were naturally dry to begin with, forest fires are the biggest threats. Forests that have been clear-cut in dry areas of the world have an additional problem: the fire-dependent systems that kept down fuels and created a lightly treed forest have been dramatically altered. For example, a natural dry forest may have spread out large pine trees as the dominant species with a variety of nondominant plants, bushes, and trees throughout. These forests have evolved with a natural fire regime that keeps tree density low while keeping the trees that survive healthy. Once a forest like this has been clear-cut, it will not return to its natural equilibrium. The mixed forest with spread out dominant species will be replaced by a crowded forest that is completely made up of the dominant tree species. A good example of this is in northern Arizona, where the dominant species is the Ponderosa pine. When forests were clear-cut for logging starting at the end of the 1800s, people assumed they would grow back the same way. The dense Ponderosa forests that came up instead were weak due to so many small trees competing for resources. These second-growth forests were prone to disease and fire. Foresters in the area worked to fight forest fires, but stopping the early forest fires only made the later ones worse. When these forests got taller and denser, low-intensity fires were replaced with crown fires that completely destroyed forests. (A crown fire is a fire that burns up to the tops or the crowns of the trees.) Crown fires can burn so intensely that they destroy the soil as well as the forests, killing the mycorrhizal fungi that live in a symbiotic relationship with Ponderosa pines. In some cases, the forest floor is left in such a state that it is unable to absorb water, which results in erosion and flash flooding. Some areas are so damaged that the forests have not been able to recover and are lost for the foreseeable future.

These are just a few examples of ways in which humans must grapple with forest management. There are of course worse cases. There are parts of the British Isles, including Ireland, where all the forests have been cut down and people have resorted to burning peat for heat and/or cooking. Not only are there no longer forests storing carbon, but carbon is being released from peat that has been stored for millennia. Finding an equilibrium between human needs and the benefits that other species (on a local level) and the atmosphere (on a global level) receive from forests is necessary to prevent the worst effects of climate change and protect the health of humans and the species with which we share the planet.

FUEL CELLS

There are many types of fuel cells, but here we focus on hydrogen fuel cells. Hydrogen fuel cells separate hydrogen into electrons and protons. The electrons create an electric current, and the protons connect with oxygen to become water, the only byproduct of a hydrogen fuel cell. The difficulty with this system is that it is expensive to extract hydrogen from water for use in fuel cells. Current hydrogen extraction techniques rely on using natural gas to recover hydrogen, but there is new technology that uses solar power. As this technology becomes cheaper and more efficient, it could be scaled up to create the basis for a hydrogen economy.

Rudimentary fuel cells were created in the early 1800s. The man given credit for inventing the first fuel cell was William Grove in 1839, but there were a number of both professional and amateur inventors that advanced the science of fuel cells. These early fuel cells demonstrated the ability to create an electric current from a chemical reaction between hydrogen and oxygen. By the 1890s, scientists were experimenting with converting coal gas into electricity using a fuel cell. However, it was not until a man named Francis "Tom" Bacon focused his considerable intelligence and energy on creating a fuel cell that the technology produced a significant amount of electricity. His work from 1932 to 1959 became the basis for the fuel cells that were used by NASA (National Aeronautics and Space Administration), the US space agency. The need for energy and water in space was solved by using hydrogen fuel cells, which powered the electronics and mechanical devices inside the spaceship. Fuel cells allowed astronauts to have power without the danger of a flame. Combustion engines and combustion (fossil fuel) generators were used in the vast majority of all transportation and energy projects on earth in the 1960s, so the use of fuel cells was revolutionary. However, it was also extremely expensive, as a nascent technology that required individual construction of each fuel cell. The future of fuel cells will require mass production and lowered costs for both the fuel cells and the hydrogen fuel required to run them.

The fuel cells used by NASA were alkaline fuel cells. An alkaline fuel cell works essentially like a battery except that it never runs out as long as it has hydrogen and oxygen. A fuel cell has an anode and a cathode. Hydrogen is fed into the cathode, and oxygen (or air that contains oxygen) is fed into the anode. The anode and the cathode are separated by an electrolyte that only allows certain ions to pass through (Burke 2003). The chemical reaction that occurs releases heat and water while it produces electricity. The next generation of fuel cells developed are called proton-exchange membrane fuel cells (PEMFC). They are more advanced than the traditional alkaline fuel cell, but they operate on the same principles. They are more efficient and can operate within a large range of temperatures.

In the future, fuel cells could replace combustion engines, diesel engines, and even jet engines, dramatically reducing the amount of CO_2 and other GHGs humans produce through the transportation sector. The question that remains is, when will green hydrogen (hydrogen made from water, using solar energy) be available in the amounts that will be necessary for a transition from fossil fuel-powered transportation? If in the near future, fuel cells will likely be an extremely important part of the carbon free energy transition. If it takes decades to scale up green hydrogen fuel cells, then it is likely that other technology will surpass green hydrogen fuel cells and they will only be a niche market, used in places in which EV or other new types of technology cannot do the job.

FUSION

Fusion is the process of combining two hydrogen atoms into one helium atom, releasing tremendous heat and light. It requires immense amount of pressure. Fusion is the process that powers the sun. Almost all energy on earth, except nuclear, is derived from fusion reactions in the sun, in at least a secondary or tertiary way. All fossil fuels, for example, are derived from long dead life forms that used the heat and light of the sun to grow. Solar energy is directly derived from sunlight, and wind energy is driven by differentials in warm and cold air, created by the sun. Fusion is important for current sources of energy, but it also has the potential to be the key to almost limitless affordable energy.

Many nations across the planet are working toward understanding fusion in the hopes of eventually harnessing it to power the planet. One of the difficulties of creating a fusion experiment is the intense heat the process releases; nothing can contain a nuclear fusion reaction. Scientists use high-powered magnets to create a space for the plasma that is created in the fusion process. The plasma appears to be levitating in the air, as it does not touch anything solid. The enormous expense that is involved in creating a space in which fusion can occur limits the rate of experimentation. Billions of dollars are invested in multinational efforts to maintain nuclear fusion reactions that create more energy than is put into the reaction, but this goal has remained elusive. If at some point a fusion reaction can be created and maintained on earth, all energy problems would be solved. There would no longer be a need for fossil fuels that lead to global warming through the greenhouse effect. The success of fusion would limit the long-term negative effects of climate change.

GEOTHERMAL ENERGY

Geothermal energy is created by using natural heat from the earth to drive a turbine to create electricity (see Chapters 1 and 3 for more information). A natural geyser such as Old Faithful in Yellowstone National Park in the United States is a visual example of geothermal energy. This is a place where the heat from deeper layers of the earth is extremely close to the surface. The core of the earth is extremely hot, 5,000–6,000 K or up to 10,340°F and 5,726°C (Alfe et al. 2002). Most of this heat comes from planetary accretion (the process of accumulation of particles due to gravity into the massive objects we call planets) and from radioactive decay. Although generally deep within the earth, in some places, this heat escapes to the surface, as our example of the Old Faithful geyser demonstrates. To use the heat of the earth to create steam that then turns a turbine and thus creates usable energy, we must be able to access that heat. In most places, this involves drilling deep into the earth, which is extremely expensive and can be dangerous. For this reason, geothermal energy makes up less than 1 percent of the electrical energy generation across the world. New advances in searching for geothermal energy pioneered by Jason Craig at the University of Nevada at Reno show promise for discovering geothermal resources without first having to drill.

Geothermal plants use either "dry" steam, which comes directly from the earth and can be used to drive a turbine, or flash steam, which comes out of the earth as hot water under pressure and only becomes steam as it nears the surface. These are the two most common forms of geothermal electric plants, with flash steam being the most common. Binary plants can use hot water from the earth to heat another fluid that has a lower boiling point. This other fluid can then turn a turbine and create electricity. This type of plant is useful in an area with natural hot water that is not hot enough to create a viable plant. Finally, there are a number of countries, including Iceland, Japan, Russia, Italy, Mexico, and the United States, that are investigating the possibility of tapping into supercritical fluid. This is water that is under pressure and hotter than 400°C. It is believed that if this supercritical fluid, which exists in a state that is between a liquid and a gas, could be controlled, it would be one of the more energy-dense materials on earth. A breakthrough in the use of this supercritical fluid could bring geothermal from producing only 1 percent of the world's energy needs toward something much more substantial.

There are a few countries that have a lot of geothermal energy available close to the surface of the earth. Some countries are able to use geothermal energy for the majority of their electricity and/or heating needs. While this is not the case in most countries, there is hope that some of the advances in drilling for oil and gas can be used to more affordably access geothermal heat in places that have not done so in the past.

The cost of building a geothermal plant and the slow return on investment makes geothermal energy similar to hydropower in that it is best suited for governments that are not looking for short-term gains. The United States produces the most geothermal power than any country on earth at 3,676 MWe (megawatts of electricity) of generating capacity. Indonesia ranks second in geothermal power at 2,133 MWe generating capacity. The Philippines, Turkey, and New Zealand are at the third, fourth, and fifth places producing, respectively, 1,918 MWe, 1,688 MWe, and 1,005 MWe of the generating capacity. No other country produces more than 1,000 MWe of geothermal energy, but Mexico and Italy are close at 963 MWe and 944 MWe, respectively. Kenya has 861 MWe of installed geothermal capacity. Iceland has 755 MWe of geothermal capacity, which is impressive given that they have a human population of less than 400,000. Japan can produce 603 MWe, while all other countries combined produce only a total of 1.025 MWe.

Geothermal electricity has plenty of room to grow, as there are many untapped geothermal resources. As countries have pledged to decrease the amount of CO_2 they produce, they have turned to geothermal electricity as a potential carbon-free energy source. A number of geothermal plants are slated to go online by 2030. These plants will be able to help many countries achieve their goals of carbon-neutrality. Technological advances could accelerate that process. It is possible that the combination of countries' desires to lower CO_2 emissions and technological advances could propel geothermal energy from a minor player in the renewable energy market to one that becomes an important mix in the global struggle to avoid the worst possible effects of climate change.

GREENHOUSE EFFECT AND GREENHOUSE GASES

A greenhouse traps heat by creating a physical barrier that limits how much heat from the inside can escape to the outside. This allows people to grow plants earlier in the spring and past the first freeze of winter. In sunny climates, additional heat is not needed because the sun heats up the air inside the greenhouse, and the glass or plastic creates a barrier that traps the warm air. This small-scale system illustrates the same process that occurs on a planetary scale, called the greenhouse effect. The gases that make up our atmosphere, along with water vapor, trap heat and keep the earth warm enough to sustain life. Without the atmosphere and water vapor, the earth would be too hot in the day and too cold at night for most forms of life. Greenhouse gases (GHGs) can be carbon dioxide (CO_2), methane (CH_4), ozone, nitrous oxide, water vapor, and chlorofluorocarbons (CFCs). So, if the greenhouse effect is a good thing, why are we worried about it? The scientific community, world leaders, and individual people are actually worried about something called the enhanced greenhouse effect. ("Enhanced" tends to be dropped from the term when people use it in casual conversation and popular science articles.) The enhanced greenhouse effect is caused by adding additional greenhouse gases to the atmosphere. The greenhouse effect keeps the planet warm and habitable, but the enhanced greenhouse effect threatens to make many parts of the earth uninhabitable. CO_2 is the GHG that is talked about the most because it is most responsible for enhancing the greenhouse effect. It is not alone, and it is not the most potent of the GHGs, although it is the most prevalent. While there are some changes in the enhanced greenhouse effect that are unlikely to be reversed in a human lifetime, there are a number of others that can be stopped if humanity takes the appropriate measures.

As mentioned, there are many types of GHGs) The most prominent and talked about is CO_2. CH_4 is the second most prominent GHG; it is pound-for-pound much more harmful in the atmosphere than CO_2 but is much less prevalent. Nitrous oxide (N_2O) is the third most dangerous GHG. We will also look at the following four industrial gases: hydrofluorocarbons (HFCs), perfluorocarbons (PFCs), sulfur hexafluoride (SF_6), and nitrogen trifluoride (NF_3). The EPA breaks down GHG emissions into five sectors, listed here with the percent of GHGs that each sector produces in the United States: transportation (29 percent), electricity (25 percent), industry (23 percent), commercial and residential (13 percent), and agriculture (10 percent) (EPA 2022).

CO_2 is produced by burning anything that is carbon-based. The vast majority of CO_2 is produced by the burning of fossil fuels for transportation, electricity production, and heat. Burning wood, trash, or dried animal waste also produces CO_2. The CO_2 created by human activities breaks down very simply: 87 percent of human-caused CO_2 comes from burning fossil fuels, 9 percent results from land use changes (such as clearing forest for farming or cattle), and 4 percent comes from industrial processes including cement manufacturing (CO_2 Human Emissions 2017). All human-created CO_2 adds up to about 32.5 gigatons, while natural sources add up to about 750 gigatons. So, human activities are responsible for a small fraction of total CO_2. Natural processes, such as animals and plants dying and decomposing in the ocean, matter decomposing on earth, forest fires, volcanoes, plants and animals releasing CO_2, and naturally occurring CO_2 deposits eroding, all are part of the earth's CO_2 cycle. The planet has been in balance for hundreds of thousands of years, absorbing the same amount of CO_2 that it produces. The problem is that by burning fossil fuels, people are digging up CO_2 that was previously absorbed by the earth and releasing it into the atmosphere on top of all the other natural sources of CO_2, throwing the natural CO_2 cycle out of balance.

Humans have only three choices regarding CO_2: find a way to pull the additional CO_2 out of the atmosphere and the ocean, replace all carbon intensive processes with carbon-neutral or carbon-negative processes, or do nothing and accept a warmer, less hospitable world with more ocean and less land that will have significantly fewer humans and other species living on it. The last choice is not really a choice; it is what happens if humanity just sticks its collective head in the sand. It will mean the dislocation of over a billion people as well as the deaths of millions if not billions. Because of these high stakes, humanity is motivated to find ways to stop the continual release of CO_2 and to find ways to sequester as much CO_2 as possible. (See the Climate Engineering or Geoengineering entry for more on the major methods for removing CO_2 through both enhanced natural processes and mechanical and chemical methods.) Unfortunately, progress had been stymied by corporate and political interests that attempt to obfuscate the truth to serve short-term financial interests. At this point in human history, all governments of the world recognize the problem. Every problem is also an opportunity. Humanity has the chance to remake itself in a manner that serves all people better. A world with less pollution will save millions of lives annually. People will have the opportunity to have healthier and thus happier lives. It is not often that humanity has such an opportunity; it is our hope that humanity will grab that opportunity and create a world that we cannot yet imagine.

Methane (CH_4) is another potent greenhouse gas. It is released by livestock, by the burning of biomass, in the process of obtaining fossil fuels, in rice cultivation, and during the decomposition of all types of waste especially in an abiotic environment such as a landfill. Methane is a more potent GHG when it comes to how effectively it increases the greenhouse effect. It is about 30 times more powerful than CO_2 on a per-ton basis over a 100-year period. So, although methane only accounts for 3 percent of the total amount of GHGs emitted, it is responsible for 23 percent of the warming of the earth (Sanois et al. 2020). On a geological timescale, however, methane is not very dangerous as it leaves the atmosphere after only twelve years, whereas CO_2 will remain in the atmosphere for thousands of years. Concentrating efforts on severely and quickly decreasing the amount of methane released in the atmosphere could have an extremely positive impact on global warming and climate change.

The vast majority of methane emissions are from agriculture, fugitive emissions, and waste. Agricultural methane is mainly from cattle and rice farming. There are good solutions to both of these problems. Most cattle are finished in concentrated animal feeding operations (CAFOs); legislation could easily require that these operations capture methane. There are simple steps that can be taken to drastically reduce the amount of methane that is released from rice farming as well, such as drawing down water to the roots and having alternating dry and wet times. Fugitive emissions is another fixable problem. Fugitive emissions are from the accidental release of natural gas, which has methane as a major component. Natural gas producers can be required to capture all natural gas, and any leaks would require remediation to capture the equivalent amount of methane from another source. Using drones with sensors, monitoring can be done easily and affordably. Waste is another easily fixable problem. Industrial-scale composting by cities such as San Francisco, California, have been able to significantly reduce methane emissions from their landfills. Compost is then sold to surrounding farms and vineyards. The wine and food produced often end up back in San Francisco, creating a closed-loop system. This system can be adopted anywhere in the world. Finally, waste that is already buried can be managed in a way that allows for the collection and use of methane. Currently in the United States only about 500 of the 2,600 landfills collect methane. That number could easily double if the right incentives and fines were implemented. Overall, methane emissions are much easier to reduce than CO_2. There are fewer changes required, and nobody is asked to change anything about their lifestyle. For these reasons it would make sense for countries, states/provinces, municipalities, and individuals to focus on reducing methane emissions as soon as possible.

Nitrous oxide (N_2O) is a naturally occurring gas that is well known as "laughing gas" and is used for medical and dental procedures as it numbs pain and creates euphoric feelings. The problem with nitrous oxide is that it is 300 times more powerful than CO_2 as a GHG. Nitrous oxide is created by overfertilization of crops using synthetic fertilizers. Farmers are always motivated to overfertilize instead of underfertilize as they do not want their crop yields to decline. Fortunately, there are two methods of decreasing nitrous oxide. The first method is simply installing more efficient watering systems. Drip irrigation can greatly decrease the amount of nitrous oxide that is released into the atmosphere. Second, if synthetic fertilizer is not overapplied, roots have a better chance of competing with the microbes that create nitrous oxide as they break down the synthetic fertilizers. Another source of nitrous oxide is the ocean. Luckily, there are natural nitrous oxide sinks that take up more nitrous oxide than is released around them. Mangrove forests are one of the best nitrous oxide sinks that exist. Coastal communities can restore and create mangrove forests on their shores, as they are beneficial for numerous reasons (see Climate Engineering or Geoengineering entry for more information). Nitrous oxide, like methane, is a dangerous GHG that can easily be contained, and humans can limit the amount that is released into the atmosphere without abrupt changes to any human systems.

Hydrofluorocarbons (HFCs), perfluorocarbons (PFCs), sulfur hexafluoride (SF_6), and nitrogen trifluoride (NF_3) are fluorinated gases that are created through industrial processes. In many cases they are used to replace ozone-depleting substances that have been banned internationally. (Ozone is technically a greenhouse gas, but it creates the ozone layer, which protects the planet from ultraviolet light, harmful to both plants and animals.) These gases are currently not very abundant, so there is not a lot of movement to ban or decrease the use of these GHGs. They are categorized as high global warming potential gases or GWP gases as they have the potential to be powerful and dangerous GHGs. Hopefully, more environmentally friendly chemical replacements will be used in the future.

Lowering the emission of GHGs is a difficult task, but there is some low-hanging fruit; there are some fairly easy ways to lower methane and nitrous oxide. CO_2 is more difficult as fossil fuels are still completely integrated into both transportation and energy production. However, renewable energy power plants are displacing fossil fuel plants and EVs are replacing combustion engines. Fluorinated gases that could become a problem are currently at reasonable levels. The task is daunting but well within the ability of human endeavors.

GREEN REVOLUTION

The Green Revolution was an agricultural revolution, sometimes referred to as the third agricultural revolution. Starting in the 1950s and picking up steam, the Green Revolution was in full swing by the 1960s. A set of new agricultural practices and advanced agricultural techniques and technology allowed farmers across the planet to greatly increase their yield. New hybrids of cereal grains that had been grown for millennia contributed, as did petroleum-based fertilizers. Technological advances in mechanized farming allowed one farmer to produce more than ever before. Pesticides became more affordable and more potent; pests that had been the scourge of farmers for generations were wiped out with a few applications. The Green Revolution meant that farming was able to keep pace with the needs of a burgeoning human population that was growing exponentially, possibly saving a billion people from starvation and malnutrition. The world eventually had more food than it needed, leading to the obesity problem many developed nations are facing. The Green Revolution allowed for continued human population growth and led to a healthier overall population of people, but it created a set of new problems as well.

The beginnings of the Green Revolution are rooted in the mechanization of agriculture that began in technologically advanced countries at the beginning of the twentieth century. As automobiles began to adapt the technology that powered trains into a more mobile technology that did not need to be connected to a track, farmers found ways to use engines to do the backbreaking work that humans and plow animals had done for time immemorial. Tractors allowed one person to plow many times more the land in one day than could be done with a team of mules. This allowed for more fields to be planted faster. Post-Second World War flight technology allowed vast tracts of land to be sprayed with pesticides from the air. By the 1960s, hybrid rice and wheat varieties had spread across the planet. Combined with inorganic fertilizers, usually derived from oil, and the powerful pesticide DDT invented by Paul Muller, yields had reached heights never dreamed of before. Beginning in the 1950s, the Rockefeller Foundation and the Ford Foundation worked to spread new agricultural practices and new higher yielding varieties of rice and wheat across the planet in the hopes of ending world hunger (Jain 2018). The combined efforts of world governments, NGOs, and foundations helped to bring the Green Revolution to every corner of the globe. By the 1980s, there were very few places left on earth that had not adopted some or all of the techniques, hybrid varieties, and technology of the Green Revolution.

As the Green Revolution spread, along with vaccines that wiped out many deadly diseases, the world population grew faster than it ever had before. Growing at over 2 percent a year from the 1950s to the 1970s, between 1950 and 2010, the world population increased from 2.5 billion to 6.9 billion, an increase of 174 percent (UN 2021). These numbers could have never been supported before the Green Revolution. Not only did human populations increase, human life expectancy and health outcomes increased as well. The human population was overall healthier than before.

There are a number of downsides to the Green Revolution, such as decreasing genetic diversity in the major cereal crops and the extinction of local crops that have been crowded out of the market by hybrids (and now genetically modified hybrids) that are less expensive to grow and produce a bigger harvest. The use of pesticides such as DDT (now banned in most countries) led to serious losses in predator species that ended up with large concentrations of DDT in their bodies. Many pesticides have been detrimental to the environment and fertilizer runoff has created hypoxic zones devoid of life in lakes and oceans (Pingali 2012). The Green Revolution has pushed many small farmers out of farming as new market conditions favor agribusinesses that can purchase huge swaths of land and, through monoculture crops, make more profit than a traditional farm with diversified crops and livestock. This process has also led to less sustainable food systems. A traditional farm may take the waste from livestock and humans and use it to grow crops, which in turn are used to feed livestock and humans. Large concentrated animal feeding operations (CAFOs) have created enormous amounts of toxic waste that is mixed with antibiotics and sometimes steroids. Pig CAFOs have created especially toxic waste that is simply held in plastic lined "ponds." One flood event can bring this toxic waste into healthy ecosystems or people's drinking water supply.

The explosive growth in the human population also comes with its own set of downsides. Large human populations crowd out other species as our need for land (housing, farming, and ranching) is in direct conflict with other species' need for habitat. There is also the risk that continued human population growth will someday outstrip agriculture's ability to adapt and advance and thus lead to widespread starvation again. However, up to this point all predictions of a starving world population have turned out to be somewhere between hyperbole and hysteria. One problem connected to human population growth that is neither hyperbole nor hysteria is the problem of global warming and climate change. Agriculture and animal husbandry are responsible for between 12 and 30 percent of all GHGs (the large discrepancy depends on how transporting food, storing food, and disposing of food and livestock waste are categorized). There are many methods that can be implemented that both store carbon and allow farmers to continue to see high yields in their fields. (See Climate Engineering or Bioengineering entry for more information.)

Overall, the Green Revolution was a huge boon to humanity, saving millions from hunger and malnutrition. These advances in farming will continue to allow more people to be pulled out of poverty. The next transition in agriculture will need to combine the advances of the Green Revolution with new knowledge about how to produce food while storing carbon as well.

HUMAN DISPERSAL

Humans evolved in Africa. They evolved from a shared genetic line that branched off of the same evolutionary line as bonobos and chimpanzees. For most of human existence, humans stayed in Africa. Modern humans or *Homo sapiens sapiens* did not venture out of Africa until after their close genetic relatives, the Neanderthals, had already been settled in Europe and Asia for hundreds of thousands of years. Neanderthals were able to interbreed with humans; most humans, outside of those of sub-Saharan African descent, carry an average of about 2 percent Neanderthal genes in their genome. Some DNA evidence suggests that female Neanderthals and human males were able to produce viable offspring and that the male Y chromosome in Neanderthals was replaced by the human Y chromosome (Gibbons 2020). This brings up the question of whether Neanderthals were simply overrun by larger numbers of humans and absorbed into the human population or if humans outcompeted Neanderthals for energy resources. Geneticists are still trying to unravel this mystery. A couple of things that are becoming clear are that Neanderthals lived in Europe and Asia from roughly 400,000 to 40,000 years ago and that they disappeared from the archeological record just as humans started to show up.

Humans likely moved to new areas in search of food sources. As local human populations grew, competition for food resources would have caused conflict between tribes or within tribes. These conflicts over energy resources would have pushed younger groups or weaker groups to search for new areas to obtain the calories they needed to survive. People traveled along the seashore where food resources were plentiful or followed herds of megafauna (animals weighing more than 45 kg) that created opportunities for plentiful caloric/energy resources. Human dispersal and the extinction of megafauna are correlated. While there is disagreement as to whether the two events are related primarily by direct causation (hunting) or indirect causation (from disease carried by human populations), it is clear that humans had a part to play in the extinction of megafauna. Megafauna represent an immense energy resource, the pursuit of which may have driven human dispersal.

Using mitochondrial DNA (which is only passed down from the mother), scientists have been able to prove that all modern humans are related to a single female ancestor who lived around 200,000 years ago. The human population at the time likely consisted of between 1,000 and 5,000 individuals. The majority of these modern humans stayed in Africa until about 60,000 years ago. Some early hominin relatives left Africa much earlier, including *Homo erectus*, which left about 1.8 million years ago. The Denisovans descended from *Homo erectus* and therefore were native to Asia. Neanderthals left Africa about 200,000 years ago. Modern humans had some small early forays into the Near East in modern-day Israel and other sites in the Levant (also called the Fertile Crescent). DNA evidence demonstrates that these groups were not related to the later waves of humans who settled across the entire planet. When modern humans settled into an area, they eventually replaced the other human species in that area. In some cases, such as in Europe, Neanderthals and modern humans shared an area for a time (15,000 years), sometimes interbreeding. DNA evidence suggests that Denisovans bred with modern humans as well.

The "Great Migration out of Africa" that took place from 60,000 to 50,000 years ago is responsible for permanently altering the planet, as humans established themselves as the apex predator on every continent (except Antarctica) and almost every inhabitable island as well. This migration forever changed the flora and fauna of the world. Everywhere humans settled in large numbers, most megafauna disappeared. Throughout the Middle East, Europe, Asia, Australia, and the Pacific Islands, as well as North and South America, permanent human settlements predated the extinction of local populations of megafauna. During the age of European exploration, Pacific Islanders were still settling the last of the uninhabited islands. European travelers took note of some of the huge flightless birds that existed on these islands, but after permanent human settlement, these species quickly decreased and in many cases went extinct. The most famous of these flightless birds is the moa, the largest species of which could be up to 12 feet tall and weigh over 500 pounds. Recent scientific studies have shown that human overhunting caused their extinction 600 years ago. While it may be impossible to prove that humans alone were responsible for the extinctions of hundreds of species of megafauna, it is fair to assume that humans played a part, either directly or indirectly, in the majority of the extinctions that correlate with the arrival of permanent human settlement.

HYDROCARBONS AND HYDROCARBON GAS LIQUIDS

Hydrocarbons are defined as organic compounds that have only hydrogen and carbon as constituent parts. Importantly, hydrocarbons are what make up the majority of what we call oil and gas. There are twelve types of hydrocarbons from a scientific perspective, defined by the number of carbon atoms that each contains: 1, methane; 2, ethane; 3, propane; 4, butane; 5, pentane; 6, hexane; 7, heptane; 8, octane; 9, nonane; 10, decane; 11, undecane; and 12, dodecane. The nomenclature for some of these hydrocarbons, 5 through 12, are taken from the Greek (pent—meaning five and so on). These hydrocarbons are used as fuel as well as in all plastics and most synthetic pharmaceuticals. Hydrocarbon gas liquids (HGLs) are hydrocarbons that occur as liquids below the ground under pressure or at especially cold temperatures. These same liquids become gases at atmospheric pressure. Each type of HGL liquefies at a different temperature and/or pressure. The nine most common types of HGLs are categorized chemically as follows:

Alkanes, or paraffins
Ethane—C_2H_6
Propane—C_3H_8
Butanes: normal butane and isobutane—C_4H_{10}
Natural gasoline or pentanes plus—C_5H_{12} and heavier
alkenes, or olefins
Ethylene—C_2H_4
Propylene—C_3H_6
Normal butylene and isobutylene—C_4H_8

Paraffin is used for a multitude of purposes, including as a lubricant, for candles, to make crayons, for electrical insulation, and, most commonly, for petroleum jelly. Ethane is used to create energy in power plants; it is also used to make plastics, in antifreeze, and in detergents. Propane is a commonly used product for backyard grills, outdoor heaters, home heating, and gas fire pits and is used in plastics. Butanes are used as a petrochemical feedstock and in motor gasoline blending, as well as butane lighters. Gasoline and pentanes are used as fuel for automobiles and other forms of transportation. Olefins/alkenes, ethylene, propylene, and normal butylene/isobutylene are all used in the petrochemical industry to create plastics, paints, solvents, resins, and artificial rubber (IEA 2018).

The uses for these hydrocarbons are almost endless. They touch every part of our lives, from the computers and phones we communicate with and learn on, to the water bottles we drink out of, to almost every product we purchase. If a product is not manufactured from a hydrocarbon, it is probably wrapped in one or put into a (hydrocarbon-derived) plastic bag and handed to us at the store. There has been a push to decrease the amount of plastics that are commonly used, especially single-use plastics. The vast majority of single-use plastic ends up in a landfill where it will take thousands of years to decompose or in the ocean where it becomes part of the food chain as plastic pieces are commonly mistaken by marine organisms as food. Current estimates are that there will soon be more plastic than fish in the ocean on a pound for pound basis. Despite the pushback on plastics, demand for plastic products continues to grow. Even as humanity transitions away from fossil fuel-based energy and transportation, a high demand will continue for the multitude of products that are derived from hydrocarbons and hydrocarbon gas liquids in particular.

HYDROELECTRIC POWER

Hydroelectric power is power created by running water and is one of the oldest forms of renewable energy. Since ancient times, peoples across the world have used rivers to transport humans and goods across vast distances and created machines that used the power of rivers to grind grain, lift things, and move things. For this book, we are most interested in how hydroelectric power is used to create electricity (see also related discussions in Chapters 1, 3, and 4).

Most electricity is created by turning a turbine, which spins an attached magnet. The magnet is surrounded by coiled insulated wires, and the spinning creates an electromagnetic field which in turn creates an electric current. Most fossil fuel-dependent power plants use heat to create steam and that steam is used to turn a turbine. Hydroelectric power plants use the force of water to turn turbines and create electricity.

Rivers generally have a seasonal ebb and flow. They are usually most powerful in the spring as winter snow melts, but water can slow to a trickle in a dry summer or fall season. In order to guarantee consistent water flows, dams are built on rivers; the newly formed lake guarantees consistent water and pressure to turn the turbines. The water intake is usually located many feet deep underwater, and the pressure of the water coming through the intake valve is enough to turn the turbine and provide consistent electricity year-round. As electricity became more common at the beginning of the twentieth century, many countries looked to their river systems as a consistent source of electricity. In some cases, hydroelectric dams have also provided a consistent source of irrigation water for farmers in drier areas. In 1882, the first hydroelectric dam was built in Appleton, Wisconsin. It was built off the work of a British-American engineer named James Francis who was the first to develop a modern water turbine some decades earlier. From that point on, a hydroelectric power rush began. Any place with a decent sized river looked to find a way to build a hydroelectric power plant. As demand for electricity increased, governments started to think bigger than small local hydroelectric projects and started designing grandiose dams that could power whole cities and, in some cases, multiple cities. These enormous dams were modern marvels showcasing human ingenuity and knowhow. The United States began planning the Hoover dam in the 1920s. Construction began in 1931, and the dam was completed in 1935. At the time it was the largest hydroelectric dam in the world, three times larger than any other dam on earth. It was also the world's largest concrete structure. It ushered in an era of state-sponsored hydroelectric power super structures. Hydroelectric power at one time looked to be the leading power source of the future, but there were not enough hydroelectric power sites to keep up with the growing demand for electricity.

Hydroelectric power is expensive up front, but one of the most affordable forms of long-term power generation. States such as Washington and Idaho that get the majority of their power from hydroelectric power stations have lower energy bills than other US states.

Due to the high upfront costs, the vast majority of all hydroelectric power plants are built by governments, who are not motivated by the idea of short-term profits. Hydroelectric power today still provides between 16 percent and 24 percent of all electricity on the planet. So, hydroelectric power is still a very important form of power. It provides the majority of the world's renewable energy but is likely to eventually be surpassed by solar and wind power.

Hydroelectric power may have another important role to play in the transformation toward a carbon-free energy system. Pumped hydroelectric storage is a form of energy storage that does not require any batteries. In a future time, in which there is more solar or wind power than is needed, the extra energy can be used to pump water from below a dam to above the dam. In this way, that water is there waiting when more electricity is needed. Hydroelectric power plants can quickly increase power to match demand. It is likely that pumped hydroelectric power will play an even more important role than it does today, so hydroelectric power plants will remain a necessary and useful component of the carbon-free electricity of the future.

HYDROGEN ECONOMY

We are carbon-based life forms, and we live in a carbon-based economy. Work takes energy. In ancient times, that energy came from the food we ate and the food we provided to our work animals. We are solar-powered in that the sun grows the plants that we and our animals eat. During the Industrial Revolution, humans invented machines that do work. Machines also need to "eat." At first, they primarily ate (burned) wood, which is also solar-powered, as the sun grows trees. Soon machines were fed coal, formed when peat (consisting of plants millions of years old) was compressed and heated by the earth until it became the flammable black rocks of coal we see today. As the Industrial Revolution continued, more fossil fuels (from the remains of ancient life forms under pressure and heat) were added to the list: natural gas and oil, along with its various derivatives like gas (petrol), kerosene, and diesel. For over 200 years people have moved things, made things, and grown things with the power of carbon-based fossil fuels, which could be called secondary energy from the sun. So, what does the sun run on? The immense gravity of the sun compresses hydrogen into helium through the process of fusion (see Fusion entry). The idea of a hydrogen economy is to use hydrogen to power the things that are now powered by fossil fuels. When hydrogen burns, it leaves only water behind. It is the cleanest possible burn, with no pollution, no GHGs, just heat and water. Hydrogen is the most abundant element in the universe. Scientists estimate that it makes up 90 percent of all that is visible in the universe. However, hydrogen is so light that it floats right out of our atmosphere and into space, so there is not a lot of pure hydrogen available. Hydrogen is abundantly available in the form of water (H_2O), but energy is needed to split water into hydrogen and oxygen.

The obstacles to a hydrogen economy are simply time, money, and political will. It takes time and money to create the infrastructure needed to supply hydrogen or hydrogen fuel cells (see Fuel Cells entry). People and their political leaders must be willing to invest in the effort and expense of becoming a hydrogen economy, which would require certain sacrifices. Money spent on building a hydrogen economy is money that cannot be spent on military or social spending that may also be a priority. People must also be willing to accept some major changes in their lives, such as new modes of transportation that align with the emergence of hydrogen and/or hydrogen fuel cells as a primary energy source.

The biggest and most important advantage of a hydrogen economy is that humans will no longer be causing additional climate change. They will also no longer be polluting, which will prevent millions of deaths per year from pollution-induced asthma attacks, lung ailments, cancers, and heart attacks. Many dangerous carcinogens will be removed from daily life. Human health will improve overall. Moving a capitalist world into a hydrogen economy will likely prove to be difficult, as many large and powerful companies know how to make money in the carbon economy and will want to protect themselves, fighting to continue this system regardless of the negative effects on human health and on the health of the planet.

There are also many unknowns in making this large of a transition. Will the use of all this hydrogen have unknown negative consequences? Will the water that is released from the process of using hydrogen for fuel be drinkable or "polluted" in some way? Will water released onto streets from hydrogen fuel cell vehicles lead to wet and slick streets, creating a safety hazard? Will a hydrogen economy work for all people, or will it exasperate the already large gaps between haves and have nots? Many other questions linger about how a hydrogen-based economy will work. The speed and feasibility of any country transitioning to a hydrogen economy are uncertain; for now, nations may consider other low-carbon technologies more affordable and easier to adopt.

MEGAFAUNA

Megafauna are large mammals, generally considered to be those over 100 lbs./45 kg. Some researchers further differentiate super megafauna as animals over 1,000 kg or 2,200 lbs. The vast majority of megafauna are extinct. When we think of megafauna today, we tend to think of the large animals of the African Serengeti (elephants, hippos, rhinoceroses, giraffes, wildebeests, lions, and crocodiles); however, at one time these animals and/or animals similar in size roamed all the continents except for Antarctica. Wooly mammoths, mastodons, and wooly rhinoceroses roamed across Eurasia and the Americas. Camels, lions, twenty-foot-tall sloths, giant cave bears, giant armadillos, and Hagerman horses roamed with bisons across American plains. Across modern-day Siberia, herds of mammoths shared space with wooly rhinoceroses (MacPhee 2018). The exact causes of megafauna extinctions are debated. There are four main theories: (1) people overhunted them in the process of feeding a rapidly growing human population; (2) an abrupt climate change event created conditions that species did not have time to adjust to, mainly drastically colder weather caused by the slowing of the Atlantic Meridional Overturning Circulation (AMOC) that was in turn caused by an influx of fresh water from the Greenland Ice Sheet; (3) an impact event caused a great die-off of plant life and a cooling of the planet, leading to the megafauna extinctions; and (4) a novel "hyperdisease" brought by immigrating human hunters led to extinctions of most of the worldwide megafauna. Other theories involve combinations of these pressures on megafauna populations; for example, the most commonly accepted idea is that changing climate and human hunting combined to cause the demise of 80 percent of the megafauna on earth.

Archeological evidence shows that humans and their hominin relatives (such as Neanderthals and Denisovans) hunted megafauna. However, while Neanderthals and Denisovans lived in near proximity to megafauna for tens of thousands of years, permanent settlement by modern humans in an area correlated with an often rapid decline in many of the megafauna species that lived locally. The decline of many megafauna species follows the settlement of humans across all continents (with the exception of Antarctica) and islands across the planet.

Paul S. Martin developed what has come to be known as his overkill hypothesis (also known as the blitzkrieg model) in the 1970s. This model was at first accepted and then later derided by many as too simplistic and not accounting for other factors. However, over time, newer research has confirmed his theory, which says that a relatively small number of hunters can have a huge effect on a naive population. If a population of megafauna (woolly mammoth, for example) did not evolved with humans, they would have little to no fear of humans. Humans would be able to take advantage of this situation and more easily hunt a population, obtaining an enormous abundance of calories. This would in turn lead to an increase in the human population. The larger human population would need to hunt even more of the prey species, eventually leading to local extinctions. Local extinctions would cause tribes of people to move in search of other megafauna, and the process would repeat itself. According to this theory, in a relatively short amount of time, as little as one to two thousand years, humans could hunt an entire species to extinction across a continent.

During the European Age of Exploration, explorers visited uninhabited islands (with no permanent human settlements) and found animals that they could walk up to and pick up. These animals demonstrated absolutely no fear of humans. Megafauna across the planet would likely have, at least at first, reacted in a nonplussed manner to the appearance of humans as well. There were times in which humans did settle into a rhythm with local megafauna; archeological records demonstrate human populations living with woolly mammoth populations for tens of thousands of years. Other species, such as bison, deer, and elk in North America, survived human hunting and continue to exist. This is not the case with the vast majority of megafauna that no longer exist on the planet. While it may never be conclusively proven that humans were responsible for the extinction of most of the megafauna, evidence produced by recent scholarship demonstrates that megafauna extinctions in the Americas are directly correlated to human migration and settlements.

MILANKOVITCH CYCLES

The earth experiences natural cycles of long-term cooling and long-term warming that help us to understand recent human influences on climate. We are fortunate that Milutin Milanković survived imprisonment in the First World War and the Nazi takeover of his native Serbia in the Second World War. Milanković's work combined and extended all previously known research on long-term cyclical climate changes caused by earth's position in relation to the sun. With this information, scientists have been able to reconstruct the historical climate of the earth and are able to predict the future climate. His work has become even more poignant and important as humans grapple with anthropogenic climate change and its long-term consequences. Having the knowledge of where earth's climate should be with respect to these natural, predictable cycles has allowed scientists to understand how far outside the norm humans have changed earth's climate.

The Milankovitch cycles describe three cyclical aspects of earth's orbit around the sun: eccentricity, axial precession, and obliquity. Eccentricity is the shape of the earth's orbit around the sun. In a 100,000-year cycle, the earth's orbit varies from 0.0034, which is an almost perfect circle, to 0.058 which is slightly elliptical (Buis 2020). Understanding this difference can explain the variation in climate over millions of years. Axial precession, which can be described as earth's wobble, is on a 26,000-year cycle. The wobble can explain why the North Pole and the South Pole are not always in the same physical location; they change in a predictable manner. Obliquity can be described as earth's tilt because it measures the angle of the earth relative to the axis of rotation as compared to the direction of the orbit around the sun. The earth's obliquity causes areas of the planet that are not close to the equator to experience seasonal changes in climate. The Milankovitch cycle describes how those seasons vary in a predictable way over a 41,000-year cycle. Taken together, these three cycles are used to understand past climate changes and to predict the future climate of the earth.

Paleoecology is the study of past climates on earth and is a field that has helped to affirm the accuracy of the Milankovitch cycles. One of the most popular methods of learning about past climates is core sampling. Scientists can take core samples from, for example, the bottom of a lake or the bottom of the ocean or through a glacier. A core is sliced into fine layers, each of which can tell scientists about a year or a number of years from the past. Using knowledge of the temperature requirements for certain plants, scientists will look for pollen from these plants. The abundance or lack of abundance of pollen indicates which time periods were warmer and which were cooler. Scientists can also discern between wet and dry periods using the same methods. To look deeper into the past, scientists must look at the geological record left behind in the rocks and the layers of the earth. Fossils allow scientists to dig even deeper into past climates.

The Milankovitch cycles have been an important tool for scientists as they unravel the mysteries of the climate past and try to understand the climate of the present. Scientists will continue to refer to these cycles as they look to bring the planet's climate closer to where it would be without the anthropogenic forcings that have changed the climate so dramatically.

MINERAL RIGHTS

Mineral rights can be a complicated matter. It is usually clear who owns the land on the surface, but who owns the riches underneath can vary from country to country, state to state, and city to city. Even within each city, who owns the rights to the minerals below your home depends on who first owned it after it was taken from the Native American people of the area (in the case of the United States). West of the Mississippi River in the mid-1800s, the US government used land grands to incentivize railroad companies to build railroads across the country. However, to prevent the railroad companies from having a monopoly on all the new cities that were sure to pop up along the railroad, the government provided land grants to the railroads in a checkerboard pattern, retaining ownership of the other half of the checkerboard. This remaining government land was then used for government purposes, sold, or made available for homesteading. According to the 1862 Homestead Act, settlers could claim land if they fenced it, built a house, and added other "improvements"; the land would become theirs after five years if all the required paperwork was filled out. Today, if you buy a home within twenty miles of one of these major railroads built between 1862 and 1871, you have a 50/50 chance of owning the mineral rights underneath your home. Many times, when railroads sold their land, they retained the right to the minerals underneath. If oil is discovered in one of these areas, the BNSF (Burlington Northern-Santa Fe) Railroad or Union Pacific Railroad might own the rights to that oil even if it is directly under the home you own. This is just one example of the complications that occur with mineral rights. On the country-wide level, some nations lay claim to all resources underground regardless of ownership of location, while other countries allow anyone to make a claim on land owned by the government and charge no fees or royalties. The United States is the only country in the world in which mineral rights are the property of the landowner (Gruy 2000) or the claimant (even on land that is owned by the federal government).

There are a number of different types of mineral rights. Surface mineral rights involve minerals on the surface that can be accessed without digging into the earth. These minerals may belong to the property owner, although historically, the royalty of some countries has laid claim to certain minerals (usually gold and precious stones) regardless of their location. Subsurface rights involve minerals under the earth that require digging to retrieve them. These rights also apply to coal, oil, and natural gas. In most countries, subsurface minerals are property of the crown or of the state. Some countries allow the property owner to retain a right to a royalty payment, meaning that the owner gets a percentage of the profit from the sale of the mineral, oil, or gas. The word "royalty" comes from the English tradition of royalty (the Crown) demanding a percentage of the gold or silver mined on any land in England. Another type of mineral rights is called a working interest, meaning that a person or business is entitled to some of the potential profits but also responsible for some of the costs of mining or drilling for the natural resources. In the United States, the federal government, through the Bureau of Land Management (BLM), offers competitive leases for mineral rights. In this case, a person or, more likely, a business can bid on the rights to explore, extract, and sell minerals on certain tracts of land or seafloor. The winning bidder must also pay a minimal land rental fee of $2 or less per acre each year. In some countries, the depth of a landowner's rights is limited; for example, they can end at 500 feet or 100 meters deep. Some countries retain the rights to all underground minerals, but they will pay the landowner a percentage or royalty payment based on the amount of natural resources that are extracted.

Even within one country the various laws and regulations vary. In the United States, land can be under federal, state, or municipal laws and regulations. For example, New York state bans fracking for oil and gas, but its neighbor Pennsylvania allows it. If the land is on tribal territory, tribal laws and regulations apply on top of the appropriate state and federal laws and regulations. The tribes retain the rights to all natural resources on their land, and each tribe can set its own royalty rates. Mineral rights can be further complicated by what each city or municipality allows. Some municipalities do not allow drilling or mining of natural resources of any kind within city limits, while others are more lenient.

There are a multitude of approaches to how to regulate and legislate mineral rights. In the United States, some believe that the tendency to allow private ownership of mineral resources allows for greater exploitation of natural resources. It also encourages individuals and corporations to innovate, as they are motivated by the potential to make profits. Some argue that this system contributes to inequality. Countries such as Norway use the profits from their national oil reserves to fund many programs that can be used by all of their people. In this way natural resources are used to create more equality and less economic disparity. Regardless of how a country has allowed for the extraction of natural resources in the past, all countries must now look at how to regulate the renewable resources that are available on their lands and decide how landowners should be compensated and how profits should be used.

MOLTEN SALT REACTOR

A molten salt reactor (MSR) is a type of nuclear reactor used to generate power. Generation IV Nuclear Reactors, today's MSR technology, are purported to be a safer alternative to the water-cooled reactors used in earlier plants, including those that experienced accidents that leaked radioactive material over large areas causing loss of life and making some areas inhabitable. Water-cooler reactors must always have a connection to the electrical grid or to backup generators, if the connection is lost, in order to cool down the core of the reactor. If something happens to the backup cooling process, then a horrible nuclear accident can occur, as was the case in the Fukushima nuclear disaster in Japan in 2011. MSRs solve that problem by using grid electricity to cool a plug. When the plant loses power, the plug melts and the molten salt drains into a cooling tank that is a heat sink, cooling the salt. MSRs are under low pressure, so a small hole somewhere would not result in an explosive amount of radioactive material pouring out (Touran n.d.). The molten salt, or fuel salt, does not have chemical reactivity with either air or water, avoiding another problem with most other types of nuclear power plants. For these reasons, there is a lot of excitement about a nuclear future centered on new MSR technology.

MSRs were first developed by the United States to fuel nuclear airplanes. The Aircraft Reactor Experiment, as it was called, was performed at Oak Ridge National Laboratory in Tennessee. It began as an offshoot of the Manhattan Project during the Second World War, which had the goal of creating the first nuclear bombs. When interest in a nuclear-powered airplane began to wane in 1956, MSRs were on a list of possible nuclear technology that could be translated into civilian uses (MacPherson 1985). After determining that MSRs were the most promising technology on the list, the US government continued to fund MSR research. Experiments demonstrated that MSRs could become breeder plants, nuclear plants that produce more fuel than they use. This is particularly useful if there is limited uranium. Breeder plants can be designed to produce weapons-grade plutonium. MSR research in the 1960s focused on molten salt breeder reactors (MSBR), and this focus eventually doomed MSR research because it soon became clear that the United States had more uranium than they needed to create nuclear weapons, making breeder plants less attractive. For four years, the United States ran the MSR experiment or MSRE; the reactor went into critical operations in 1965 and was producing 8 MW of power during that time. Experiments demonstrated that it could create additional nuclear fuel and operate on uranium-233 or U-233, which no other nuclear plant had done before. U-233 is bred from thorium, which is many times more common in the earth than uranium. Most nuclear plants operate using U-235, which makes up less than 1 percent of total uranium deposits. The MSRE was ended in 1969 despite its success; it was an expensive experiment and monies were moved into other types of nuclear power.

The possible benefits of MSR are clear, as the technology involves a safer nuclear reaction that is less likely to end in disaster. Nuclear power creates carbon-free electricity. MSRs can be relatively small, so they may be able to power ships, planes, or spacecraft in the future. Like all nuclear power possibilities, they come with potential problems. Despite their touted safety, MSRs could have breakdowns, leaving the possibility for nuclear disasters. Nuclear reactors are extremely expensive to build, and because MSRs have only rarely been used to build large nuclear power plants, costs will likely be extremely high for at least the first few MSRs. Russia is the only country that is successfully operating an MSR. They have two MSRs, the BN-600 and BN-800, that were specifically designed to use weapons-grade plutonium so that it is destroyed and will not get into the hands of the wrong people. The United States was planning to send their weapons-grade plutonium over to the Russian BN-800, but deteriorating trust and cooperation between the countries led to the United States pulling out of the agreement.

The future of MSRs is unclear. There are many countries around the world planning on creating MSR plants. However, the continually decreasing costs of renewable energy are making all nuclear power plants look like overpriced dinosaurs from another era. The future of MSRs will depend on each country's commitment to creating safer nuclear power and on whether MSR technology is truly the most affordable and practical way to decrease CO_2 emissions and/or collect CO_2 from the air.

NATURAL GAS

Natural gas was at one point a niche product. Most power plants ran on hydropower or coal, and almost every vehicle (planes, trains, automobiles, tanks) ran on oil. Humans have been aware of natural gas since ancient times; it was used to light ancient Greek temples and to boil salt water to create drinking water and salt in China as early as 500 BCE. However, it did not have a mainstream use until London used it to light the streetlamps at the end of the eighteenth century. Other cities followed suit, and it soon became common practice; throughout the nineteenth century natural gas lit up cities. A new invention called the Bunsen burner, a precursor to the gas furnace, established and inspired new uses for natural gas at the end of the nineteenth century. As the twentieth century began, cities created gas pipeline systems that could bring natural gas into people's homes for cooking and heating. Natural gas was not used for generating electricity due to the affordability and ease of using coal. While there had been experimental plants and research into the feasibility of creating natural gas power plants, it was not until the 1960s that countries started building plants at a size and scope to power the grid.

Natural gas in its natural state is made up of methane (CH_4) and smaller amounts of hydrocarbon gas liquids. Both natural gas and hydrocarbon gas liquids have a number of uses in addition to fuel (for more information, see the Hydrocarbons and Hydrocarbon Gas Liquids entry). Natural gas is a key component (sometimes referred to as a feedstock) in the creation of synthetic fertilizers; it is combined with nitrogen, which is taken from the air, to create ammonia and urea. Natural gas is also used for heat and electricity during the fertilizer production process. Without natural gas, there would never have been the Green Revolution (responsible for increasing agricultural yields at a faster pace than human population growth (see Green Revolution entry)). Without the Green Revolution, predictions about mass starvation due to overpopulation would have come true. Agricultural production had to grow exponentially to keep up with exponential population growth. A transition away from fossil fuel power plants may end up being a boon for agriculture as it will not have to compete with power plants for natural gas. It does not appear likely that synthetic fertilizers will go away anytime soon. Currently, organic methods do not always have the same amount of year over year field stability as conventional crops, meaning that an organic crop (one that does not use synthetic fertilizers) has a greater chance of having a disappointing year than a conventional crop. The difference is only 15 percent, but that could prove significant if there was a worldwide transition back to traditional (organic) agriculture. Until current agricultural practices can close that gap, synthetic fertilizers will be preferred by farmers. Farmers are always looking at ways to minimize their risks, whether they be related to natural disasters, pests, or the commodities market. Current practices that use genetically modified organisms (GMOs) along with heavy pesticide use and synthetic fertilizers are likely to be the preferred method for the near future. Natural gas will be needed to produce that fertilizer.

Natural gas is drilled in the same manner as oil. Traditionally, vertical drilling was used to find and extract the majority of all oil and gas deposits. Today, fracking and horizontal drilling are used to extract more oil and gas out of tighter deposits and old deposits that appeared to be no longer producing. Natural gas was formed millions of years ago when plants and animals were buried under layers of soil, sand, or silt., Fossil fuels were formed over millions to hundreds of millions of years. Depending on the amount of pressure and heat involved in the process, either coal, oil, or natural gas was formed.

The first natural gas power plants were experimental plants in the United States and Switzerland in the late 1940s. It was not until the 1960s that electricity providers started to build natural gas plants to scale. At that time, coal was usually less expensive, so natural gas plants were only used in certain situations. That usually meant that natural gas was inexpensive in areas that had local supply. This reduced the costs significantly in places that had natural gas. There was no international market for natural gas, at this time, as it is expensive to create infrastructure for transporting it.

The growth of natural gas power plants was at first a slow process, as natural gas was still expensive compared to coal. Coal prices more than quadrupled in the 1970s, which resulted in an uptick in natural gas power plants. While there was steady growth in the 1970s and 1980s, it was not until the fracking boom began in the early 2000s that natural gas power plants really began to replace coal power plants. Government regulations on pollution also put pressure on coal plants as coal can be many times dirtier than natural gas, especially in regard to particulate matter. Many countries have also looked to lower their CO_2 emissions, and natural gas creates roughly half as much CO_2 emissions as coal. These three forces—lower gas prices, pollution regulations, and efforts to limit CO_2 emissions—have combined to make natural gas the number one source of electricity generation in the United States (EIA 2017.

The future of natural gas is complicated. To move forward quickly toward a carbon-neutral economy, no new natural gas plants should be built because natural gas still produces copious amounts of CO_2. However, many leaders see natural gas as a bridge energy between coal and renewable energy. The problem is that a natural gas power plant has an expected life cycle of about thirty years on average, so investing in building new plants means that they are likely to remain in use for some time, delaying the transition to a carbon-neutral or carbon-negative future. Natural gas, like oil, is not evenly distributed across the planet. The United States, Russia, and Iran are the top exporters of natural gas, and many countries do not want to be beholden to the politics of these countries or others that they may not agree with. Local energy sources such as sun, wind, hydro, nuclear, and even coal can help limit energy dependency on other nations. Coal is generally seen as untenable due to worldwide agreements to reduce CO_2 emissions; that leaves nuclear and renewables. Many countries are moving away from nuclear due to lack of public support over safety and nuclear waste storage issues, which leaves renewables as the biggest competition for natural gas power plants. If the rest of the world follows the same track as the United States and the European Union, renewable power plants will quickly start to replace natural gas power plants. In 2020, 78 percent of new installed power in the United States was renewable energy. Natural gas still produces the most electricity, but it does not appear it will be the leader for long. Natural gas power plants will likely continue to be built in some countries but will not be able to compete with the lower prices of renewable energy in the long-term. Coupled with a worldwide drive to lower carbon emissions, natural gas power plants will decline, but natural gas will continue to be used to create synthetic fertilizers and other products.

NET ZERO

Net zero means that something does not add CO_2 to the atmosphere. For example, a tree absorbs CO_2 during its lifetime, but when it dies a natural death or burns in a forest fire, it releases CO_2. The tree is a net-zero CO_2 admitter, as it absorbs as much CO_2 as it releases. (Even a dead tree will store carbon in whatever is left of the tree and the root system, so in some ways trees can be carbon negative, especially if they are in areas where decomposition is difficult, such as a swamp or a bog.) The term is most often used to refer to human endeavors, in businesses, governments, cities, states, countries, and homes. Human entities can become net zero by aiming to take in as much CO_2 as they create, with all activities and resources attributable to an entity adding up to carbon-neutrality. When people devise methods to judge if something is carbon-neutral, they are usually looking at a simpler equation: renewable energy "in" (or produced) is equal to or greater than the energy "out" (or consumed).

For example, an average home in Southern California consumes 6,000 KWh (kilowatt hours) of electricity. For this household to be (electrically) carbon-neutral it would have to produce 6,000 KWh of renewable energy, from solar panels, for example. The reason why this home is described as "net" zero and not carbon free or carbon zero is that during the day the solar panels produce more electricity than the household needs and send it out to other homes, but at night the home pulls electricity from the grid. The grid electricity could be "clean" (produced by renewable energy) or "dirty" (produced by coal or gas electric power plants). So, overall this home creates enough "clean" energy during the day to counteract the "dirty" energy used at night, making it a net-zero home. A simple math problem can determine if an entity is net zero, net positive, or net negative. The vast majority of homes in the United States are net positive in their creation of CO_2, meaning they create CO_2 and do not produce enough renewable energy to offset it. There are a few homes that are net negative, which means they create more renewable energy than the total amount of energy they consume. If cities, states, and countries want to achieve net-zero status, they will have to create more renewable energy than they need. The additional energy could be used to power carbon capture machines that could take carbon out of the atmosphere.

Energy use is not the only cause of CO_2, so it is not the only sector that needs to consider getting to net zero. Transportation is the second most carbon intensive sector after energy (electricity) production. Transportation methods would all have to switch to renewable energy to operate a net-zero rate. For some people and industries, this should not be very difficult. For example, an electrified bus, train, or EV car would just need to be charged up using renewable energy. This would be especially easy for a person who works a day job. During the day, while they are working, their EV car could be hooked up to a rooftop solar system that charges the car. For industries such as cross-ocean shipping and flight, the transition is more daunting and may require transition to a hydrogen-based fuel system. Green hydrogen is the process of creating hydrogen using only water and solar power. Hydrogen can be burned in the case of rockets, or it can create electricity in a much more controlled process through fuel cells (see Fuel Cells entry). The only things that a hydrogen fuel cell produces are electricity, heat, and water (which is where the hydrogen came from in the beginning). In the near future, the transition to electric vehicles for all forms of ground transportation will move the transportation sector as a whole much closer to net zero. Internal combustion engines (ICE) and diesel engines will become less and less desirable for a multitude of reasons. They are more expensive to run and to maintain. They have 100 times the number of moving parts as an electric vehicle; EVs have an average of 20 moving parts compared to 2,000 for ICE vehicles. ICE vehicles are more polluting; they literally stink in comparison to EVs. Once people start to switch, it will be strange to smell like gas and exhaust fumes. EVs also outperform ICE vehicles by almost every measure. Once total lifetime repairs and fuel costs are calculated, they are less expensive to own. At this point it is only a matter of people getting used to the change, just as they adapted to using smart phones. It seemed strange at first to have a computer/camera/phone in our pockets, but now many of us could not live without one.

In order for travel industries such as air lines to become carbon-neutral in the near future, they will have to either purchase carbon offsets or create carbon-negative projects to offset the CO_2 they put into the atmosphere. One of the most popular and simplest carbon offsets is to plant trees that over their lifetime will consume more carbon than the flight created. The problem with this type of carbon offset is that it is good for the long-term, but it is not very effective in the short-term (see Carbon Offsets entry). An example of a short-term effective strategy would be for an airline to buy a coal power plant that has an additional twenty years in its lifespan and retire it early, replacing the electricity it produced with renewable energy.

For container ships (as well as other types of ships), there are some interesting options. Hydrogen fuel cells can be used alongside solar panels covering the shipping containers. Some plans involve using large sails on routes with favorable winds; after all, it was wind power that powered all ocean travel until the Industrial Revolution introduced the steamship. The difficulty with these new technologies is that it will take a long time for them to get to market and take even longer to bring down the price. The speed of this transition could be accelerated with national or international prices on carbon (see Carbon Tax/Carbon Market/Carbon Credits/Cap and Trade entry).

Energy and transportation are not the only places for creating net-zero systems. Food systems tend to be carbon intensive. There are a wide variety of options available for farmers to become net zero or to run a carbon-negative farm (see Climate Engineering or Geoengineering entry). Food, clothing, and everything else we buy can be created in a carbon-intense way or a net-zero (even a net negative) way. It is up to governments and consumers to create regulations and demand to encourage a more net-zero carbon world.

NUCLEAR FISSION, NUCLEAR FUEL, AND NUCLEAR POWER PLANTS

Nuclear fission is the process of splitting one heavy and unstable nucleus (usually from a uranium-235, uranium-238, or plutonium-239 atom) into smaller and lighter nuclei by colliding a neutron into that atom at extremely high speed. Fission creates two (and in rare cases, three) completely different elements. The process of splitting the atom releases massive amounts of energy, often in the form of gamma rays, which release heat that can be used to create steam to drive a turbine to produce electricity. The splitting of the atom also sets off a chain reaction. The aim of a fission nuclear power plant is to harness the energy produced by fission without letting the chain reaction get out of control. Different types of moderators are used to control the reaction by absorbing neutrons that are moving through the reactor at high speeds. These moderators allow the reactor to continue producing heat without overheating and leading to a meltdown.

The original research on fission was done by Lise Meitner and Otto Hahn in Germany and Enrico Fermi in Italy, among others. Lise Meitner deduced that atoms were fissioning apart during these early experiments and, in the process, releasing energy. Meitner was born Jewish (though raised Protestant), so she had to flee to Sweden to escape from the Nazis. Hahn, still in Germany, could not credit Meitner for her contributions for fear of Nazi reprisals; he eventually received the Nobel Prize for work she contributed to. Fermi was married to a Jewish woman in Italy. After he accepted a Nobel Prize in 1938, they fled to New York to escape the Nazis and became American citizens. Fermi went on to lead the Manhattan Project, which produced the first nuclear bomb and was also deeply involved in the creation of the first functioning nuclear reactors in the United States.

The work of these early researchers and hundreds of others led to the proliferation of nuclear power plants across the planet. Today, nuclear power plants produce about 10 percent of the world's electricity. The United States produces the most nuclear power, but France receives the greatest percentage of its power from nuclear power plants at over 70 percent. Nuclear fuel in its natural state (uranium) is not evenly dispersed across the planet. Some countries have much more than they could ever use. Others have none. Having raw nuclear fuel, however, is not enough. A country must have the knowledge and the technology to process uranium into a form that is usable to start a nuclear reaction. Once this process is completed, nuclear fuel is the most efficient fuel on earth. It can produce 320,000 times the power of coal on a gram-by-gram basis. Despite this incredible power, current nuclear plants do not consume most of their fuel. Enormous amounts of spent fuel, which is still radioactive and dangerous, are left behind. In most cases, this spent fuel will remain dangerous for 10,000 years or more. Newer nuclear reactors are looking to both use new fuel and reduce the radioactive lifespan of the waste they produce. Nuclear power has held the promise of limitless free energy without pollution or CO_2 emissions. It has not lived up to that promise for a number of reasons, the most important of being the risk (and fear) of nuclear accidents. The disasters at Three Mile Island, Chernobyl, and Fukushima eroded public trust in the safety of nuclear power to the point that there are hardly any country building new nuclear power plants. Most democratic countries are either holding steady with the amount of nuclear power plants they have (as in the United States) or are actively decommissioning them (as in Germany). Some autocratic countries, such as Russia and China, where public opinion does not hold as much weight, continue to expand the number of nuclear power plants that they have as part of their energy grid.

However, next-generation nuclear power plants appear to be safer and should produce less radioactive waste and less waste overall. Generation IV nuclear plants could help stop the worst effects of climate change by creating carbon-free electricity. They could also be used in climate engineering, to cleanly power machines that extract CO_2 from the atmosphere. There are currently very few Generation IV nuclear power plants, so it is unclear if that new technology will expand across the world as water-cooled power plants did. Plants can be expensive to build and impossible to insure, so in most cases only governments will be able to build them. Nuclear power will probably remain an important part of the energy mix, but it is unclear if nuclear power will be used to augment renewable energy in the future or if it will simply hold its place as renewable energy slowly replaces all other energy sources.

OCEAN CURRENTS

Ocean currents (and wind) have been important in many conflicts over energy resources. They have moved people, ships, and armies across the world. They were the engine that drove human migration across the Pacific Ocean as Pacific Islanders settled in new lands. Europeans used ocean currents to explore and conquer much of the world and then to establish shipping routes with their colonies, enabling the exchange of raw materials and finished products. Centuries before European colonization, the great Chinese empires used ocean currents to travel all around Asia and as far as Africa to demand tribute and fealty.

Ocean currents move nutrients across the oceans, which effects the location and abundance of human food sources. They also regulate temperature across the planet. Without the moderating effects of ocean currents, specifically the Atlantic meridional overturning circulation (AMOC), temperatures in Europe would plummet, significantly altering the ability of Europeans to grow their own food and the energy needed to heat their homes. A drop in temperature may also affect the amount of solar and wind power that is created in Europe.

The AMOC is just one part of a large worldwide network of global ocean currents that circulate warm and cold waters. The AMOC brings warm saltier (less dense) water that travels near the surface north from equatorial waters and then returns cooler denser water that travels back at a deeper level. The warm water releases heat as it passes by Western and Northern Europe, creating warmer weather year-round for these areas.

Paleo climatologists have discovered that ocean currents have changed over the millennia. That is a reason for concern because weather patterns and seasonal temperatures depend on these currents. While observational data is limited on a timescale, ocean currents appear to be slowing down. The AMOC has decreased by 15 percent according to studies (Praetorius 2018). If the AMOC decreases significantly or stops altogether, this could lead to drastic drops in temperatures, as much as 5 °C. The two driving factors in these changes are the fresh water melting off of the Greenland Ice Sheet and a warming atmosphere. The waters between Nova Scotia, Greenland, and Ireland are already colder and the waters of the Atlantic coast of the United States and in the Caribbean are warmer.

The possible long-term effects of these changes could be quite extreme, drastically warming already warm areas and freezing temperate areas that have enjoyed mild weather for millennia. While there are many effects of climate change that can still be stopped, it is unknown if the slowing of the AMOC is reversible or if the process will continue unabated regardless of human intervention to reduce the emissions of GHGs. The breakdown of worldwide ocean currents would leave a few places much colder but make most places much warmer. Hopefully, these processes are in an early enough stage that changes in human behavior can slow or even stop them, thus preserving these ancient ocean currents for generations to come.

PEAT MEGAFIRES AND COAL SEAM FIRES

Peat is a material made of many layers of dead plants It is usually formed in a bog, fen, moor, muskeg, or swamp. As each successive generation of plants dies, the plants fall into the water underneath them. Fairly stagnant water creates a low-oxygen environment, so the dead plants just pile on top of each other, compacting before they can decompose. This spongy, partially decomposed material sits in water that is generally fairly acidic and not well drained. The plant material accumulates over thousands or even millions of years, resulting in compacted, wet peat that can be up to 25 feet deep. However, people will sometimes drain these wetlands for a variety of reasons. They may need the water for farming, or the water might be diverted for a growing city. The water may also be drained specifically so that the peat can be used for fuel. Peat can be used for local heating and cooking, or it can be used to create electricity, as a power plant can run on almost anything that is flammable. Unfortunately, peat is one of the few substances that is dirtier than coal when it burns, meaning it releases more fine particulate matter. That particulate matter is dangerous for people with asthma and other lung ailments. It causes increased deaths locally and also pollutes the atmosphere at a higher level globally, producing more GHGs than any other form of fuel. There is also greater danger in draining a peat wetland, as it can catch on fire, resulting in a slow smoky burn that is incredibly hard to put out. Some fires have burned for decades.

Underground peat fires alone are estimated to be responsible for 15 percent of the planet's annual GHG production (Rein 2015). Many areas across the globe have drained natural peat bogs for a wide variety of reasons, including using the water for agriculture or municipal fresh water systems. In other cases, bogs have been drained specifically so that the dried peat could be used for heat and/or electricity production. Drought conditions tied to climate change can also precipitate peat fires; for example, the melting of permafrost due to warming climate has led to peat fires across the Arctic. Peat fires have occurred all over the planet, from the Arctic to Southeast Asia, from Scotland to the Southeastern United States.

The exact process of a smoldering peat fire is not as well understood as that of a classic combustion forest fire. A smoldering fire moves much slower and creates much more smoke, haze, and toxic compounds, so a peat fire is more likely to cause sickness and death for people who have preexisting lung conditions or those with asthma (Rein 2015). Once these fires start, they are extremely difficult to extinguish. Rain is usually not sufficient, as peat fires can be up to 60 feet below the surface. If these fires are also in remote and rugged terrain, it can be nearly impossible to bring in the immense amounts of water necessary to extinguish them.

In some areas, peat fires are started by local people practicing slash and burn agriculture. The long-burning fires can completely destroy the soil underneath these areas. The toxic gases also lead to local, national, and global health problems. The Indonesian peat megafire of 2015 produced enough smoke to affect neighboring countries. There was so much toxic smoke that airplanes had to be rerouted and flights had to be canceled.

In some instances, there is little that can be done to stop peat fires. This is true in places near the Arctic that were once permafrost and are now defrosting. However, in many places, public policy could limit or eliminate peat fires by restoring natural water flows into peat bogs and banning slash and burn agricultural practices. For some governments, enforcing these types of regulations would be difficult with limited resources. A concerted political effort would be the best way to foment positive change and limit the number of peat megafires.

Coal seam fires are another underground fire problem. Many of these fires are man-made, accidentally started during the mining process. Others, such as the "Burning Mountain" in New South Wales, Australia, have been burning for what scientists estimate to be 6,000 years (Monroe 2019). These fires are extremely difficult to extinguish as they have almost endless amounts of fuel and oxygen and are located deep in the earth. There are several methods to put these fires out, but they are all expensive and difficult; the seam can be packed with tons of clay to suffocate the fire, flooded with copious amounts of water, or completely dug out to deny the fire fuel. Another solution for reducing the number of coal seam fires is to stop mining for coal. While that is not something that can happen overnight, the movement away from coal may eventually make man-made coal fires a thing of the past.

PETROLEUM

Humans have used petroleum for many thousands of years. It was used to seal things, from pitchers to canoes. It was used to pave and seal stone roads in ancient Mesopotamia. It was burned for heat. Although there were surface seeps providing access during early history, most petroleum was underground and inaccessible. When people realized that petroleum could replace whale oil used in oil lamps in cities and homes across the world, digging and drilling increased enormously. (The transition to petroleum inadvertently saved many whale species from extinction.)

Petroleum, also called crude oil or oil, consists of compressed dead organisms, mostly algae, phytoplankton, and zooplankton. These organisms are abundant in the oceans. Algae and phytoplankton convert sunlight and CO_2 into oxygen and carbon-based life forms. Zooplankton eat phytoplankton and algae. This is the basis of the ocean food chain. The algae, phytoplankton, and zooplankton that are not eaten by fish, corals, crustaceans, and whales sink to the bottom of the ocean when they die. Here, it is extremely cold, and there is enormous pressure. Layer after layer of dead microscopic sea organisms are compressed into the seabed, as millions of years pass; almost everything besides carbon is squeezed out. The carbon-rich fluid becomes trapped in rock, remaining underground until it is removed through geological activity, volcanic activity, or human activity. The world's oceans and landmasses have moved around considerably over the earth's history. For this reason, oil is found in many places that are no longer sea beds.

Humans first obtained petroleum through surface availability along the Euphrates River (modern-day Iraq), on Ionian islands (modern-day Greece), in Cessia (modern-day Iran), Baku, Azerbaijan, and in China. As surface seeps were exhausted, hand-dug wells allowed continued access. Eventually, mechanical wells were dug, first in China and eventually everywhere oil was known to exist. The oil rush of the early 1800s found oil in Pennsylvania in the United States and in Baku, Azerbaijan (then part of the Russian Empire). These two areas remained the two largest oil-producing regions for many years. In the Second World War, the Battle of Stalingrad (1942), between Nazi Germany and the Soviet Union, was fought to secure access to the Baku oil fields that lay past Stalingrad. (Other battles were fought across the Caucasus as well to access Baku from the south.) Eventually, oil was found in Ohio, Oklahoma, Texas, and California in the United States. The Dutch East Indies (modern-day Indonesia), Persia (modern-day Iran), Galicia (modern-day Ukraine and Poland), Peru, Venezuela, and Mexico were all producing oil at the beginning of the First World War in 1914. Oil production began in Iraq and Saudi Arabia between the two great wars and soon almost all the countries that were adjacent to Iraq and Saudi Arabia were producing as well. These countries include modern-day United Arab Emirates, Kuwait, and Qatar, as well as smaller producers such as Egypt, Bahrain, and Yemen.

Oil grew in importance as it was converted to diesel and gasoline/petrol. Oil was the fuel source for cars, ships, trains, and planes, and it became impossible to operate a modern world without it. The countries that controlled access to sufficient oil prospered and grew. Countries that shipped out oil for exports grew rich. Desert countries that had little resources such as Saudi Arabia became economic and political powerhouses because of the influence, power, and money that their oil brought them.

Oil is mostly used for fuel, but an increasing large percentage is used for plastic manufacturing and the production of fertilizer. When the need for oil as a fuel eventually diminishes and disappears, these uses will likely remain. The use of plastic products continues to increase, but there are concerns regarding plastic pollution in the soil and the waters of the world. Petroleum-based fertilizers have revolutionized agriculture across the planet, increasing yields and land productivity, but fertilizer runoff causes toxic algal blooms that kill sea life. While there may be regulatory pushback to limit the negative effects of plastics and fertilizers, it is likely that they will both be used for the foreseeable future.

Oil deposits, known as proven oil reserves, and oil production are not always correlated. The United States produces the most oil of any nation but barely ranks in the top ten in proven oil reserves. Venezuela has the most proven oil reserves of any country, but it is currently only the twelfth largest producer due to mismanagement by its government and international sanctions led by the United States (a response to the Venezuelan government suppressing the will of the voters). The only other countries that produce oil close to the amount in the United States are Saudi Arabia and Russia. Today, there are 125 countries producing over 100,000 bpd (36,500,000 barrels per year). A barrel of oil is 42 gallons or 159 liters. The entire world uses about 100 million barrels of oil a day.

Proven reserves are measured in billions of barrels or Gbbl. Venezuela is number one with 303.3 Gbbl; Saudi Arabia is second with 297.7 Gbbl; Canada is third with 167.8 Gbbl; Iran is fourth with 155.6 Gbbl; Iraq is fifth with 147.2 Gbbl; Russia is sixth with 106.2 Gbbl; Kuwait is seventh with 101.5 Gbbl; the UAE is eighth with 97.8 Gbbl; the United States is ninth with 61.2 Gbbl; Libya comes in tenth with 48.4 Gbbl (BP 2019). However, as oil demand decreases over the following decades, a more important statistic is annual production. The countries that produce the most oil in the short term are likely to receive the most monetary benefits. These numbers are skewed, however, because members of OPEC (Oil Producing and Exporting Countries), which banded together in the 1970s to limit supply in order to control prices, agree to quarterly production limitations to keep prices higher. Now that the United States and Canada are producing high amounts of oil, OPEC has found it more difficult to control oil prices. It was a very effective method until 2015 when North American production reached a level to sway prices by producing record amounts of oil. Prices are unlikely to reach the highs seen in the first twenty years of the twenty-first century, when $100+ prices were the norm. The United States, Saudi Arabia, Russia, Canada, China, Iraq, the United Arab Emirates, Brazil, Iran, and Kuwait are the top ten oil-producing countries, listed in descending order. Together, they produce

71 percent of the world's oil and will continue to produce the majority of oil in the declining oil demand years.

The IEA estimates that pollution from oil and gas production and consumption is responsible for 55 percent of all CO_2 released into the atmosphere. Humanity must move away from petroleum, as well as natural gas and coal. If trends continue, coal and oil usage will continue to decline. Coal is already declining significantly, and it appears that oil is on a precipice as well. The transition to EVs, and possibly hydrogen fuel cell–powered vehicles as well, will continue until petroleum-powered vehicles are about as common as horse drawn carriages. The timeline of this transition will depend on government policies on pollution, the build out of charging infrastructure, and comparative pricing of EVs compared to their petroleum-powered counterparts.

The end of oil is not near, but the beginning of the end of oil may have already occurred. It seemed quite possible that peak oil demand occurred shortly before the onset of the worldwide Covid-19 pandemic. Demand slowed at the end of 2019 and beginning of 2020. Demand has continued to grow ever since. In the meantime, carmakers across the world are offering 134 EV models as of 2024. If demand is equal to production, then there will be a significant drop in demand for gasoline/petrol and diesel. The online shopping behemoth, Amazon, has committed to becoming fully electric. Many governments are switching their public buses from combustion engines to EVs. While there are many variables in play, including public buy-in of EVs and government infrastructure for the charging of EVs, oil-based transportation is going to decrease significantly in the next decade. It is feasible that oil demand could recover before entering its permanent decline. Regardless of the exact year of peak oil demand, demand will be in permanent decline from the 2030s into the foreseeable future. This will have enormous effects on the geopolitical landscape, and countries that derive their power and wealth almost exclusively from oil proceeds, such as Saudi Arabia, must find a way to pivot to other industries before they face economic collapse and social upheaval. Humanity has been in conflict over energy resources in one manner or another for almost its entire existence. It is possible that the renewable energy revolution may help put an end to some energy resource conflicts.

PHOTOVOLTAICS

Photovoltaic energy is created by light from the sun, which travels as photons and collides with a semiconductor, usually silicon. The thin wafer of silicon has a positive and negative charge on each side. This is a solar cell. The photons knock electrons loose as the medium absorbs the photon, and the electrons combine to create an electronic current. A solar panel is made of multiple solar cells and usually combined with other solar panels to create a solar power system. The electrons create a DC current that is then converted to an AC current for use in the home or put back into the grid for distribution to other homes and businesses. Photovoltaic paint is a form of thin film solar that can be applied like a paint.

Solar power was first theorized by Albert Einstein in 1905; the work eventually won him the Nobel Prize (Knier 2008). A working prototype was eventually created by Bell Labs, but due to its expense, it was not used to create electricity at a significant scale. The US space program, now NASA (National Aeronautics and Space Administration), eventually became interested in solar power for its work in space. The space program developed solar power to the point that it was functional, but it was still expensive. It took another fifty years of innovation and scaling for solar to be truly affordable. As solar power gained popularity, the price of solar panels decreased. As more people, businesses, and governments purchased solar panels, companies began to scale up and that decreased the price as well. Solar panels are now so affordable that solar power is less expensive than almost any other form of energy.

Solar power has been growing exponentially. At its current rate of growth, solar may eventually become the most used form of energy, although wind power is also growing at an exponential rate. Solar power has an average growth rate of 42 percent; the amount of solar power available in the world doubles on average every two years. Currently, solar power creates less than 3 percent of all the electric power in the world, so even with its fast rate of growth, it will be another ten years before solar is supplanting some of the traditional fossil fuel power plants.

Photovoltaic power plants (generally referred to as solar power plants) are fairly easy to build. All that is needed is some open land. When they are paired with pumped hydro or battery systems, they can continue to produce even in the absence of sunlight. Photovoltaics have the added advantage of being able to be attached to the roofs of existing buildings, so in cities and towns, photovoltaic systems do not need new land at all.

Agrivoltaics are a simple idea with huge possibilities. The idea is to grow crops under solar panels that are high enough for plants to absorb the light that is not absorbed by the panels and for farm machinery to drive under the panels when planting and harvesting the crops. Studies by the University of Arizona have demonstrated that some plants grow better under solar panels. Tomatoes produced twice as much fruit and chiltepins (a spicy pepper) produced three times as much as fruit as control groups in full sun. Agrivoltaics is a potential large growth area for solar power, as there are millions of square miles of farmland.

Photovoltaics or solar panels will be an enormous part of the renewable energy transition. Portability and adaptability set solar panels apart, and their continually lowering prices will make them an indispensable part of every country's renewable energy portfolio.

PROVEN RESERVES

Proven reserves are the amount or quantity of natural resources in an area that a country or company expects to extract with 90 percent profitability. To qualify as a proven reserve, a resource must be proven through data from geologists and engineers. That data must be gathered by seismic testing or test drilling and/or digging sites, depending on the resource. Proven reserves can grow with continued exploration. As the world persists in consuming fossil fuels, companies and governments continue to explore and discover new deposits. While valuable minerals, metals, and gems have proven reserves, the discussion here focuses on coal, natural gas, oil, and uranium deposits.

Coal is available in large quantities all over the planet, but five countries hold 75 percent of all reserves: the United States (22 percent), Russia (15 percent), Australia (14 percent), China (13 percent), and India (10 percent). Coal demand decreased in 2019 by 2 percent and in 2020 by 5 percent. While there was a bounce in coal demand in 2021, coal demand has been leveling out since. Long-term coal is likely to continue to decrease in use due to the local pollution it causes, the CO_2 it emits into the atmosphere, and the affordability of renewable energy. Renewable energy costs have made coal power impossibly expensive in areas where energy companies can bid on producing energy. Natural gas prices have also put additional pressure on coal-powered plants. Across the United States, energy companies continue to announce closing of coal-powered plants years and sometimes decades before their scheduled decommissioning. Further pressure is coming from state governments controlled by the Democratic Party; these states are putting end dates on both coal plants and fossil fuel plants in general. This trend is likely to continue as renewables that were once more expensive than coal continue to see decreases in costs.

The world has plenty of coal; if it was not polluting, CO_2 producing, and relatively expensive, it could be used for the next 300 years. It is likely that very few countries will be using coal for electricity in 2050. European countries have already closed half of their power plants in the five years between 2016 and 2021 (Taylor 2021). There will be only a few remaining coal plants across the continent by 2030. US coal consumption decreased by 50 percent from 2010 to 2020 (EIA 2022). The rest of the world will likely begin to decrease its coal use at the same rate as the United States and Europe over the next twenty years as each country moves away from fossil fuel energy, specifically coal. China is the largest consumer of coal and plans to level out coal use by 2025 and quickly move away from coal starting in 2030. If China follows through with these plans, the last major market for coal will be extinguished.

Natural gas is the one fossil fuel for which usage is currently growing; it will continue to grow into the foreseeable future due to low prices and advances in extraction processes. It is a cleaner alternative to coal, although it still produces CO_2 and other GHGs. Natural gas production and consumption has increased every year since at least 1973, except for 2009 due to the Great Recession and 2020 due to Covid-19.

Natural gas reserves, like all other proven reserves, are likely to continue to grow if demand keeps growing and prices rise to levels that encourage expensive exploration and the development of ingenious extraction techniques. The United States is an excellent example of this phenomenon; US proven reserves for natural gas have grown every year since 2000, despite increasing production levels. Scientists and engineers have estimated the total reserves on the planet, but these estimates are based on existing knowledge and existing technology, so they are likely to be proven wrong as knowledge grows and technology advances. All that can be known with confidence is what proven reserves are at this moment, looking at today's prices and today's technology.

Natural gas proven reserves are concentrated in four countries, listed by percentage of known worldwide proven reserves: Russia (19.1 percent), Iran (16.1 percent), Qatar (12.4 percent), and Turkmenistan (9.8 percent) (BP 2019). However, while the United States has a distant fifth most proven reserves totaling 6.5 percent, it produced more natural gas than any other country, totaling 23.1 percent in 2019. Russia produces 17 percent and Iran produces 6.1 percent. Canada, Algeria, Qatar, Norway, China, Saudi Arabia, and the United Arab Emirates round out the top ten producers, but all of them produce less than 5 percent of the world's natural gas.

Using a Reserves-to-Production (R/P) ratio, the world will run out of natural gas in the next 50–5 years depending on whose numbers are being used. As discussed, more reserves will likely be found if demand continues at present rates of growth. If new technology finds an economically affordable way to capture carbon at the source of natural gas power plants, the demand for natural gas will continue to grow. Conversely, if affordable carbon capture remains elusive and the world continues to move toward renewable energy, then reserves will not increase due to a lack of financial incentive for exploration and drilling technology research and development. There may be an uptick in the use of natural gas as a "bridge fuel," a fuel that bridges the gap between fossil fuels and renewables with energy storage. If countries honor their promises from the Paris Climate Accords, then demand will not increase past 2030. The future of energy will continue to move away from fossil fuels toward carbon-free energy sources, primarily renewables. Decreasing demand will not mean the end of natural gas. There will likely continue to be uses for natural gas even as it declines in use. New natural gas plants are being built, and most of these plants are expected to run for at least thirty years. Natural gas plants may run into the same economics and politics that are confronting coal; gas prices could reach a point at which gas power plants can no longer compete with renewable energy. If countries put a price on CO_2 emissions, gas power plants will become increasingly expensive, in which case gas power plants are likely to decrease in use regardless of proven reserves.

Proven oil reserves, like all proven reserves, are not fixed. Higher prices driven by demand can push forward exploration and drilling technology, leading to more proven reserves. In 1956, M. King Hubbard projected that US oil production would reach its peak in 1970. His predictions, based on contemporary known oil reserves and drilling technology, appeared to be correct until the fracking boom that began around 2005 combined with horizontal drilling technology to completely revitalize the US oil industry. Figure II.1 demonstrates how close his predictions were. He had also predicted a worldwide oil peak in 2005, which appeared possible until new technology changed that.

The reality for oil markets is that peak oil demand (after which demand for oil will decrease every year) is more likely to develop before peak oil production. The main driver of peak oil demand is likely to be the mass adoption of electric vehicle (EV). The transition from combustion engines to EVs has already begun and is moving at a quick pace. EV sales make up a larger percentage of vehicles sold every year. The first quarter of 2021 in the United States saw an 81 percent growth in the sales of EVs and hybrid EVs. If this trend continues the oil production landscape will change to adjust to diminishing demand. Worldwide demand for oil decreased in 2019 and 2020, but that is likely due to the economic shutdowns in the effort to halt the spread of the Covid-19 virus. Demand for oil increased significantly in 2021 due to pent-up demand for air travel and road trips after a year and a half of lockdowns and limits on travel. There has been incremental demand increases in 2022 and 2023. Increased adoption of electric vehicles and energy efficiency will lead to the eventual decline in oil demand. The actual date of peak oil demand is not yet clear, it is estimated to be in the late 2020s.

Proven oil reserves are sometimes controversial due to disagreement over how "easy" oil and "tight" oil are counted. Some would argue that the easy oil that is in Saudi Arabia

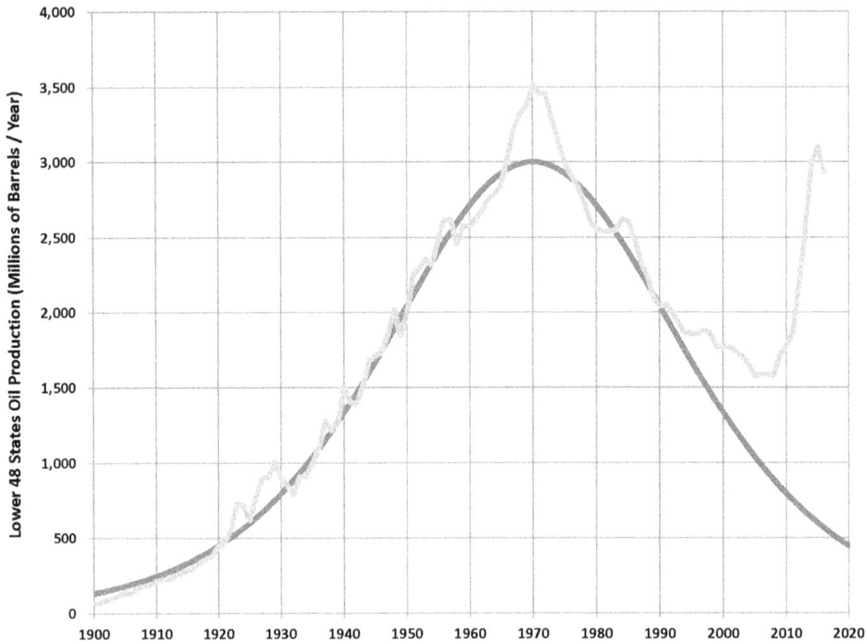

FIGURE II.1 M. King Hubbert's upper-bound prediction for US crude oil production (1956) and actual lower-48 states production through 2016. Plazak, Wikimedia Commons, https://commons.wikimedia.org/wiki/File:Hubbert_Upper-Bound_Peak_1956.png.

is not equivalent to the tight oil that is in Venezuela, the majority of which is located miles below the ocean floor in deep water.

RENEWABLE ENERGY

Renewable energy is defined as an energy source that does need to use a finite fuel. Wood, peat, coal, trash, oil, natural gas, and nuclear energy are not renewable energy. Carbon-free energy should not be confused with renewable energy. Nuclear energy is carbon free and produces no other GHGs, but it is not renewable because it requires fuel. It uses fuel the most efficiently of any process that requires fuel, but it is not endlessly renewable. Nuclear fuel could theoretically run out, although it would take thousands of years. The most common types of renewable energy are hydropower, solar power, wind power, geothermal power, wave power, and tidal power. This is not an exhaustive list, but these six types of renewable power cover 99 percent of all renewable energy production that does not create any GHGs. Burning wood and/or farm waste can be considered renewable, but it is also polluting and can lead to an increase in global warming and add to the growing problems associated with climate change (see Biofuels entry).

Hydroelectric power produces more electricity than any other single source of renewable energy. It is the most mature form of renewable energy; there have been massive hydropower projects since the beginning of the twentieth century. Over 100 years of history have created an industry that is able to efficiently generate electricity with running water. These hydroelectric dams store water behind them to guarantee a steady supply of energy and to guarantee a steady supply of water for cities and farms. This dual purpose has made hydropower dams an attractive investment for cities, states/provinces, and, especially, federal governments that are yearning to modernize their countries. Hydropower is an established energy source, but growth opportunities are limited. Most of the "low-hanging fruit" has been plucked, in that almost every "easy" hydropower dam site has been built on. There is also a movement to remove or modify dams that block salmon runs or the runs of other fish. The future of hydropower is likely to be a combination of improving efficiency at existing hydropower stations through more advanced generators and building smaller hydropower stations or run-of-river (ROR) hydropower that does not require dam building and allows for the passage of fish and watercraft without interruption. ROR has the downside of having to deal with seasonal fluctuations in river flow, making ROR hydroelectricity more intermittent than a traditional hydroelectric dam power station. Although it is unlikely that hydroelectric power will be an area of massive growth, it will remain an important part of the world's renewable energy portfolio.

Solar power capacity has been increasing at an exponential rate. If its rate of increase continues, solar power will soon produce more electricity than any other form of renewable energy. The two most prevalent forms of solar-powered electricity generation are photovoltaic (see Photovoltaics entry) and CSP (see Concentrated Solar Power entry).

Most solar power comes from photovoltaic electricity production (solar panels), which differs from most other forms of electricity production. Almost every other form of electricity is produced by spinning a turbine that is lined with magnets, which creates an electric current. Hydroelectric power is created by having water spin a turbine. Wind turbines are directly spun by the wind. All fossil fuels, geothermal, biomass, and nuclear sources of energy use heat to create steam to turn a turbine that creates electricity. The photovoltaic process uses the sun, which sends out photons. The medium in the solar cell absorbs a photon and releases an electron. The electrons from many photons combine to make an electric current. A photovoltaic system can be small and portable, but it also can be scaled up to enormous size. The only additional equipment required is a transformer. Photovoltaic systems create direct current (DC), but many countries, including the United States, use alternating current (AC). Photovoltaics in the form of solar panels also have other advantages. Solar panels can shade parking lots and can help cool homes by absorbing solar energy. They can be used above certain agricultural products that grow better while shaded during the hottest hours of the day (see Photovoltaics entry). There are other experimental or emergent types of photovoltaic systems as well, among them are paint-on solar thin film, solar roadways, and floating solar arrays. With the continually falling prices of solar photovoltaics of all types, photovoltaic solar power will likely continue to be integrated into many more parts of daily human activity.

Concentrated solar power (CSP) is also a growing renewable energy source. It uses mirrors to reflect the sun toward a target area, to heat water or another fluid. The fluid is either directly converted to steam or used to secondarily heat water and create steam. CSP is more expensive than photovoltaics and cannot be used on a small scale. It also creates bright light which can attract bugs that in turn attract birds; both bugs and birds can be incinerated in the intense heat emanating off a central tower. An advantage of a CSP plant is that it can store heat, which can then create electricity after the sun sets. The heated liquid, which could be a type of molten salt with large systems or simply water with smaller systems, is used to create steam that spins a turbine, creating an electric current. CSP systems are an important piece of a renewable energy plan as they can create a continual amount of electricity, especially useful during the evening hours in which electricity providers see peak demand. CSP plants can provide electricity until dawn in some cases, although most CSP projects only produce electricity for 4–6 hours after sunset. Demand for electricity drops significantly after 10:00 p.m., so a twenty-four-hour CSP plant is not a good investment unless CSP is the only form of electricity available. CSP is a useful yet expensive (compared to other renewables) form of renewable energy. It will likely be used to bridge the gap between sunset and bedtime, but unless the costs decrease significantly and CSP builders create better ways to protect bird populations, CSP will continue to be a niche form of electricity generation.

Wind power generation has many advantages and is growing at an exponential pace along with solar. One of the biggest advantages of wind power is that there are places on earth that are almost always windy. Windy places can produce electricity day and night, although even windy places can have days where there is not enough wind or there is too much wind. Wind turbines are also scalable. A wind turbine can be used for an individual home or combined with other wind turbines to power an entire city. Wind turbines continue to grow in size, durability, and efficiency. There are offshore wind turbines that are taller than the Statue of Liberty. Another advantage of wind power is that it can be installed both onshore and offshore. Offshore wind turbines have the advantage of being out of site (or nearly so). Most of the pushback against wind power comes from local NIMBYism (not in my backyard). Many people do not want large wind turbines blocking their view or potentially affecting the value of their property. Others are concerned about noise, bird impacts, deforestation when turbines are located on windy mountain ridges, negative effects on tourism, the risk of fire, and the costs to local communities in decommissioning turbines. Many projects have been scuttled over the years due to community pushback. It is usually easier for the business trying to build the wind farm, even if they will eventually win court, to move on to another project that does not have well-heeled opposition. Despite some opposition to the site of wind turbines, wind power continues to grow both onshore and offshore.

In Northern Europe, offshore wind power is driving the growth in wind power. In the United States, wind power largely comes from onshore production. States such as Texas and Oklahoma, which have deregulated wind power production, have seen some of the largest and quickest gains in wind power projects. Texas produces more wind power than any other state. The rise of Texas as a wind power superstar is a surprise because it has long been known as the heart of oil production in the United States. Today, oil production and wind power are both proud parts of Texas' energy economy.

Like solar power, wind power projects do well when paired with energy storage systems such as pumped hydro storage or large battery storage systems. As battery storage costs decline, it is easier for utilities to incorporate large amounts of wind power into their energy portfolios. Wind power growth will likley accelerate as countries, states/provinces, municipalities, corporations, and individuals continue to strive toward carbon-free energy production.

Wood and waste have been used to generate heat for millennia, but today they are also used to create electricity. The downside of this technology is the pollution created in the burning process. Wood, agricultural waste, and trash create high amounts of particulate matter when burned. A cleaner way to create electricity using these materials would involve an additional process that both "scrubs" the emissions of all particulate matter and captures CO_2 at the site for either long-term storage or repurposed for fuel (see Waste-to-energy entry). If the process of burning waste for energy can be done cleanly and CO_2 can be captured, then usable fuel supplies are nearly unlimited. Wood and agricultural waste are renewable resources that can be grown (every year in the case of agricultural waste and every 10–20 years for wood). Digging up and burning buried trash would produce electricity for decades.

Although many countries still burn wood and/or trash to produce electricity, other entities (countries, states/provinces, and municipalities) have discontinued the use of these highly polluting energy sources to protect the health and well-being of their citizens. Areas that still use these materials vary in how much they limit the pollution released. Some do not scrub for particulates or capture emissions. Other places, specifically in Northern Europe, have found ways to both burn these fuels and limit the pollution that they cause. Currently, there is movement away from using these resources for electricity, but if affordable processes can be developed for limiting emissions and capturing CO_2, this could be a major growth area in the world of energy resources.

Wave energy is a nascent technology and up until this point has only been used experimentally to create electricity. There have been small-scale demonstrations of the effectiveness of this technology (see Wave Energy entry). It is truly unknown if the massive potential energy of waves can be accessed and used at an affordable price and brought up to scale. There are many different proposed designs. Some designs have turbines that are anchored to the seafloor, and the force of a wave pulling up on the tether rotates the turbine and creates electricity. Other designs allow for and harness more types of movement, in addition to vertical motion. Until the emergence of a scalable technology that produces electricity at reasonable rates, it is unlikely that wave energy will produce enough energy to be a serious part of any country's energy portfolio. Wave energy may eventually be an important energy resource but not in the foreseeable future.

Using tidal energy for electric power generation is old technology, as it is basically a modified version of a hydroelectric dam. Tidal power plants are built in areas that have especially big tides. A great difference between high tide and low tide allows a tidal energy plant to use the water either as it is flowing in or as it is flowing out. While some plants have been operating for decades, the creation of new plants is limited by geography (see Tidal Energy entry). Not all areas have appropriate places to build tidal power plants. This is further complicated by the fact that many, if not most, areas in which a tidal plant would be appropriate are areas that are also used by fishermen and outdoor recreationists. In these cases, public opposition to using these areas for a tidal power plant would make its building difficult or impossible. For these reasons, tidal energy is unlikely to play a large role in the renewable energy transition.

Geothermal electric power generation may be one of the most underused renewable energy resources. Geothermal power is either extracted from naturally occurring steam or created by injecting water into the ground in places where the heat from inside the earth is near the surface. These sources of energy are also completely renewable as the intense heat at the center of the earth is unlikely to cool anytime in the next few billion years. The difficulty with using geothermal energy is finding it. In most places, the earth is not hot enough to produce steam near the surface. Drilling miles into the earth is expensive and creating a water heating system that is miles below the earth is unfeasible. Finding heat that is nearer the surface in order to build a geothermal energy plant requires drilling test holes, which is expensive, and there is no guarantee that a usable geothermal site will be found. This financial risk deters many national governments and private electricity providers from creating new geothermal power plants.

However, there are some countries that have taken advantage of their obvious near-surface geothermal energy resources. The United States is a leader in geothermal energy but because of the country's large size, geothermal is a very small percentage of US electricity production. There are many other countries that have substantial geothermal resources, including those located around the Pacific Ring of Fire (a volcanically active path around the Pacific Ocean). Some countries use their ample geothermal resources to create electricity and, as in the case of Iceland, to heat their homes as well (see Geothermal Energy entry). If proper resources are put toward geothermal energy exploration, including technology that has advanced oil and gas drilling, it is possible that geothermal energy will be a more important part of the renewable energy transition.

STEAM POWER (STEAM ENGINES)

Steam power is power generated by using pressurized steam to turn or rotate a shaft, which can then be used to do mechanical work. In the Roman Empire, the Greek engineer Hero (d. 70 CE) of Alexandria (in what is now Egypt), created an early steam engine, which was found preserved underground. Similar devices appear in the historical record in Ottoman Egypt and Renaissance Italy. There is even a disputed record of a Flemish mathematician/astronomer who lived in China using steam power to create the first car in the 1600s. However, it was not until steam power was used to propel vehicles, including cars and trains, in the 1800s that steam power began to change the world.

Using steam is simply a way to convert heat into power. Water is heated by fire and turned into steam, which is funneled into a device that narrows and pressurizes the steam. The pressurized steam is then focused on a turbine, pushing the turbine blades and causing rotation. The rotation can be used to perform work, for example, to pull a rope or raise a weight.

Historically, people used animal power or manpower for most operations, but this started to change with the invention of efficient and reliable steam machines in the late 1800s. While many inventors were working on similar steam turbines, Charles Parsons is credited with the first patent. James Watt's steam engine design was instrumental in bringing steam engines to the masses. These machines were primarily wood-powered in the beginning. As their popularity increased, wood became more expensive and harder to find. Coal became an excellent substitute. Steam engines accelerated the transition to coal because they could be used to pump water out of coal mines and to transport dirt and the coal itself.

Steam engines were soon put to work in the transportation sector. Steamboats were able to travel upriver and across lakes, and they no longer needed the right trade winds to journey across the ocean. This new technology allowed people in more technologically advanced cultures to go to places that had always been inaccessible. Steam engines powered trains (locomotives), and trains transformed the world. Until the invention of the steam engine, travel along rivers was the main method of moving people and goods across continents and sailing ships was the only way to travel across the ocean. Most cities were built near either a seaport or a major river, and often both. With the invention of the steam-powered trains, cities could be built almost anywhere, as a train could bring any goods that people needed.

Steam locomotives and train tracks were also used by European powers to access and conquer peoples across Africa. European advanced technology in the form of steam engine trains, firearms, and artillery decimated African populations. The United States used trains and firepower to decimate Native American populations, completing its 300-year-old war on the native peoples of North America. By the early 1900s most of Africa was conquered by European powers, and most Native American territories, except for reservation lands, were subsumed by the United States.

Steam engines were also used to create the first cars. The Stanley Steamer is an example of one of the many successful brands of steam cars that were popular from the 1890s to the 1920s. Eventually, Henry Ford's Model A and Model T combustion engine cars pushed most of the steam and electric cars out of the market due to their low prices. Steam engines initiated mechanized travel but were eventually supplanted by combustion engines that run on gasoline (petrol) and diesel.

Steam engines were part of the design for early power plants, and eventually the steam engine became the steam turbine. The steam turbine is the heart of electric generators that convert heat energy into electric current. These turbines are used in about the vast majority of power plants that provide electricity to people across the planet.

SUSTAINABILITY

Sustainability refers to any system that can operate indefinitely without exhausting resources or destroying the environment that it operates within. A sustainable system is one that does not need additional outside inputs. It is a self-supporting system. There are entire books written on how to define sustainability across different disciplines. Sustainable energy systems are generally the ones that create electricity using renewable energy, although nuclear energy holds some promise of becoming a sustainable energy source. Currently, nuclear energy is not sustainable because it creates radioactive waste that needs to be stored indefinitely. Even renewable energy systems are not completely sustainable, as they do need resources to create power plants and devices used to collect and transport renewable energy.

Sustainable energy sources are inexhaustible energy sources that do not destroy anything. These sources do not use a finite energy source. Solar power is sustainable because the sun will shine for billions of years. Wind energy is sustainable because the wind will continue day after day. Geothermal energy is sustainable because the heat from the earth will exist for billions of years. Wave and tidal energy are sustainable because the moon will continue to pull on the earth for as long as humans live on this planet. As mentioned, nuclear power is not sustainable because its dangerous waste must be stored for tens of thousands of years. Fossil fuels are unsustainable for several reasons: they are limited, they pollute, and they are warming the planet.

The earth has enough sustainable energy types that it is unnecessary to use unsustainable energy sources in the future. Sustainable transportation will run on sustainable energy. EVs that are powered by renewable energy are one form of sustainable transportation. Hydrogen fuel cells will be used to move people and goods around the planet. As long as hydrogen is produced using renewable energy, it is a sustainable form of transportation. Sustainable food systems are ones that use inputs that are not poisonous or dangerous for the environment and the plants and animals that live in it. These are closed-loop systems: animal manure is used to fertilize fields, and the chaff left over from the harvest is fed to the animals. Natural forms of pest control are sustainable, including encouraging the establishment of insects that eat common pests. Food is then transported using sustainable transportation to where people live. Sustainable consumer products are ones that are made to last and are recyclable when their life cycle finishes. They are not made with chemicals that pollute, and they are produced using renewable energy. They are transported with sustainable forms of transportation.

Sustainability sounds simple, but it is difficult to achieve because governments have incentivized profits without incentivizing the environment. Destruction of natural resources does not cost a company. In most places, producing GHGs does not cost a company. People create governments, and governments create regulations for people and companies. By creating a financial system that disincentivizes unsustainable behavior, transition to a sustainable society will be driven by businesses and consumers.

TAR SANDS/OIL SANDS

Tar sands, oil sands, or crude bitumen are names given to a geological formation that contains sand, clay, sandstone, and water mixed with bitumen. Bitumen is a form of thick oil. The process of recovery of this type of oil is resource intensive, and the oil created in the process is full of more particulate matter than most other types of oil. Because of this, tar sands oil is often referred to as the dirtiest form of oil. The use of tar sands has been a controversial topic both locally (at the sites of tar sands oil collection) and internationally, as some environmental groups have fought to have governments stop companies from mining for this type of oil.

Mining tar sands usually requires the removal of massive amounts of material, a process akin to strip mining. The forest and plants are cleared from an area, and then the earth underneath is removed until the tar sands are reached. The tar sands must then be moved and processed to separate the oil from the rest of the material. This is energy-intensive and requires a lot of water, so the cost per barrel of tar sands oil is much higher than other types of oil. It is the most polluting form of oil and is filled with impurities and releases more particulate matter when burned. From beginning to end, the mining and burning of tar sands release three to five times as much GHGs as conventional oil.

While tar sands exist in dozens of countries around the world, the majority are located in Canada and Venezuela. Venezuela's oil production has decreased in recent years due to political volatility, nationalization of oil resources, and a mass exodus of skilled labor. Nevertheless, Venezuela still produces half of a million barrels of oil from tar sands every day. Canada, on the other hand, is increasing its production of tar sands oil and is producing over three million barrels of tar sands oil every day. The United States purchases most Canadian tar sands oil.

As mentioned, mining tar sands near the surface is energy-intensive and destructive. Some tar sands are too deep to dig out and haul to a facility for processing. In those situations, steam is injected deep into the ground to separate the bitumen from the sand. This process is called Steam-assisted gravity drainage (SAGD). Two wells are dug: one above and one below. The upper well is pumped with hot steam, which liquifies the bitumen. The bitumen drips down to the bottom well and is extracted to the surface (Government of Canada 2021). In both near-surface mining and SAGD, the bitumen must be processed so that it can be turned into synthetic crude oil. This involves adding hydrogen and/or removing carbon while also removing impurities so that the material is "clean" enough to be processed by oil refineries. It takes about 300 kWh of energy to turn one barrel of bitumen into one barrel of crude oil.

As long as oil prices remain high, Canada will likely continue to increase its production of tar sands oil. The country has an almost limitless amount of tar sands sitting under 130,000 square miles of mostly pristine forests. If the political situation improves in Venezuela, Venezuela will likely do the same. As soon as EV adoption hits a tipping point, demand for oil will decrease rapidly and tar sands oil will no longer be profitable. At that point, tar sands oil will have reached the end of its useful life. The question that remains is when the world will make that transition.

TIDAL ENERGY

Tidal energy is captured by building power plants located in an area that has a high tidal range. As a general rule, power plants cannot operate efficiently with less than 10 feet of tidal range (EIA 2022). These tidal energy power plants work on the same concept as hydroelectric power plants on a river. As the tide comes in and out, the water is pulled through the tidal power plant, where it turns a turbine. The spinning turbine transforms the physical energy of the moving water into electricity. Tidal energy is a niche energy type because there are not many places in which tidal energy plants can be built and successfully used to create electricity for an area.

Some successful tidal energy plants have existed for decades and continue to produce electricity today. The two largest plants are the Rance Tidal Power Plant in France, built in 1966, which puts out 240 MW, and the Sihwa Lake Tidal Power Plant in South Korea, built in 2011, which produces 254 MW of electricity. That makes the Sihwa plant the largest in the world. Canada has the Annapolis Royal Generating Station, which produces 20 MW of electricity. There are larger tidal power plants planned for the future, including two in Great Britain. The MeyGen tidal power plant on the northern tip of Scotland has the potential to produce up to 398 MW of electricity (Calder 2021). It is in the beginning stages and currently produces only 6 MW of electricity. MeyGen is different from many other types of tidal power plants because it works more like an underwater wind turbine than a hydroelectric dam. Currently, each water turbine produces 1.5 MW of power, so with only four turbines MeyGen is able to produce 6 MW of power. Construction goals aim to upgrade to 86 MW of electric production by 2025/26 (Calder 2021). Another large project in development is the Swansea Tidal Lagoon project in Wales, which could one day produce up to 240 MW of electricity. This project will operate with similar technology as hydroelectric dams; four times a day, the tide will pull water through the turbines as the tide comes in and goes out.

There are many possible tidal energy sites left in the world; the UK, Russia, Korea, and India all have major plans to install large tidal power stations (Ramos and Ringwood 2016). However, the growth of tidal energy can be stymied by competing interests in coastal areas, as ocean waters that are near cities or towns are used for recreation and/or fishing. Remote locations have the added difficulties of finding staff that will relocate and of creating the infrastructure to transfer the electricity long distances.

While some countries have plans for building new, large tidal power plants, most countries with coastal areas do not. Tidal power makes up only a tiny fraction of the world's renewable energy and will likely continue to play only a minor part due to limited potential sites and competing uses for those sites. Russia has proposed a 24,000 MW project, although some studies have indicated a lower projection of 8,000 MW. If constructed, it would be the largest tidal power plant in the world, almost one hundred times larger than the Sihwa and the Rance tidal power stations. If large projects like this can be built and proven successful, they could inspire other nations to invest in their own plants.

WASTE-TO-ENERGY

Waste-to-energy (WTE), the process of burning waste or trash to create energy, has had mixed success over the years. It has been hailed as a solution to all of humanity's trash problems and vilified as one of the most polluting energy sources in existence. WTE power plants work similarly to wood- or coal-powered plants. Trash that is dry enough to burn is used to create heat to boil water and generate steam. Steam turns a turbine that produces an electric current. Burning trash to produce electricity was common practice until countries started measuring pollutants. Many countries then started burying their trash and burning fossil fuels instead, because although fossil fuels pollute and add CO_2 to the atmosphere, they are less polluting than WTE plants.

Historically, trash was often burnt for heat if it was flammable. In ancient times, trash mainly consisted of compostable and/or flammable materials. Plastics, e-waste, and batteries did not exist. Most things humans used were from nature and could easily return to nature. In the early 1900s, advanced economies began producing things out of materials like metal and glass that did not easily decompose. After the Second World War, plastic became a normal part of life and continue to make up a larger and larger part of the human waste stream. Most plastic products are not easily recycled. Using WTE power plants is one way in which plastic and other waste can be reused and disposed of.

One advantage of WTE power plants is that the amount of ash to be disposed of after incineration is about one-tenth of the original amount of trash. This means that WTE power plants could potentially reduce landfill waste by 90 percent and produce electricity at the same time. These plants pollute much less than coal or oil power plants but a tiny fraction more than natural gas power plants. In Norway, more than half of all waste is burned for electricity. In Sweden, 52 percent of trash is burned, 47 percent is recycled, and 1 percent is put in landfills (Yee 2018). Sweden also imports trash to recycle and burn at 43 USD per tonne. By burning waste that would have been put into a landfill, millions of tons of GHGs are averted annually; burning waste replaces coal or natural gas that would have been burned, and methane that would have been created in the landfill is not produced. These success stories demonstrate that WTE power plants can be done right and responsibly.

One of the reasons more countries are not rushing to build WTE plants is that burning trash still emits GHGs. Replacing coal- and gas-powered plants with renewable energy makes more sense than switching to WTE plants, at least in terms of limiting human effects on climate change. Burning waste also creates pollution. Creating new WTE power plants is a large investment, and on their own they are not profitable. Most of their income comes from the fees they charge waste management companies to "dispose" of their waste. Building a WTE power plant means that cities or states are locked into a thirty-year commitment because it takes that long to reclaim the investment in the plant. There also is a hope that, over the next thirty years, society can move away from single-use plastics and other cheap plastic products that are not made to last, thereby reducing waste.

The future of WTE power plants will depend on public perception and political will. Modern WTE plants are cleaner (less polluting) and better smelling than they were in the past. However, most communities are not interested in the idea of burning trash in their neighborhoods, so considerable public outreach and education would be required for the not-in-my-backyard problem. Countries that are doing well with WTE plants will likely continue to do so, and they may inspire their neighbors to do the same. Countries that are not short on landfill space are unlikely to invest in WTE power plants, but WTE does appear to be a good option for countries that want to decrease the footprint of their landfills or get rid of old landfills.

WAVE ENERGY

The potential energy of waves is estimated at two terawatts or two million megawatts, which would be enough power to cover all the world's energy needs twice over. While the potential is incredibly large, the amount of wave power that people harness is incredibly small. At this time there are many experimental wave energy projects taking place, but it will take a long time to get these projects from experimental to implemented and scaled up. Wave energy is not likely to be a major part of the renewable energy transition, although there is hope that within twenty years there will be wave power plants that can serve the electrical needs of coastal communities across the planet.

People have ridden waves for sport and for transportation for centuries. Using waves to create electricity is a much newer endeavor. The technology that exists today is varied, and no one particular type of wave energy extractor has come to prominence. Scientists and inventors have organized the various prototypes into roughly seven different categories. (1) Point absorbers are the most common type of wave energy converters. These are devices that are generally connected to the ocean floor and have a floating mechanism. As the wave pulls the floating mechanism away from the piece attached to the ocean floor, energy is created. The pulling motion can spin a generator that produces electricity. (2) Attenuators consist of two floating sections that can create energy by using the "flapping" motion that is generated by a wave lifting one section and then the other as the first section goes back down. (3) Terminators, or overtopping devices, let a wave break over the device, and the falling wave turns a turbine before it is released back into the ocean. (4) An oscillating wave surge device is mounted to the seafloor and basically is swung back and forth like an upside-down pendulum. This back-and-forth motion is converted to energy. (5) An oscillating water column is a hollow structure that is open on one side. The waves come in and push up a device that converts that energy to electricity. (6) A submerged pressure differential device uses the changing pressure of the water above to lower and raise the device, creating motion that is converted into electricity. (7) Finally, rotating mass devices use the power of waves to force a round-bottomed device to wobble. Inside the device, a ball-like structure rolls around, and that movement is converted to electricity (EMEC n.d.). All these devices and methods show promise but must demonstrate that they can be large-scale and long-term, providing continual energy.

There are some startup companies that are looking to be the first to bring wave power into the mainstream. Most of these companies are in Europe as a result of significant investment from the European Union. Promising demonstrations may mean that coastal communities will soon begin to augment their power with wave power. If these projects are successful, before long, most coastal communities will see a large portion of their power coming from wave power. At that point, wave power will be an important part of the low-carbon economy and help humans avoid the worst effects of climate change.

WIND POWER

Wind power in one form or the other is an ancient form of energy. Humans harnessed wind to travel across the world's oceans on sailing vessels, settling on distant islands. Windmills used wind power to grind down, or mill, grain into flour. While wind power has been used for millennia, it has only been used to create electricity since the late 1800s. The natural energy of the wind is used to mechanically spin a magnet inside of coiled wire to create an electric current, which can directly power a home or business or be put into the grid to be a part of the total electric energy generation for an area. Wind power has moved from a localized form of manual labor to an impressive and growing source of electrical generation across the planet (see also Chapters 1, 2, and 4).

Wind power is growing at an incredibly fast rate, doubling every five to six years worldwide. There was 177,000 MW of installed wind power in 2010 and 698,000 MW by the end of 2020 (IRENA 2021); that is almost four times as much wind power in only ten years. At that pace, wind power will likely overtake hydropower as the largest renewable energy source near 2027. With the commitments to renewable energy that almost every country on earth has signed on to, the rate of increase will likely continue to grow aggressively. The second quarter of 2021 saw a sixfold increase in the number of wind power capacity awarded (GWEC 2021). This process continues unabated. In 2023, the world added 50 percent more wind capacity than they added in 2022, and 2024 and 2025 are set to be record years in new wind power installed. Wind and solar are set to make up 96 percent of all new energy added to the grid until 2030 at least.

Wind power does have the drawback of inconsistency. Even in windy areas, there may be times that the wind does not blow. For this reason, it is important to pair wind power with pumped hydro or battery systems that can back up the wind turbines on non-windy days. As natural gas power plants are supplanted by renewable energy, some of them may find a use as what is called a peaker plant. Peaker plants are kept online at a low rate and ready to fire up during peak electricity demand, which can help to supplement wind power.

China has more installed wind power plants than any other country. The United States is second in wind power, unless the European Union is counted as one block. China has double the wind power capacity of the United States, and the United States has double the wind capacity of the next closest country, Germany. Therefore, there is considerable room for improvement. Most countries are woefully behind in building enough wind turbines to catch up to their wind potential. However, rapid growth rates mean that it will not be long before many other countries are producing enough wind power to be worth mentioning.

The future of wind power is almost limitless, as there is enough potential wind power to provide adequate electricity for the entire world many hundreds of times over. The central plains in the United States alone could produce enough power for the entire planet. Wind power will continue to grow until it is either the most common or second most common (after solar) form of electricity generation.

ZERO EMISSIONS ENERGY

Zero emissions energy refers to an energy source that does not produce CO_2 or any other type of GHG emissions during the process of energy generation. Zero emissions energy differs from net-zero energy in that net-zero systems may produce GHG emissions but balance them out by sequestering carbon or exporting clean electricity from zero emission sources (see Net Zero entry). The energy produced from zero emission sources is usually used to create electricity or to propel a vehicle, but it could be used for anything.

Most zero emission energy sources create electricity by spinning a turbine. Hydroelectric dams use the force of water to turn large turbines. Nuclear plants create steam that turns turbines. Wind turbines are powered by the wind. Geothermal energy uses the heat and steam from the earth to turn turbines. Wave and tidal energy tend to turn turbines as well. Photovoltaic solar power is the one exception, as it does not require a turbine or a generator. Solar cells produce electricity directly from the sun when photons collide with the cell and release electrons (see Photovoltaics entry). There is no electricity generation process that does not require hardware. Every system requires a certain amount of energy and resources to build it, and at present, this means that emissions were likely produced during its creation. Even zero emissions energy requires energy to begin the process of its generation. Therefore, most of the power sources discussed in this entry are technically considered low-carbon energy sources rather than zero emissions, although they may be considered net zero after some time in operation. However, if humans switch to zero emissions (or net zero) energy for all electricity and transportation used in constructing these systems, then they will indeed be completely zero emissions systems. These sources combine to create about 40 percent of the electricity on the planet.

Hydroelectric power is the number one zero emissions energy source (putting aside the energy and resources used to build dams and plants). It is the oldest technology discussed here and creates 17 percent of the electricity in the world (IEA 2021), meaning that hydroelectric power produces 42.5 percent of all zero emissions energy in the world. Hydroelectric technology is mature, with few ideal sites remaining for growth. While hydroelectric power will remain an important part of a net-zero energy mix, it will not expand significantly.

Nuclear power has always held the potential to completely free the world from the need to use fossil fuels. That potential has never been realized for three reasons: (1) it is much more expensive than traditional fuel sources; (2) it leaves waste that remains radioactive for tens of thousands of years; and (3) it remains dangerous. Despite nuclear power's generally safe record, the three major disasters at Three Mile Island, Chernobyl, and Fukushima have turned public perception against the continual use of nuclear power and made it nearly impossible to build new reactors within democracies. Public resistance to nuclear power has forced some countries to shutter functioning plants well ahead of their planned lifecycle. New generations of nuclear technology promise safer nuclear energy that creates less waste. Public pressure will dictate that continual testing of new technology is likely to be limited to countries with authoritarian governments, such as Russia and China. If these countries can prove that nuclear power can be safe and create significantly less radioactive waste, then there will be an opportunity for nuclear power to grow once again. Nuclear power currently provides 10 percent of the world's power, which is 25 percent of the world's zero emissions energy (World Nuclear Association 2021). The number of reactors has not increased worldwide since the late 1980s; while a few countries are still building nuclear reactors, other countries are shutting them down and decommissioning them. If trends continue, nuclear power will continue to be an important, yet shrinking part of the worldwide energy mix. Unless new technology can assure a nervous public that it is both safe and reliable to operate nuclear power plants and to store the waste they create, nuclear power will not have the opportunity to grow.

Wind power produces the third most zero emissions electricity worldwide at 12.5 percent; it generates 5 percent of worldwide power overall. Wind power is growing at an incredibly fast rate. Finding appropriate sites and gaining local community support will be important to the growth of wind power, as will connecting it with backup systems that can cover periods when wind speed is too low or high for turbines to operate (see Wind Power entry).

Solar power is another renewable and zero emissions energy resource that is growing at an exponential rate. It currently makes up 5 percent of all zero emissions electricity, and

2 percent of worldwide power overall. Solar is poised to eventually overtake wind power, as it is available anywhere with adequate sunshine. Solar power is easy to implement locally and to scale up. Used in combination with systems that store or make energy when there is no sunlight, solar energy can provide a constant energy source around the world.

Geothermal energy is for heating homes and to create electricity. While geothermal is extremely important in countries like Iceland, it only produces about 1 percent of all electricity or 2.5 percent of zero emissions electricity. There is room for growth, but the expense involved will likely prohibit significant long-term investments.

Wave and tidal are nascent technologies. They both have enormous potential, but neither is likely to contribute much to the planet's clean energy needs in the near future. They will need time to develop and prove their worth.

The transition to net zero or zero emissions is not as difficult as it may appear. The United States decreased their emissions in the power generation sector by 50 percent in just fifteen years (between 2005 and 2020). This was achieved by attacking the problem from multiple angles. Solar and wind power has significantly increased. Nuclear power has remained steady. Energy efficiency has increased across many platforms and in many industries. Natural gas has replaced coal plants that pollute more and emit more CO_2; this transition has led to fewer deaths from pollution (usually from asthma and other respiratory problems as well as heart attacks). Lowering CO_2 emissions can help to slow the process of climate change. Even with all these advancements, prices for energy in the United States continue to decrease. Jobs in the energy field continue to grow. This demonstrates that energy systems can be changed quickly and affordably while increasing health, saving money, and providing jobs. All that is needed for this transition is the political will to stand up against entrenched fossil fuel interests. Once the power sector is based on net-zero and zero emissions energy, all electric transportation will also run on net-zero energy.

FURTHER READING AND REFERENCES

Alfe, D., M. J. Gillan, L. Vočadlo, J. Brodholt, and G. D. Price. 2002. "The Ab Initio Simulation of the Earth's Core." *Philosophical Transactions of the Royal Society of London*. Series A: Mathematical, Physical and Engineering Sciences, 360, no. 1795: 1227–44. https://doi.org/10.1098/rsta.2002.0992.

Alongi, Daniel M. 2012. "Carbon Sequestration in Mangrove Forests." *Carbon Management* 3, no. 3: 313–22. https://doi.org/10.4155/cmt.12.20.

Arenas-Corraliza, M. G., V. Rolo, M. L. López-Díaz, and G. Moreno. 2019. "Wheat and Barley Can Increase Grain Yield in Shade through Acclimation of Physiological and Morphological Traits in Mediterranean Conditions." *Scientific Reports* 9, no. 1: 1–10.

Blue Carbon Institute. n.d. "About Blue Carbon." Accessed September 18, 2021. https://www.thebluecarboninitiative.org/about-blue-carbon.

BP Statistical Review of World Energy. 2019. 68th edition. London: BP. Accessed October 30, 2022. https://www.bp.com/content/dam/bp/business-sites/en/global/corporate/pdfs/energy-economics/statistical-review/bp-stats-review-2019-full-report.pdf.

Briguet Bastos, Milton. 2007. "Brazil's Ethanol Program: An Insider's View." *Energy Tribune,* June 20, 2007. https://web.archive.org/web/20110710193215/http://www.energytribune.com/articles.cfm?aid=534.

Buis, A., 2020. Milankovitch (orbital) cycles and their role in earth's climate. *NASA Climate*, 27, p.8. Accessed July 12, 2024. https://science.nasa.gov/science-research/earth-science/milankovitch-orbital-cycles-and-their-role-in-earths-climate/.

Burke, K., 2003, November. Fuel cells for space science applications. In *1st International Energy Conversion Engineering Conference (IECEC)* (p. 5938). Acessed July 12, 2024.https://ntrs.nasa.gov/api/citations/20040010319/downloads/20040010319.pdf.

Calder, Gordon. 2021. "MeyGen Electricity Production to Increase from Six to 86 Megawatts by 2025/26, Say Simec Atlantis Energy." *John O'Groat Journal and Caithness Courier*, September 10. https://www.johnogroat-journal.co.uk/news/meygen-electricity-production-to-increase-from-six-to-86-meg-250474/.

Cimons, Marlene. 2017. "Discarded Oyster Shells Can Help Us Grow Food, Make Cement, and Fight Climate Change: Solutions on the Half Shell." *Popular Science*, July 19. https://www.popsci.com/oyster-shells-alternate-uses/.

CO_2 Human Emissions. 2017. "Main Sources of Carbon Dioxide Emissions." December 13. https://www.che-project.eu/news/main-sources-carbon-dioxide-emissions.

Donev, J.M.K.C. et al. n.d. "Energy Education: Coal Types." *Energy Education*. Accessed November 17, 2020. https://energyeducation.ca/encyclopedia/Coal_types.

Energy Information Administration (EIA). 2017. "Natural Gas Generators Make Up the Largest Share of Overall U.S. Generation Capacity." December 18, 2017. https://www.eia.gov/todayinenergy/detail.php?id=34172.

Energy Information Administration (EIA). 2022a. "Coal Explained: How Much Coal Is Left." Updated October 21, 2022. https://www.eia.gov/energyexplained/coal/how-much-coal-is-left.php.

Energy Information Administration (EIA). 2022b. "Coal Explained: Use of Coal." Updated June 6, 2022. https://www.eia.gov/energyexplained/coal/use-of-coal.php.

Energy Information Administration (EIA). 2022c. "Hydropower Explained: Tidal Power." Updated August 10, 2022. https://www.eia.gov/energyexplained/hydropower/tidal-power.php.

Environmental Protection Agency (EPA). 2022a. "Understanding Global Warming Potentials." Updated May 5, 2022. https://www.epa.gov/ghgemissions/understanding-global-warming-potentials.

Environmental Protection Agency (EPA). 2022b. "Sources of Greenhouse Gas Emissions." Updated August 5, 2022. https://www.epa.gov/ghgemissions/sources-greenhouse-gas-emissions.

The European Marine Energy Centre LTD (EMEC). n.d. "Wave Devices." Accessed September 12, 2021. https://www.emec.org.uk/marine-energy/wave-devices/.

Gibbons, Ann. 2020. "How Neanderthals Lost Their Y Chromosome: *Homo sapiens'* Y Chromosome Spread through Neanderthal Populations after Early Mating." *Science*, September 24. Accessed July 13, 2024. https://www.sciencemag.org/news/2020/09/how-neanderthals-lost-their-y-chromosome.

Government of Canada. 2021. "Energy Facts." Updated December 23, 2021. https://www.nrcan.gc.ca/science-and-data/data-and-analysis/energy-data-and-analysis/energy-facts/20061.

Gruy, H. J., and A. J. Gruy. 2000. "History of the Ownerships of Mineral Rights." *SPE Annual Technical Conference and Exhibition*. OnePetro, October 1–4, Dallas, TX. https://onepetro.org/SPEATCE/proceedings-abstract/00ATCE/All-00ATCE/132038

Global Wind Energy Council (GWEC). 2021. "Global Wind Power Capacity Awarded in Q2 2021 Six Times Larger than Q2 2020." September 3. Accessed July 13, 2024. https://gwec.net/global-wind-power-capacity-awarded-in-q2-2021-six-times-larger-than-q2-2020/

International Energy Association (IEA). 2018. "The Future of Petrochemicals—Analysis." .Accessed July 12, 2024. www.iea.org/reports/the-future-of-petrochemicals.

International Energy Association (IEA). 2021. *Hydropower Special Market Report*, IEA, Paris https://www.iea.org/reports/hydropower-special-market-report, Licence: CC BY 4.0.

International Renewable Energy Agency (IRENA). N.d. "Wind Energy." Accessed August 22, 2021. https://www.irena.org/wind.

Jain, H. K. 2018. *Green Revolution: History, Impact and Future*. Houston: Stadium Press LLC.

Knier, Gil. 2008. "How Do Photovoltaics Work?" *Nasa Science*, August 6. https://science.nasa.gov/science-news/science-at-nasa/2002/solarcells.

MacPhee, Ross D. E. 2018. *End of the Megafauna: The Fate of the World's Hugest, Fiercest, and Strangest Animals*. New York: WW Norton.

MacPherson, H. G. 1985. "The Molten Salt Reactor Adventure." *Nuclear Science and Engineering* 90, no. 4: 374–80.

Maestas, A. "Mexican Wolf Population Grows for Eighth Consecutive Year: U.S. Fish & Wildlife Service." *FWS.Gov*, March 5, 2024. Accessed July 12, 2024. www.fws.gov/press-release/2024-03/exican-wolf-population-grows-eighth-consecutive-year.

Munroe, Thailynn. 2019. "Embers Under the Earth: The Surprising World of Coal Seam Fires." *Global Fire Watch*, January 30. https://www.globalforestwatch.org/blog/fires/embers-under-the-earth-the-surprising-world-of-coal-seam-fires/.

National Park Service (NPS). 2022. "Wolf Restoration." Updated February 16. Accessed July 12, 2024. https://www.nps.gov/yell/learn/nature/wolf-restoration.htm.

Nelson, Max. 2005. *The Barbarian's Beverage: A History of Beer in Ancient Europe*. London: Routledge.

Pingali, P. L. 2012. "Green Revolution: Impacts, Limits, and the Path Ahead." *Proceedings of the National Academy of Sciences* 109, no. 31: 12302–8.

Praetorius, Summer K. 2018. "North Atlantic Circulation Slows Down." *Nature* 556: 180–1. https://www.nature.com/articles/d41586-018-04086-4/.

Ramos, V., and J.V. Ringwood. 2016. "Implementation and Evaluation of the International Electrotechnical Commission Specification for Tidal Stream Energy Resource Assessment: A Case Study. *Energy Conversion and Management* 127: 66–79. https://doi.org/10.1016/j.enconman.2016.08.078.

Rein, G. 2015. "Smoldering-peat Megafires." In *Coal and Peat Fires: A Global Perspective, Volume 4: Peat – Geology, Combustion, and Case Studies*, 1st edition, edited by Glenn B. Stracher, Anupma Prakash, and Guillermo Rein, 1–11. Amsterdam: Elsevier Science.

Saunois, M., Stavert, A.R., Poulter, B., Bousquet, P., Canadell, J.G., Jackson, R.B., Raymond, P.A., Dlugokencky, E.J., Houweling, S., Patra, P.K. and Ciais, P., 2020. The Global Methane Budget 2000–2017, *Earth Syst. Sci. Data* 12: 1561–623.

Taylor, Kira. 2021. "Europe Halfway towards Closing All Coal Power Plants by 2030." *EuraCTIV*, March 23. https://www.euractiv.com/section/climate-environment/news/europe-halfway-towards-closing-all-coal-power-plants-by-2030/.

Tisserand, O. 2014. "Introduction to Oil & Gas Offshore Drilling." *Indelac Controls, Inc.*, June 24. Accessed July 12, 2024. www.indelac.com/blog/introduction-to-oil-gas-offshore-drilling.

Tourand, N. n.d. *Molten Salt Reactors. whatisnuclear*. Accessed July 12, 2024. https://whatisnuclear.com/msr.html.

UNEP, F., 2020. The State of the World's Forests 2020. *Forests, Biodiversity and People*, p. 214. Accessed July 12, 2024. https://openknowledge.fao.org/items/f77312ae-67bc-4769-96a1-2b6f8a72d88c.

United Nations. n.d. "Population." Accessed August 21, 2021. https://www.un.org/en/global-issues/population.

US Department of Agriculture. n.d. "Conservation Energy Estimators." *Natural Resources Conservation Service*. Accessed August 21, 2021. https://www.nrcs.usda.gov/wps/portal/nrcs/detailfull/national/energy/conservation/?cid=nrcs143_023637.

US Fish and Wildlife Service. n.d. "Conserving the American Wolf." Accessed August 12, 2021. https://www.fws.gov/southwest/es/mexicanwolf/.

World Health Organization. n.d. "Air Pollution." Accessed August 21, 2021. https://www.who.int/health-topics/air-pollution.

World Nuclear Association. 2022. "Nuclear Power in the World Today." Updated October 2022. https://world-nuclear.org/information-library/current-and-future-generation/nuclear-power-in-the-world-today.aspx.

Yee, A., 2018. "In sweden, Trash Heats Homes, Powers Buses and Fuels Taxi Fleets." *New York Times*, 21. Accessed July 13, 2024. https://www.nytimes.com/2018/09/21/climate/sweden-garbage-used-for-fuel.html.

ABOUT THE AUTHORS AND CONTRIBUTORS

AUTHORS

Adrah N. Parafiniuk teaches at Northern Arizona University, United States. He is the author of *108 Steps to Fix the Planet: A Guide to a Healthier, Happier, and More Joyous Life* (2013). He has published work on energy, green gilding, motivations for sustainable behaviors, pro-environmental behaviors, and a variety of other topics.

Zachary A. Smith was Regents' Professor at Northern Arizona University, United States. Smith is widely published, producing such works as *The Environmental Policy Paradox,* 8th Edition (2022), *Globalization* (2008) (with Justin Ervin), and *Protecting Our Environment: Lessons from the European Union* (2005) (with Janet R. Hunter).

CONTRIBUTORS

Haley N. Anderson works at Princeton University as a program coordinator. Her research interests are in the areas of nonprofit administration, organization sustainability, and organization management. She is also a PhD student.

Vivian Ike is a political scientist who focuses in global government and the environment. Dr. Ike is an assistant professor at Syracuse University in New York. Areas of publication include a journal article on the impact of veto players on Australia's carbon tax and repeal, a book review on the environmental health effects of fracking in the United States, and a chapter in a book about different countries' national policy commitments regarding climate change ethics and justice.

Laine Munir is a political and legal anthropologist examining the nexus of gender, environmental conflict, and law in sub-Saharan Africa. She is Teaching Assistant Professor in the School of Politics and Global Studies at Arizona State University, United States. In addition to contributing as a Senior Research Fellow at the University of Rwanda's Center of Excellence in Biodiversity and Natural Resource Management, she is a gender equality consultant with international organizations, nonprofits, and government agencies. She holds an interdisciplinary PhD in Law and Society from New York University (2015) and an MA in Human Rights from Columbia University (2008).

Ayesha Sadiqa received her master's degree in Solid State Physics from the University of the Punjab, Pakistan. She has worked as a research assistant and lecturer in Pakistan. She is currently pursuing her doctoral degree at Lappeenranta University of Technology, Finland. She is working in the Solar Economy group and her area of research is 100 percent renewable energy-based system optimization, where she is working on 100 percent renewable energy system transition research for Pakistan. Her areas of interest are renewable energy, sustainability, and the role of women in renewable energy transition.

Helena Solman is a PhD researcher at the Environmental Policy Group at Wageningen University and Research, the Netherlands. Her research focuses on the process of public engagement with wind energy. Her expertise lies in investigation of innovative approaches to wind energy design, planning, and management that incorporate public concerns, values, and practices.

INDEX

Note: Page numbers in *italics* refer to figures.

www.ingramcontent.com/pod-product-compliance
Ingram Content Group UK Ltd.
Pitfield, Milton Keynes, MK11 3LW, UK
UKHW050946290926
15470UKWH00005B/1670

9 781440 871801